D1651201

PCR Primer Design

METHODS IN MOLECULAR BIOLOGY™

John M. Walker, SERIES EDITOR

419. **Post-Transcriptional Gene Regulation,** edited by *Jeffrey Wilusz, 2008*
418. **Avidin–Biotin Interactions:** *Methods and Applications,* edited by *Robert J. McMahon, 2008*
417. **Tissue Engineering, Second Edition**, edited by *Hannsjörg Hauser and Martin Fussenegger, 2007*
416. **Gene Essentiality:** *Protocols and Bioinformatics,* edited by *Andrei L. Osterman, 2008*
415. **Innate Immunity,** edited by *Jonathan Ewbank and Eric Vivier, 2007*
414. **Apoptosis in Cancer:** *Methods and Protocols*, edited by *Gil Mor and Ayesha Alvero, 2008*
413. **Protein Structure Prediction, Second Edition,** edited by *Mohammed Zaki and Chris Bystroff, 2008*
412. **Neutrophil Methods and Protocols,** edited by *Mark T. Quinn, Frank R. DeLeo, and Gary M. Bokoch, 2007*
411. **Reporter Genes for Mammalian Systems**, edited by *Don Anson, 2007*
410. **Environmental Genomics,** edited by *Cristofre C. Martin, 2007*
409. **Immunoinformatics:** *Predicting Immunogenicity* In Silico, edited by *Darren R. Flower, 2007*
408. **Gene Function Analysis,** edited by *Michael Ochs, 2007*
407. **Stem Cell Assays,** edited by *Vemuri C. Mohan, 2007*
406. **Plant Bioinformatics:** *Methods and Protocols,* edited by *David Edwards, 2007*
405. **Telomerase Inhibition:** *Strategies and Protocols,* edited by *Lucy Andrews and Trygve O. Tollefsbol, 2007*
404. **Topics in Biostatistics,** edited by *Walter T. Ambrosius, 2007*
403. **Patch-Clamp Methods and Protocols**, edited by *Peter Molnar and James J. Hickman, 2007*
402. **PCR Primer Design**, edited by *Anton Yuryev, 2007*
401. **Neuroinformatics,** edited by *Chiquito J. Crasto, 2007*
400. **Methods in Lipid Membranes,** edited by *Alex Dopico, 2007*
399. **Neuroprotection Methods and Protocols,** edited by *Tiziana Borsello, 2007*
398. **Lipid Rafts,** edited by *Thomas J. McIntosh, 2007*
397. **Hedgehog Signaling Protocols,** edited by *Jamila I. Horabin, 2007*
396. **Comparative Genomics,** *Volume 2*, **edited by** *Nicholas H. Bergman, 2007*
395. **Comparative Genomics,** *Volume 1*, **edited by** *Nicholas H. Bergman, 2007*
394. **Salmonella:** *Methods and Protocols,* edited by *Heide Schatten and Abe Eisenstark, 2007*
393. **Plant Secondary Metabolites**, edited by *Harinder P. S. Makkar, P. Siddhuraju, and Klaus Becker, 2007*
392. **Molecular Motors:** *Methods and Protocols*, edited by *Ann O. Sperry, 2007*
391. **MRSA Protocols,** edited by *Yinduo Ji, 2007*
390. **Protein Targeting Protocols, Second Edition,** edited by *Mark van der Giezen, 2007*
389. ***Pichia* Protocols, Second Edition,** edited by *James M. Cregg, 2007*
388. **Baculovirus and Insect Cell Expression Protocols, Second Edition,** edited by *David W. Murhammer, 2007*
387. **Serial Analysis of Gene Expression (SAGE):** *Digital Gene Expression Profiling,* edited by *Kare Lehmann Nielsen, 2007*
386. **Peptide Characterization and Application Protocols**, edited by *Gregg B. Fields, 2007*
385. **Microchip-Based Assay Systems:** *Methods and Applications,* edited by *Pierre N. Floriano, 2007*
384. **Capillary Electrophoresis:** *Methods and Protocols,* edited by *Philippe Schmitt-Kopplin, 2007*
383. **Cancer Genomics and Proteomics:** *Methods and Protocols,* edited by *Paul B. Fisher, 2007*
382. **Microarrays, Second Edition:** *Volume 2, Applications and Data Analysis,* edited by *Jang B. Rampal, 2007*
381. **Microarrays, Second Edition:** *Volume 1, Synthesis Methods,* edited by *Jang B. Rampal, 2007*
380. **Immunological Tolerance:** *Methods and Protocols,* edited by *Paul J. Fairchild, 2007*
379. **Glycovirology Protocols**, edited by *Richard J. Sugrue, 2007*
378. **Monoclonal Antibodies:** *Methods and Protocols,* edited by *Maher Albitar, 2007*
377. **Microarray Data Analysis:** *Methods and Applications,* edited by *Michael J. Korenberg, 2007*
376. **Linkage Disequilibrium and Association Mapping:** *Analysis and Application,* edited by *Andrew R. Collins, 2007*
375. **In Vitro Transcription and Translation Protocols:** *Second Edition,* edited by *Guido Grandi, 2007*
374. **Quantum Dots:** *Applications in Biology,* edited by *Marcel Bruchez and Charles Z. Hotz, 2007*
373. **Pyrosequencing® Protocols**, edited by *Sharon Marsh, 2007*
372. **Mitochondria: Practical Protocols,** edited by *Dario Leister and Johannes Herrmann, 2007*
371. **Biological Aging:** *Methods and Protocols,* edited by *Trygve O. Tollefsbol, 2007*
370. **Adhesion Protein Protocols**, *Second Edition,* edited by *Amanda S. Coutts, 2007*
369. **Electron Microscopy:** *Methods and Protocols, Second Edition,* edited by *John Kuo, 2007*
368. **Cryopreservation and Freeze-Drying Protocols**, *Second Edition,* edited by *John G. Day and Glyn Stacey, 2007*
367. **Mass Spectrometry Data Analysis in Proteomics**, edited by *Rune Matthiesen, 2007*
366. **Cardiac Gene Expression:** *Methods and Protocols,* edited by *Jun Zhang and Gregg Rokosh, 2007*
365. **Protein Phosphatase Protocols:** edited by *Greg Moorhead, 2007*
364. **Macromolecular Crystallography Protocols:** *Volume 2, Structure Determination,* edited by *Sylvie Doublié, 2007*
363. **Macromolecular Crystallography Protocols:** *Volume 1, Preparation and Crystallization of Macromolecules,* edited by *Sylvie Doublié, 2007*
362. **Circadian Rhythms:** *Methods and Protocols,* edited by *Ezio Rosato, 2007*
361. **Target Discovery and Validation Reviews and Protocols:** *Emerging Molecular Targets and Treatment Options, Volume 2,* edited by *Mouldy Sioud, 2007*

METHODS IN MOLECULAR BIOLOGY™

PCR Primer Design

Edited by

Anton Yuryev

Application Science Department, Ariadne Genomics Inc., Rockville, MD

HUMANA PRESS ✱ TOTOWA, NEW JERSEY

999 Riverview Drive, Suite 208
Totowa, New Jersey 07512

www.humanapress.com

This publication is printed on acid-free paper. ♾
ANSI Z39.48-1984 (American Standards Institute) Permanence of Paper for Printed Library Materials

Cover illustration: Fig. 2 from Chapter 1 (Reprinted with permission from the Annual Review of Biophysics and Biomolecular Structure, Volume 33 (c) 2004 by Annual Reviews- www.annualreviews.org), Fig. 2 from Chapter 5, Fig. 2 from Chapter 2, and Fig. 6 from Chapter 18.

Production Editor: Rhukea J. Hussain
Cover design by: Nancy K. Fallatt

For additional copies, pricing for bulk purchases, and/or information about other Humana titles, contact Humana at the above address or at any of the following numbers: Tel.: 973-256-1699; Fax: 973-256-8341; E-mail: humana@humanapr.com; or visit our Website: www.humanapress.com

Printed in the United States of America. 10 9 8 7 6 5 4 3 2 1

ISBN 13: 978-1-58829-725-9

eISBN 978-1-59745-528-2

Library of Congress Control Number: 2007925517

Preface

In the past decade, Molecular Biology has been transformed from the art of cloning a single gene to a statistical science measuring and calculating properties of entire genomes. New high-throughput methods have been developed for genome sequencing and studying the cell at different systematic levels such as transcriptome, proteome, metabolome and other "...omes". At the heart of most high-throughput methods is the technique of polymerase chain reaction (PCR). PCR allows amplification of specific DNA sequences from sub-picomole concentrations to amounts sufficient for gene detection and quantification. The gene expression microarray experiments, the construction of cDNA libraries for two-hybrid experiments for studying protein-protein interaction, and the genome-wide genotyping of single nucleotide polymorphism (SNP) are all impossible without PCR. The performance and accuracy of these methods directly depend on the efficiency of the PCR reaction. Therefore, the improvement of the PCR has been a focus of much attention among molecular biologists.

The principal ingredients of the PCR reaction are DNA template, reaction buffer, DNA polymerase, and two primers that determine the specificity of the amplification. All of these ingredients have been thoroughly studied and optimized in the last few years. This book focuses on primer design, which is critical to both the efficiency and the accuracy of the PCR. The necessity of simultaneously amplifying a large variety of DNA sequences for high-throughput experiments yielded novel PCR approaches that are described in this book. Ultimately, primer design strategy is determined by the goal of the PCR method. However, there are basic oligonucleotide properties for which optimal combination is important for the success of any method. These properties are now well-understood and predictable with great accuracy. The availability of the whole-genome sequences allowed the development of highly sophisticated mathematical methods to calculate thousands of primers in order to maximize the efficiency of the amplification. This book contains the description of basic approaches for PCR primer design in addition to specialized methods. They can be used for both genome-scale experiments and for small-scale individual PCR amplifications. This book will be useful for organizations performing whole genome studies, for companies designing instruments that utilize PCR, as well as for individual scientists who routinely use PCR in their research.

Dr. Anton Yuryev
Ariadne Genomics Inc.

Contents

Contributors

REIDAR ANDRESON • *Department of Bioinformatics at University of Tartu, Tartu, Estonia, and Estonian Biocentre, Tartu, Estonia*

ARUN APTE • *Premier Biosoft International, Dove St. San Diego, CA*

TAMÁS ARÁNYI • *Institute of Enzymology, BRC, Hungarian Academy of Sciences, Karolina, Hungary*

JEREMY BUHLER • *Department of Computer Science and Engineering, Washington University in St. Louis, St. Louis, MO, and Department of Genetics, Washington University in St. Louis, St. Louis, MO*

PETER DE RIJK • *Department of Molecular Genetics, Flanders Interuniversity Institute for Biotechnology (VIB), University of Antwerp, Belgium*

JURGEN DEL-FAVERO • *Department of Molecular Genetics, Flanders Interuniversity Institute for Biotechnology (VIB), University of Antwerp, Belgium*

ILKA DETMER • *Konstanz University, Konstanz, Germany*

ROHAN FERNANDES • *Computer Science Department, Rutgers, The State University of New Jersey, Hill Center for the Mathematical Sciences, Frelinghuysen Road, Piscataway, NJ*

JENS GASTER • *Konstanz University, Konstanz, Germany*

WIM GLASSEE • *Department of Molecular Genetics, Flanders Interuniversity Institute for Biotechnology (VIB), University of Antwerp, Belgium*

JIANPING HUANG • *New Jersey Department of Health, Trenton, NJ*

LARS KADERALI • *German Cancer Research Center (dkfz), Theoretical Bioinformatics, Heidelberg, Germany*

THOMAS KÄMPKE • *Forschungsinstitut für anwendungsorientierte Wissensverarbeitung/n FAW/n, Lise-Meitner-Str., Ulm, Germany*

LAURIS KAPLINSKI • *Department of Bioinformatics at University of Tartu, Tartu, Estonia, and Estonian Biocentre, Tartu, Estonia*

DARIO LEISTER • *Department Biologie I, Botanik, Ludwig-Maximilians-Universität München, München, Germany*

LONG-CHENG LI • *Department of Urology, Veterans Affairs Medical Center and University of California, San Francisco, CA*

CHAIM LINHART • *School of Computer Science, Tel Aviv University, ISRAEL*

YVES LE LOIR • *Laboratoire de Microbiologie, UMR1253 STLO INRA Agrocampus Rennes, Rennes Cedex, France*

OWEN MARSHALL • *Chromosome Research, Murdoch Childrens Research Institute, Royal Children's Hospital, Flemington Road, Parkville Victoria, Australia*

ANDREAS MARX • *Konstanz University, Konstanz, Germany*

KENJI ONODERA • *RIKEN Genome Sciences Center, Tsurumi-ku, Yokohama, Japan*

MAIDO REMM • *Department of Bioinformatics at University of Tartu, Tartu, Estonia and Estonian Biocentre, Tartu, Estonia*

WOJCIECH RYCHLIK • *Molecular Biology Insights, Inc., Cascade, CO*

JOHN SANTALUCIA, Jr. • *Department of Chemistry, Wayne State University, Detroit, MI; DNA Software, Inc., Ann Arbor, MI*

RON SHAMIR • *Head, School of Computer Science, Tel Aviv University, Tel Aviv, ISRAEL*

SIDDHARTH SINGH • *Premier Biosoft International, Dove St. San Diego, CA*

STEVEN SKIENA • *Department of Computer Science, State University of New York at Stony Brook, Stony Brook, NY*

RICHARD SOUVENIR • *Department of Computer Science and Engineering, Washington University in St. Louis, St. Louis, MO*

GYAN PRAKASH SRIVASTAVA • *Digital Biology Laboratory, Computer Science Department University of Missouri-Columbia, East Rollin Road, Columbia, MO*

GARY STORMO • *Department of Computer Science and Engineering, Washington University in St. Louis, St. Louis, MO, and Department of Genetics, Washington University in St. Louis, St. Louis, MO*

MICHAEL STRERATH • *Konstanz University, Konstanz, Germany*

GÁBOR E. TUSNÁDY • *Institute of Enzymology, BRC, Hungarian Academy of Sciences, Karolina, Hungary*

CHRISTINE VAN BROECKHOVEN • *Department of Molecular Genetics, Flanders Interuniversity Institute for Biotechnology (VIB), University of Antwerp, Belgium*

CLAUDIO VAROTTO • *Center for the Study of Biodiversity – Trentino, Istituto Agrario di San Michele all'Adige, San Michele all'Adige (TN), Italy*

STEFAN WECKX • *Department of Molecular Genetics, Flanders Interuniversity Institute for Biotechnology (VIB), University of Antwerp, Belgium*

DONG XU • *James C. Dowell Associate Professor, Director, Digital Biology Laboratory, Computer Science Department, Engineering Building West University of Missouri-Columbia, Columbia, MO*

ANTON YURYEV • *Application Science Department, Ariadne Genomics Inc., Rockville, MD*

NOURI BEN ZAKOUR • *Laboratoire de Microbiologie, UMR1253 STLO INRA Agrocampus Rennes, Rennes Cedex, France*

WEIXIONG ZHANG • *Department of Computer Science and Engineering, Washington University in St. Louis, St. Louis, MO, and Department of Genetics, Washington University in St. Louis, St. Louis, MO*

I

Basic Principles and Software for PCR Primer Design

1

Physical Principles and Visual-OMP Software for Optimal PCR Design

John SantaLucia, Jr.

Summary

The physical principles of DNA hybridization and folding are described within the context of how they are important for designing optimal PCRs. The multi-state equilibrium model for computing the concentrations of competing unimolecular and bimolecular species is described. Seven PCR design "myths" are stated explicitly, and alternative proper physical models for PCR design are described. This chapter provides both a theoretical framework for understanding PCR design and practical guidelines for users. The Visual-OMP (oligonucleotide modeling platform) package from DNA Software, Inc. is also described.

Key Words: Thermodynamics; nearest-neighbor model; multi-state model; Visual-OMP; secondary structure; oligonucleotide design; software.

1. Introduction

Single-target PCR is generally regarded as a robust and reliable technique for amplifying nucleic acids. This reputation is well deserved and is a result of the inherent nature of PCR technology, the creativity of a wide variety of scientists and engineers, and the huge financial investment of private industry as well as government funding. An incomplete list of some of the important innovations includes a variety of engineered thermostable polymerases, well-engineered thermocycling instruments, hot-start PCR, exonuclease-deficient polymerases, addition of dimethylsulfoxide (DMSO), buffer optimization, aerosol-blocking pipette tips, and use of uracil DNA glycosylase to minimize contamination artifacts. Despite these innovations and the large investment, there are many

From: *Methods in Molecular Biology, vol. 402: PCR Primer Design*
Edited by: A. Yuryev © Humana Press, Totowa, NJ

aspects of PCR that are still not well understood (such as the detailed kinetic time course of reactions that occur during thermocycling). These gaps in our knowledge result in less-than-perfect design software; the human experts are not perfect either. Nonetheless, there is a series of widely believed myths about PCR that result in poor designs. This chapter is devoted to stating explicitly some of these myths and providing explanations and guidelines for improved PCR design. These principles are fully implemented in the commercial package from DNA Software, Inc. (Ann Arbor, MI, USA) called Visual-OMP (oligonucleotide modeling platform) *(1,2)*. I co-founded DNA Software in year 2000 to implement the advanced thermodynamic prediction methods that were discovered in my academic laboratory as well as the best of what was available in the literature from other laboratories *(2)*. This chapter is organized into a series of sections that provide the background for understanding DNA thermodynamics and sections that specifically address each of seven myths about PCR design.

2. Background: DNA Thermodynamics

The detailed methods for predicting the thermodynamics of DNA folding and hybridization were recently reviewed *(2)*. A full description of solution thermodynamics is beyond the scope of this chapter, but a brief description is given. Review articles on the details of solution thermodynamics of nucleic acids have also been published *(3–5)*. This topic can be difficult and confusing for non-experts and can be the source of many misconceptions about PCR design. However, the serious molecular biologists should be familiar with these topics and should make the effort to educate themselves. This chapter will serve to demystify the topic of DNA thermodynamics and make it clear why thermodynamics is important for PCR design. Such knowledge is crucial for effective use of available software packages.

2.1. Solution Equilibrium and Calculation of the Amount Bound

The process of duplex hybridization for a forward bimolecular reaction is given by

$$A + B \rightarrow AB \tag{1}$$

where A and B imply strands A and B in the random coil state and AB implies the ordered AB duplex state. This is called the two-state approximation (it is assumed that there are no intermediate states). The reverse reaction is unimolecular and is given by

$$A + B \leftarrow AB \tag{2}$$

If enough time elapses, the forward and reverse reaction rates will be equal and equilibrium will be achieved as shown in Eq. 3:

$$A + B \rightleftharpoons AB \tag{3}$$

The equilibrium constant, K, for the reaction is given by the law of mass action:

$$K = \frac{[AB]}{[A][B]} \tag{4}$$

The equilibrium constant is independent of the total strand concentrations, $[Atot]$ and $[Btot]$, which is why it is called a "constant." However, K depends strongly on temperature, salt concentration, pH, [DMSO], and other environmental variables.

Even though K is a constant, if you change the total concentration of one or both of the strands, $[Atot]$ and/or $[Btot]$, then the system will respond to re-achieve the equilibrium ratio given by Eq. 4, which in turn means that the individual concentrations $[A]$, $[B]$, and $[AB]$ will change. This is called "Le Chatelier's Principle." Simply stated, the more the strand A added, the more $[AB]$ will increase. Let us illustrate how Eq. 4 is used. Assume that the equilibrium constant is 1.81×10^6 (we will see later how to predict K at any temperature) and that $[Atot] = [Btot] = 1 \times 10^{-5}$ M (both of which are easily measured by UV absorbance or other technique).

$$K = 1.81 \times 10^6 = \frac{[AB]}{[A][B]} \tag{5}$$

$$[Atot] = 1 \times 10^{-5}\text{M} = [A] + [AB] \tag{6}$$

$$[Btot] = 1 \times 10^{-5}\text{M} = [B] + [AB] \tag{7}$$

Equation 5 is called the "equilibrium equation," and Eqs. 6 and 7 are the "conservation of mass" equations. Notice that there are three equations with three unknowns (namely, $[A]$, $[B]$, and $[AB]$). Such a system of equations can be solved analytically if it is quadratic or numerically if it is higher order ***(6)***. More complex cases are discussed in the section 2.4 concerning multi-state equilibrium. Visual-OMP sets up the equations, solves them automatically and outputs the species concentrations.

Let us consider a simple example to demystify the process of solving the simultaneous equations. Substituting Eqs. 6 and 7 into Eq. 5 so that everything is expressed in terms of [*AB*], we get

$$K = \frac{[AB]}{([Atot]-[AB]) \times ([Btot]-[AB])} \tag{8}$$

$$1.81 \times 10^6 = \frac{[AB]}{(1 \times 10^{-5} - [AB]) \times (1 \times 10^{-5} - [AB])} \tag{9}$$

Equation 8 is then rearranged into the familiar form of the quadratic equation:

$$0 = K\,[AB]^2 - (K\,[Atot] + K\,[Btot] + 1)\,[AB] + K\,[Atot]\,[Btot]$$

$$0 = aX^2 + bX + C \tag{10}$$

As $[AB] = X$, we can solve for [*AB*] using the familiar analytical solution to the quadratic equation from high-school mathematics (concentration must be positive and only one root is positive):

$$X = \frac{-b \pm \sqrt{b^2 - 4ac}}{2a} \tag{11}$$

Note, however, that this equation is sometimes numerically unstable (particularly at low temperatures) ***(6)***. Plugging in the numbers for *K, Atot*, and *Btot* into Eq. 11 gives $[AB] = 7.91 \times 10^{-6}$ M, and using Eqs. 6 and 7, we get $[A] = [B] = 2.09 \times 10^{-6}$ M. This means that 79% of [*Atot*] is in the bound duplex structure, [*AB*]. In performing such calculations, it is useful to qualitatively verify the reasonableness of the result by asking whether the temperature of interest is above or below the melting temperature, T_m (discussed in section 2.3). In this particular example, T_m is 43.4 °C, whereas *K* was determined at 37 °C. As 37 °C is less than T_m, we expect that the amount bound should be more than 50%; indeed 79% bound at 37 °C is consistent with that expectation.

This is an important result because it shows that if the equilibrium constant and total strand concentration are known, then Eqs. 8–11 can be used to compute the amount of primer bound to target [*AB*], which is the quantity that matters for hybridization in PCR and is also directly related to the amount of "signal" in a hybridization assay. How can we predict the equilibrium constant and how does it change with temperature? This leads us to the next section on $\Delta G°$, $\Delta H°$, and $\Delta S°$ and the field of thermodynamics in general.

2.2. The Meaning of ΔH°, ΔS°, and ΔG°_T Parameters

In thermodynamics, there is a crucial distinction between the "system" and the "surroundings." For PCR, the "system" is defined as the contents of the test tube that contains the nucleic acid strands, solvent, buffer, salts, and all the other chemicals. The "surroundings" is defined as the rest of the entire universe. Fortunately, the discoverers of the field of thermodynamics have provided a means by which we do not need to keep track of what is going on in the whole universe, but instead we only need to determine the changes in certain properties of the system alone (namely, ΔH°, ΔS°, and $\Delta G^\circ{}_T$) to determine whether a process is spontaneous and to determine the equilibrium. The process of reaching equilibrium results in the release of heat from the system to the surroundings when strands change from the random coil state to the duplex state. At constant pressure, this change in heat of the system is called the enthalpy change, ΔH. The naught symbol, °, is added (e.g., ΔH°) to indicate that the energy values given are for the idealized "standard state," which simply means that the energy change refers to the amount of energy that would be released if a scientist could prepare each species in 1 M concentration (i.e., $[A] = 1\,\text{M}$, $[B] = 1\,\text{M}$, and $[AB] = 1\,\text{M}$, which is a non-equilibrium condition), mix them, and then allow them to come to equilibrium. The more heat that is released from the reaction to the surroundings, the more disordered the surroundings become and thus the more favorable the reaction is (because of the second law of thermodynamics). As a result of the hybridization reaction, the amount of order in the system also changes (a duplex is more ordered than a random coil because of conformational entropy). In addition, solvent molecules and counterions bind differently to the duplexes and random coils. These effects are all accounted for in the entropy change of the system, ΔS°. The ΔH° and ΔS° are combined to give the Gibb's free energy change for going from random coil to duplex:

$$\Delta G^\circ{}_T = \frac{\Delta H^\circ \times 1000 - T \times \Delta S^\circ}{1000} \qquad (12)$$

where T is the Kelvin temperature, ΔH° is given in kcal/mol, and ΔS° is given in cal/mol K. A slightly more accurate version of this equation would account for the change in heat capacity, ΔC_p, which has been described in detail ***(4,5)***. Importantly, there is a relationship between the Gibb's free energy change at temperature T and the equilibrium constant at temperature T:

$$\Delta G^\circ_T = -RT \times \ln(K) \qquad (13)$$

where R is the gas constant (equals 1.9872 cal/mol K). A good rule of thumb for the qualitative meaning of $\Delta G°_T$ is: "At 25 °C, every −1.4 kcal/mol in $\Delta G°_{25}$ results in a change in the equilibrium constant by a multiplicative factor of 10" (due to Eq. 13). Thus, a $\Delta G°_{25}$ of −4.2 kcal/mol equals 3*-1.4 and thus K equals $10^3 = 1000$.

Equation 13 provides the critical link that allows for the equilibrium constant to be computed from $\Delta G°_T$. Next, $\Delta G°$ can be computed at any temperature T, if $\Delta H°$ and $\Delta S°$ are known by using Eq. 12. Thus, if $\Delta G°$ is known at all temperatures, then K can be computed at all temperatures and thus the concentrations of all species can be computed at all temperatures as described in Eqs. 5–11. All these statements mean that given $\Delta H°$ and $\Delta S°$, we can compute the concentration distribution for all species at all temperatures. This is illustrated in **Fig. 1**. Certainly, this is more powerful than simple computation of T_m! Where do we get the $\Delta H°$ and $\Delta S°$? They are accurately predicted from the strand sequences involved in the duplex by applying the nearest-neighbor (NN) model. The details of how to practically apply the NN model has been presented elsewhere *(2,7)*.

In addition to the NN predictions of Watson–Crick base-paired duplexes, my laboratory has published the empirical equations that allow the NN model to be extended to include salt dependence *(7)*, terminal dangling ends *(8)*, and all

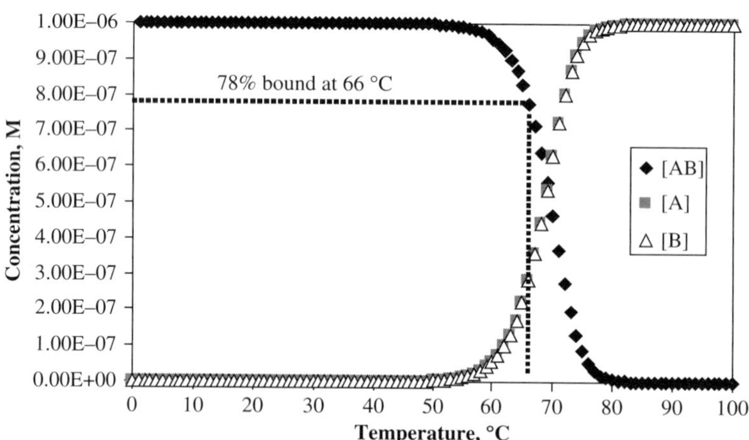

Fig. 1. Simulation of the hybridization profile for a simple non-self-complementary two-state transition, given only $\Delta H°$ and $\Delta S°$ and using Eqs. 5–13, as described in the text. Note that the percent bound can be determined at any temperature. The random coil concentrations of strands A (squares) and B (triangles) are superimposed.

possible internal *(9,10)* and terminal mismatches (S. Varma and J. SantaLucia, unpublished results). These motifs are shown in **Fig. 2**. The availability of the dangling-end parameters is important, because in PCR, the short primers bind to the longer target DNA and the first unpaired nucleotides of the target sequence next to the 5′ and 3′ of the primer-binding site contributes significantly to binding and cannot be neglected. In some instances, a dangling end can contribute as much as a full AT base pair. The mismatch parameters are important because they allow for the T_m of mutagenic primers to be accurately accounted for and for the specificity of hybridization to be computed. The availability of the salt dependence allows for the accurate prediction of

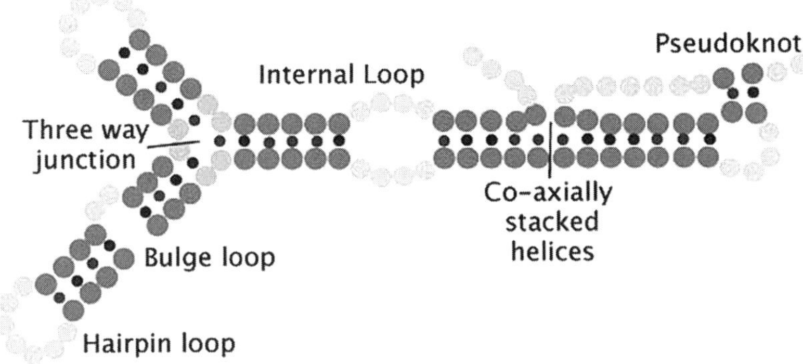

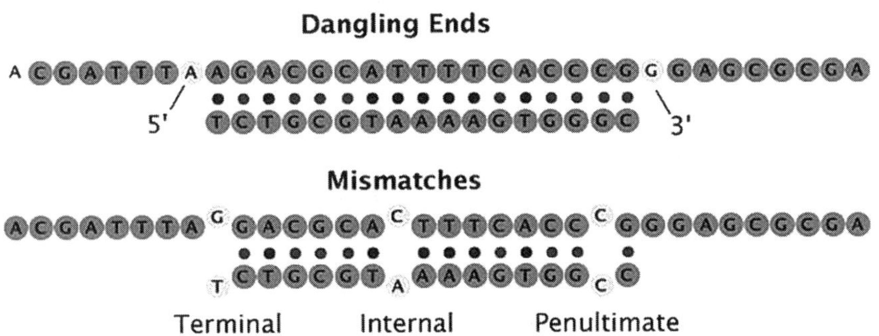

Fig. 2. Structural motifs that occur in folded DNA (top) and in bimolecular duplexes (bottom).

thermodynamics under a wide variety of solution conditions that occur in biological assays, including PCR. At DNA Software, Inc., further empirical equations have been measured (under NIH SBIR funding) for magnesium, DMSO, glycerol, formamide, urea, many fluorophores, and many modified nucleotides including PNA, LNA, morpholino, phosphorothioate, alkynyl pyrimidines, the universal pairing base inosine *(11)*, and others (S. Morosyuk and J. SantaLucia, unpublished results). For several PCR applications, the parameters for PNA, LNA, and inosine (among others) are important and unique to Visual-OMP. We have also determined complete parameters for DNA–RNA hybridization including mismatches, salt dependence, and dangling ends (M. Tsay, S. Morosyuk, and J. SantaLucia, unpublished results), which is useful for the design of reverse-transcription PCR and hybridization-based assays.

2.3. Computation of T_m from ΔH° and ΔS°

By combining Eqs. 12 and 13, one can derive the following expression:

$$T = \frac{\Delta H^\circ \times 1000}{\Delta S^\circ - R\ln(K)} \quad (14)$$

At the T_m, the equilibrium constant is determined by the fact that half the strands are in the duplex state and half are in random coil. For unimolecular transitions, such as hairpin formation or more complex folding (as observed in single-stranded PCR targets), $K = 1$ at the T_m, and Eq. 14 reduces to

$$T_m = \frac{\Delta H^\circ \times 1000}{\Delta S^\circ} - 273.15 \quad (15)$$

For self-complementary duplexes, $K = 1/[Atot]$ at the T_m, and T_m is given by

$$T_m = \frac{\Delta H^\circ \times 1000}{\Delta S^\circ + R\ln([Atot])} - 273.15 \quad (16)$$

For non-self-complementary duplexes in which $[Atot] \geq [Btot]$, $K = 1/([Atot] - [Btot]/2)$, and thus T_m (in Celsius degrees) is given by

$$T_m = \frac{\Delta H^\circ \times 1000}{\Delta S^\circ + R\ln\left([Atot] - \frac{[Btot]}{2}\right)} - 273.15 \quad (17)$$

where $[Atot]$ is the total molar strand concentration of the strand that is in excess (typically the primer) and $[Btot]$ is the molar concentration of the strand that is lower in concentration (typically the target strand). In Eq. 17, if $[Atot] = [Btot]$, then it is easy to derive that the $[A] - ([B]/2)$ term equals $Ct/4$, where $Ct = [Atot] + [Btot]$.

Importantly, all the T_m equations above apply only to "two-state transitions" (i.e., the molecules that form only random coil and duplex states), and they do not apply to transitions that involve intermediate partially folded or hybridized structures. For such multi-state transitions, the definition of the T_m changes to: The temperature at which half of a particular strand (usually the lower concentration strand, which is the target in PCR) forms a particular structure (e.g., duplex hybrid) and the remainder of the strands of that limiting strand form all other intermediates and random coil. Sometimes, the T_m is undefined because there is no temperature at which half the strands form a particular structure.

2.4. Multi-State Coupled Equilibrium Calculations

The principle of calculating the amount bound for a two-state transition was described in **Subheading 2.1**. The two-state model (*see* Eq. 3), however, can be deceptive because there are often many equilibria that can compete with the desired equilibrium (*see* **Fig. 3**). In addition to target secondary structure folding, other structural species can also form folded primer, mismatch hybridization, and primer homodimers (and primer heterodimers when more than one primer is present as is typical in PCR). It is desirable to compute the concentrations of all the species for such a coupled multi-state system. This can be accomplished by generalizing the approach described above for the two-state case (*see* Eqs. 5–11).

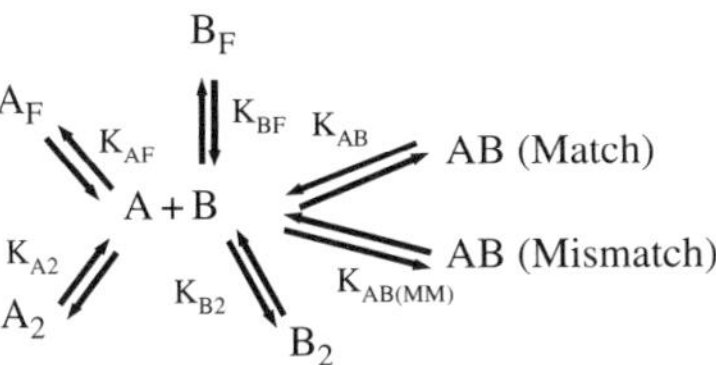

Fig. 3. Seven-state model for hybridization (*AB* match) with competing equilibria for unimolecular folding (A_F and B_F), homodimers (A_2 and B_2), and mismatch hybridization (*AB* mismatch). By Le Chatelier's principle, the presence of the competing equilibria will decrease the concentration of *AB* (match). To compute the concentrations of all the species, a numerical approach is used (described in the text).

We now consider the computation of the amount bound for the case of non-two-state transitions, a situation that is typical in PCR.

$$K_{AB} = \frac{[AB]}{[A][B]} \tag{18}$$

$$K_{AB(MM)} = \frac{[AB(MM)]}{[A][B]} \tag{19}$$

$$K_{AF} = \frac{[A_F]}{[A]} \tag{20}$$

$$K_{BF} = \frac{[B_F]}{[B]} \tag{21}$$

$$K_{A2} = \frac{[A_2]}{[A]^2} \tag{22}$$

$$K_{B2} = \frac{[B_2]}{[B]^2} \tag{23}$$

$$[Atot] = [A] + [A_F] + 2[A_2] + [AB] + [AB(MM)] \tag{24}$$

$$[Btot] = [B] + [B_F] + 2[B_2] + [AB] + [AB(MM)]. \tag{25}$$

Notice that Eqs. 18–25 give a total of eight equations with eight unknowns (A, B, A_F, B_F, A_2, B_2, AB, and *AB(MM)*), which can be solved numerically to give all the species concentrations at equilibrium. Furthermore, we can predict the $\Delta H°$ and $\Delta S°$ for each of the reactions in **Fig. 3** using the NN model and loop parameters *(2)*, and thus, we can compute $\Delta G°$ at all temperatures, and thus all the K's at all temperatures, and those can be solved to give the concentrations of all species at all temperatures. This allows us to produce a multi-state graph of the concentration of all the species as a function of temperature (*see* **Fig. 4**). Two other concepts that arise from the coupled multi-state equilibrium formalism are "net T_m" and "net $\Delta G°$" [also known as $\Delta G°$(*effective*)], which were described elsewhere *(2)*. The net T_m is simply the temperature at which half the strands form a desired species (which is sometimes undefined). Qualitatively, the net $\Delta G°$ is simply the value of $\Delta G°$ that would give the observed equilibrium [*AB*], if all the other species were lumped together and called random coil (to make the process appear to be two state). This can be visualized with the following expression:

$$XA + XB \rightleftharpoons AB \tag{26}$$

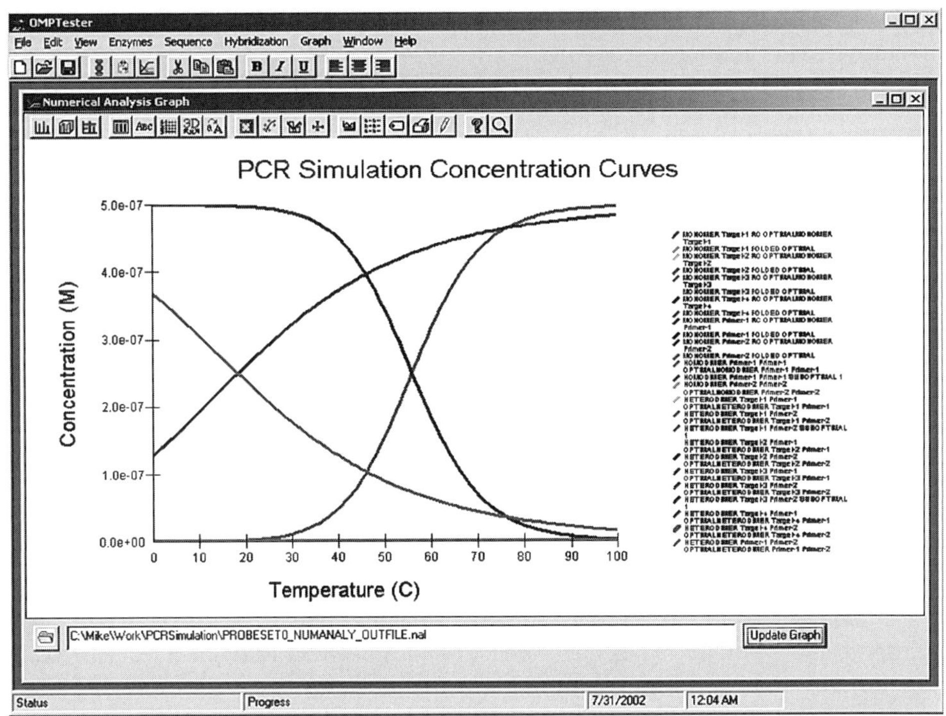

Fig. 4. Graphic output of Visual-OMP of the multi-state numerical analysis results for a PCR. Primer-target duplexes, primer homodimers, primer heterodimers, primer hairpins, and random coil concentrations are given at all temperatures. In this particular simulation, one of the primers hybridizes with a net T_m of 59°C, whereas the other primer has a net T_m of 20°C. Both primers have two-state T_m's above 60°C, but one primer fails due to competing target secondary structure. All the other species (primer dimers and suboptimal structures) were calculated but found to have very low concentration in this example (on the baseline).

where *XA* is the sum of concentrations of all species involving strand *A* except *AB*. The value of [*XA*] is equal to [*Atot*] – [*AB*]. This results in expressions for *K*(*effective*) and Δ*G*°(*effective*):

$$K_{AB}(effective) = \frac{[AB]}{([Atot]-[AB]) \times ([Btot]-[AB])} \tag{27}$$

$$\Delta G^{\circ}_{T}(effective) = -RT \times \ln(K(effective)) \tag{28}$$

The method to compute net Δ*G*° is to perform a special sum of the individual Δ*G*° for all the species, which are each weighted by their concentration values.

Such a procedure is equivalent to a partition function approach. The usefulness of the net $\Delta G°$ is that it is related to what would be observed experimentally if we were to make a measurement of a non-two-state system and yet fit the binding curve with the assumption that the system is two state, which would yield an "observed $\Delta G°$." We demonstrated the accuracy of this approach for molecular beacons that have competing hairpin, random coil, and duplex structures. The results completely validate this approach (*see* Table 6 in **ref. 2**).

The important concept here is that the solution method given is totally general (there are just more equations analogous to those given in Eqs. 18–25), so that the Visual-OMP software is scalable and can handle complex reaction mixtures, as occurs in multiplex PCR and still effectively compute the equilibrium concentrations of all species at any desired temperature.

3. Myths and Improved Methods for PCR Design

3.1. Myth 1: PCR Nearly Always Works and Design Is Not that Important

It might come as a surprise to many that despite the wide use and large investment, PCR in fact is still subject to many artifacts and environmental factors and is not as robust as would be desirable. Many of these artifacts can be avoided by careful oligonucleotide design. Over the last 10 years (1996–2006), I have informally polled scientists who are experts in PCR and asked: "What percentage of the time does a casually designed PCR reaction 'work' without any experimental optimization?" In this context, "work" means that the desired amplification product is made in good yield with a minimum of artifact products such as primer dimers, wrong amplicons, or inefficient amplification. By "casually designed," I mean that typical software tools are used by an experienced molecular biologist. The consensus answer is 70–75%. If one allows for optimization of the annealing temperature in the thermocycling protocol (e.g., by using temperature gradient optimization), magnesium concentration optimization, and primer concentration optimization, then the consensus percentage increases to 90–95%. What is a user to do, however, in the 5–10% of cases where single-target PCR fails? Typically, they redesign the primers (without knowledge of what caused the original failure), resynthesize the oligonucleotides, and retest the PCR. Such a strategy works fine for laboratories that perform only a few PCRs. Once a particular PCR protocol is tested, it is usually quite reproducible, and this leads to the feeling that PCR is reliable. Even the 90–95% of single-target PCRs that "work" can be improved by using good design principles, which increases the sensitivity, decreases the background amplifications, and requires less experimental optimization. In a high-throughput industrial-scale environment, however, individual optimization of each PCR, redesigning failures, performing individualized thermocycling and buffer conditions, and tracking all these is a nightmare logistically and leads to non-uniform success. In multiplex PCR, all the targets are obviously amplified under the same solution and temperature cycling conditions, so there is no possibility of doing individual optimizations. Instead, it is desirable to have the capability to automatically design PCRs that work under a single general set of conditions without any optimization, which would enable parallel PCRs (e.g., in 384-well format) to be performed under the same buffer conditions and thermocycling protocol. Such robustness would further improve reliability of PCR in all applications but particularly in non-laboratory settings such as hospital clinics or field-testing applications.

Shortly after the discovery of PCR, software for designing oligonucleotides was developed ***(12)***. Some examples of widely used primer design software (some of which are described in this book) include VectorNTI, OLIGO ***(12)***, Wisconsin GCG, Primer3 ***(13)***, PRIMO ***(14)***, PRIDE ***(15)***, PRIMERFINDER (http://arep.med.harvard.edu/PrimerFinder/PrimerFinderOverview.html), OSP ***(16)***, PRIMERMASTER ***(17)***, HybSIMULATOR ***(18)***, and PrimerPremiere. Many of these programs do incorporate novel features such as accounting for template quality ***(14)*** and providing primer predictions that are completely automated ***(14,15)***. Each software package has certain advantages and disadvantages, but all are not equal. They widely differ in their ease-of-use, computational efficiency, and underlying theoretical and conceptual framework. These differences result in varying PCR design quality. In addition, there are standalone Web servers that allow for individual parts of PCR to be predicted, notably DNA-MFOLD by Michael Zuker (http://www.bioinfo.rpi.edu/applications/mfold/old/dna/) and HYTHER by my laboratory (http://ozone3.chem.wayne.edu).

3.1.1. Why Is There a Need for Primer Design Software?

DNA hybridization experiments often require optimization because DNA hybridization does not strictly follow the Watson–Crick pairing rules. Instead, a DNA oligonucleotide can potentially pair with many sites on the genome with perhaps only one or a few mismatches, leading to false-positive results. In addition, the desired target sites of single-stranded genomic DNA or mRNA are often folded into stable secondary structures that must be unfolded to allow an oligonucleotide to bind. Sometimes, the target folding is so stable that very little probe DNA binds to the target, leading to a false-negative test. Various other artifacts include probe folding and probe dimerization. Thus, for DNA-based diagnostics to be successful, there is a need to fully understand the science underlying DNA folding and match versus mismatch hybridization. Achieving this goal has been a central activity of my academic laboratory as well as DNA Software, Inc.

3.2. Myth 2: Different Methods for Predicting Hybridization T_m Are Essentially Equivalent in Accuracy

The melting temperature, T_m, of duplex formation is usually defined as the temperature at which half the available strands are in the double-stranded state (or folded state for unimolecular transitions) and half the strands are in the "random coil" state. We will see later (i.e., myth 3) that this definition is not general and that the T_m itself is not particularly useful for PCR design. Over the past 45 years (1960–2005), there have been a large number of alternative methods for predicting DNA duplex T_m that have been published. The simplest equation based on base content is the "Wallace rule" ***(19)***:

$$T_m = 4(G + C) + 2(A + T) \tag{29}$$

This equation neglects many important factors: T_m is dependent on strand concentration, salt concentration, and base sequence. Typical error for this simple method compared with experimental T_m is greater than 15 °C, and thus this equation is not recommended. A somewhat more advanced base content model is given by ***(20,21)***:

$$T_m = 81.5C + 16.6 \times \log10[Na^+] + 0.41(\%G + C) – 0.63(\%\text{formamide}) – 600/L \tag{30}$$

where L is the length of the hybrid duplex in base pairs. Maxim Frank-Kamenetskii provided a more accurate polymer salt dependence correction in 1971 ***(22)***. Nonetheless, Eq. 30 was derived for polymers, which do not include bimolecular initiation that is present in oligonucleotides, does not account for sequence dependent effects, and does not account for terminal end effects that are present in oligonucleotide duplexes ***(7)***. Thus, this equation works well for DNA polymers, where sequence-dependent effects are averaged out, and long duplexes (greater than 40 base pairs) but breaks down for short oligonucleotide duplexes that are typically used for PCR. Both these simple equations are inappropriate for PCR design. Further work suggested (wrongly) that the presence of mismatches in DNA polymers can be accounted for by decreasing the T_m by 1 °C for every 1% of mismatch present in the sequence ***(20,23)***. Based on a comprehensive set of measurements from my laboratory ***(9)***, we now know that this is highly inaccurate (mismatch stability is very sequence dependent), and yet some commercial packages continue to use it.

The appropriate method for predicting oligonucleotide thermodynamics is the NN model *(2,7)*. The NN model is capable of accounting for sequence-dependent stacking as well as bimolecular initiation. As of 1996, there were at least eight sets of NN parameters of DNA duplex formation in the literature, and it was not until 1998 that the different parameter sets were critically evaluated and a "unified NN set" was developed *(7)*. Several groups ***(7,24)*** came to the same conclusion that the 1986 parameters ***(25)*** are unreliable. Unfortunately, the wrong 1986 parameters are still present in some of the most widely used packages for PCR design (namely, Primer3, OLIGO, and VectorNTI). **Table 1** compares the quality of predictions for different parameter sets.

The results in **Table 1** clearly demonstrate that the 1986 NN set is unreliable and that the PCR community should abandon their use. The fact that many scientists have used these inaccurate parameters to design successful PCRs is a testament to the robustness of single-target PCR and the availability of optimization of the annealing temperature in PCR to improve amplification efficiency despite wrong predictions (*see* myth 1). However, as soon as one tries to use the old parameters to design more complicated assays such as multiplex PCR, real-time PCR, and parallel PCRs, then it is observed that the old parameters fail badly. The use of the "unified NN parameters," on the contrary, results in much better PCR designs with more predictable annealing behavior and thereby enables high-throughput PCR applications and also multiplex PCR.

Table 1
Average T_m Deviation (Experiment-Predicted) for Different Software Packages (Delta T_m Given in °C)

Database	OMP	Vector NTI 7.0	Oligo 6.7
46 sequences 1 M NaCl	1.78	8.99	6.10
20 sequences 0.01–0.5 M NaCl	2.29	15.30	8.32
16 sequences with mismatches	1.44	NP	7.27
4 sequences with competing target structure	3.10	NP	22.1[a]

NP, calculation not possible with Vector NTI.

[a] Oligo cannot predict target folding so the number given is for the hybridization neglecting target folding.

Furthermore, the unified NN parameters were extended by my laboratory to allow for accurate calculation of mismatches, dangling ends, salt effects, and other secondary structural elements, all of which are important in PCR *(2)*.

3.3. Myth 3: Designing Forward and Reverse Primers to Have Matching T_m's Is the Best Strategy to Optimize for PCR

Nearly all "experts" in PCR design would claim to believe in myth 3. Most current software packages base their design strategy on this myth. Some careful thought, however, quickly reveals the deficiencies of that approach. The T_m is the temperature at which half the primer strands are bound to target. This provides intuitive insight for very simple reactions, but it does not reveal the behavior (i.e., the amount of primer bound to target) at the annealing temperature. The PCR annealing temperature is typically chosen to be 10°C below the T_m. However, different primers have different $\Delta H°$ of binding, which results in different slopes at the T_m of the melting transition. Thus, the hybridization behavior at the T_m is not the same as the behavior at the annealing temperature. The quantity that is important for PCR design is the amount of primer bound to target at the annealing temperature. Obtaining equal primer binding requires that the solution of the equilibrium equations as discussed in **Subheading 2.1**. If the primers have an equal concentration of binding, then they will be equally extended by DNA polymerase, resulting in efficient amplification. This principle is illustrated in **Fig. 5**. The differences in primer binding are amplified with each cycle of PCR, thereby reducing the amplification efficiency and providing opportunity for artifacts to develop. The myth of matched T_m's is thus flawed. Nonetheless, as single-target PCR is fairly robust, such inaccuracies are somewhat tolerated, particularly if one allows for experimental optimization of the temperature cycling protocol for each PCR. In multiplex and other complex assays, however, the design flaws from matched T_m's become crucial and lead to failure.

An additional problem with using two-state T_m's for primer design is that they do not account for the rather typical case where target secondary structure competes with primer binding. Thus, the two-state approximation is typically invalid for PCR, and thus the two-state T_m is not directly related to the actual behavior in the PCR. The physical principle that does account for the effects of competing secondary structure, mishybridization, primer dimers, and so on is called "multi-state equilibrium," as described in **Subheading 2.4**.

Below an alternative design strategy is suggested in which primers are carefully designed so that many PCRs can be made to work optimally at a single PCR condition, thereby enabling high-throughput PCR without the need

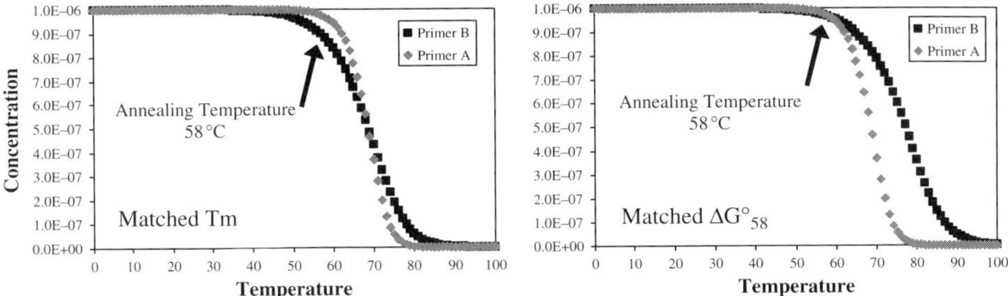

Fig. 5. Illustration of hybridization profiles of primers with two different design strategies. In the left panel, the T_m's are matched at 68.6 °C, but at the annealing temperature of 58 °C, primer B (squares) binds 87% and primer A (diamonds) binds 97%. This would lead to unequal hybridization and polymerase extension, thus reducing the efficiency of PCR. In the right panel, the $\Delta G°$ at 58 °C of the two primers is matched by redesigning primer B. The result is that both primers are now 97% bound, and thus optimal PCR efficiency would be observed. Notice that the T_m's of the two primers are not equal in the right panel.

for temperature optimization. This robust strategy also lays the foundation for designing multiplex PCR with uniform amplification efficiency in which one must perform all the amplification reactions at the same temperature.

3.3.1. Application of the Multi-State Model to PCR Design

A typical single-stranded DNA target is not "random coil" nor do targets form a linear conformation as cartoons describing PCR often show (*see* **Fig. 6**). Instead, target DNA molecules (and also primers sometimes) form stable secondary structure (*see* **Fig. 7**). In the case of RNA targets, which are important for reverse-transcription PCR, the RNAs may be folded into secondary and tertiary structures that are much more stable than a typical random DNA sequence. If the primer is designed to bind to a region of the target DNA

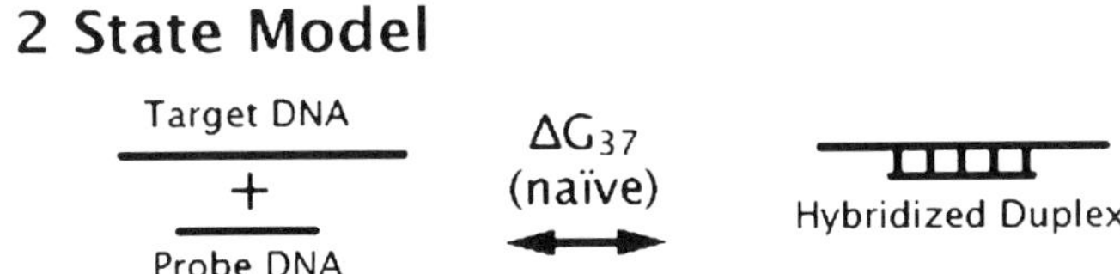

Fig. 6. The two-state model for duplex hybridization. The single-stranded target and probe DNAs are assumed to be in the random coil conformation.

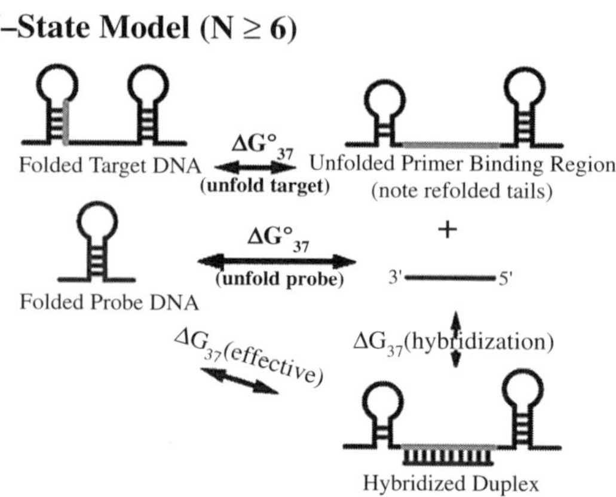

Fig. 7. The multi-state model for the coupled equilibrium involved in DNA hybridization. Most software only calculates the two-state thermodynamics (vertical transition). The competing target and primer structures, however, significantly affect the effective thermodynamics (diagonal transition). Note that $\Delta G°(effective)$ is not the simple sum of $\Delta G°(unfolding)$ and $\Delta G°(hybridization)$, but instead the sum must be weighted by the species concentrations, which can only be obtained by solving the coupled equilibria for the given total strand concentrations. Note that a more precise model would also include competing equilibria for primer dimerization and mismatch hybridization.

or RNA that is folded, then the folding must be broken before the primer can bind (*see* **Fig. 7**). This provides an energetic barrier that slows down the kinetics of hybridization and also makes the equilibrium less favorable toward binding. This can result in the complete failure of a PCR or hybridization assay (a false-negative test). Thus, it is desirable to design primers to bind to regions of the target that are relatively free of secondary structure. DNA secondary structure can be predicted using the DNA-MFOLD server or using Visual-OMP as described in our review *(2)*. Simply looking at a DNA secondary structure does not always obviously reveal the best places to bind a primer. The reason why the hybridization is more complex than expected is revealed by some reasoning. Primer binding to a target can be thought to occur in three steps: (1) the target partially unfolds, (2) the primer binds, and (3) the remainder of the target rearranges its folding to accomplish a minimum energy state. The energy required to unfold structure in **step 1** can sometimes be partially compensated by the structural rearrangement energy from **step 3** (as shown in **Fig. 7**). Such rearrangement energy will help the equilibrium to be more favorable toward hybridization, but the kinetics of hybridization will still be slower than what would occur in a comparable open-target site. Note that the bimolecular structure shown in **Fig. 7** shows the tails of the target folded. Visual-OMP allows the prediction of such structures, which is accomplished by a novel bimolecular dynamic programming algorithm.

To compute the equilibrium binding is also not obvious in multi-state reactions. We recommend solving the coupled equilibria for the concentrations of all the species. This is best done numerically as described in **Subheading 2.4**. The only PCR design software currently available that can solve the multi-state coupled equilibrium is Visual-OMP. Some recent work by Zuker *(26)* with partition functions is also applicable to the issue of multi-state equilibrium but to date has not been integrated into an automated PCR design software package. These considerations make the choice of the best target site non-obvious. Mismatch hybridization to an unstructured region can sometimes be more favorable than hybridization at a fully match site that is folded, thereby resulting in undesired false priming artifacts in PCR. Perhaps, it will come as a surprise to some that secondary structure in the primer is beneficial for specificity but harmful to binding kinetics and equilibrium. A practical way to overcome the complexity problem is to simply simulate the net binding characteristics of all oligos of a given length along the target—this is called "oligo walking." Oligo walking is automatically done in the PCR design module of Visual-OMP and is also available in the RNA-STRUCTURE software from David Matthews and Douglas Turner *(27,28)*.

3.4. Myth 4: "Primer Dimer" Artifacts Are Due to Dimerization of Primers

A common artifact in PCR is the amplification of "primer dimers." The most common conception of the origin of primer dimers is that two primers hybridize at their 3′-ends (*see* **Fig. 8**). DNA polymerase can bind to such species and extend the primers in both directions to produce an undesired product with a length that is slightly less than the sum of the lengths of the forward and reverse primers. This mechanism of primer dimerization is certainly feasible and can be experimentally demonstrated by performing thermocycling in the absence of target DNA. This mechanism can also occur when the desired amplification of the target is inefficient (e.g., when one of the primers is designed to bind to a region of the target that is folded into a stable secondary structure). Therefore, most PCR design software packages check candidate primers for 3′-complementarity and redesign one or both of them if the thermodynamic

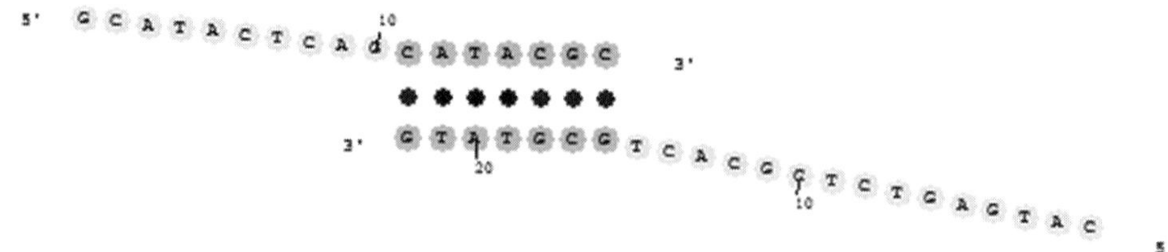

Fig. 8. Primer dimer hybridized duplex. Note that the 3′-ends of both primers are extensible by DNA polymerase.

stability of the hybrid is above some threshold. Another practical strategy to reduce primer dimer formation is to design the primer to have the last two nucleotides as AA or TT, which reduces the likelihood of a primer dimer structure with a stable hybridized 3′-end ***(29)***. For single-target PCR, two primers are present (FP and RP), and there are three different combinations of primer dimers that are possible FP–FP, RP–RP, and FP–RP. For multiplex PCR with *N* primers, there are NC2 pairwise combinations that are possible, and it becomes harder to redesign the primers so that all of them are mutually compatible. This becomes computationally challenging for large-scale multiplexing. However, such computer optimization is only partially effective at removing the primer dimer artifacts in real PCRs. Why?

3.4.1. An Alternative Mechanism for Primer Dimer Artifacts

There are some additional observations that provide clues for an alternative mechanism for primer dimerization.

1. Generally, homodimers (i.e., dimers involving the same strand) are rarely observed.
2. Primer dimer artifacts typically occur at a large threshold cycle number (usually > 35 cycles), which is higher than the threshold cycle number for the desired amplicon.
3. Primer dimers increase markedly when heterologous genomic DNA is added.
4. Primer dimers are most often observed when one or both of the primers bind inefficiently to the target DNA (e.g., due to secondary structure of the target or weak thermodynamics).
5. When the primer dimers are sequenced, there are often a few extra nucleotides of mysterious origin in the center of the dimer amplicon.

Observations 1 and 2 suggest that DNA polymerase does not efficiently bind to or extend primer duplexes with complementary 3′-ends. Observation

2 could also be interpreted as meaning that the concentration of the primer duplex is quite low compared with the normal primer-target duplex. In the early stages of PCR, however, observations 3 and 5 suggest that background genomic DNA may play a role in the mechanism of primer dimer formation. Observation 4 suggests that primer dimerization needs to occur in the early rounds of PCR to prevent the desired amplicon from taking over the reactions in the test tube. **Figure 9** illustrates a mechanism that involves the genomic DNA in the early cycles of PCR and that provides an explanation for all five observations.

The mechanism presented in **Fig. 9** can also be checked for by computer, but searching for such a site in a large genome can be quite computationally demanding. The ThermoBLAST algorithm developed by DNA Software, Inc. can meet this challenge (*see* myth 5).

3.4.2. Additional Concerns for Primer Dimers

Two primers can sometimes hybridize using the 5′ end or middle of the sequences. Such structures are not efficiently extensible by DNA polymerase. Such 5′-end primer hybrids, however, can in principle affect the overall equilibrium for hybridization, but generally, this is a negligible effect that is easily minimized by primer design software (i.e., if a primer is predicted to form a significant interaction with one of the other primers, then one or both of the primers are redesigned to bind to a shifted location on the target). If a polymerase is used that has exonuclease activity (e.g., *Pfu* polymerase), then it is possible that hybridized structures that would normally be non-extensible might be chewed back by the exonuclease and create an extensible structure. Indeed, it is observed that PCRs done with enzymes that have exonuclease

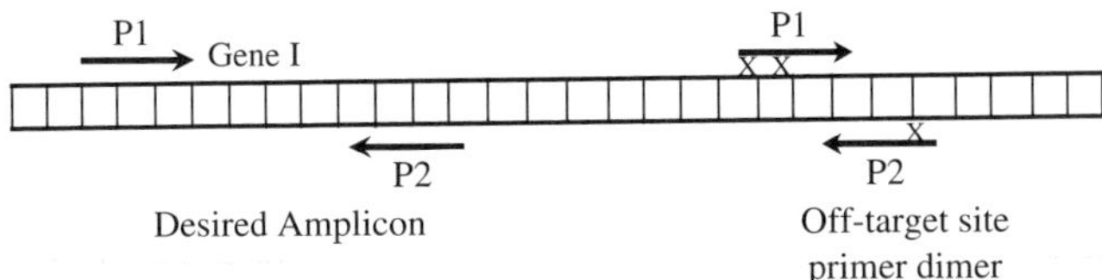

Fig. 9. Genomic DNA can participate in the creation of both the desired amplicon and the primer dimerization artifact. Notice that despite the presence of a few mismatches, denoted by "x," the middle and 5′-ends of the primers are able to bind to the target stronger than they would bind to another primer molecule. Note that this mechanism does not require very strong 3′-end complementarity of the primers P1 and P2. Instead, this mechanism requires that sites for P1 and P2 are close to each other.

activity have a much higher incidence of primer dimer formation and mishybridization artifacts. Thus, for PCR, "proofreading" activity can actually be harmful.

3.5. Myth 5: A BLAST Search Is the Best Method for Determining the Specificity of a Primer

To minimize mispriming, several PCR texts suggest performing a BLAST search, and such capability is a part of some primer design packages such as GCG and Vector NTI and Visual-OMP. However, a BLAST search is not the appropriate screen for mispriming because sequence identity is not a good approximation to duplex thermodynamics, which is the proper quantity that controls primer binding. For example, BLAST scores a GC and an AT pair identically (as matches), whereas it is well known that base pairing in fact depends on both the G + C content and the sequence, which is why the NN model is most appropriate. In addition, different mismatches contribute differently to duplex stability. For example, a G − G mismatch contributes as much as −2.2 kcal/mol to duplex stability at 37 °C, whereas a C − C mismatch can destabilize a duplex by as much as +2.5 kcal/mol. Thus, mismatches can contribute ΔG° over a range of 4.7 kcal/mol, which corresponds to factor of 2000 in equilibrium constant. In addition, the thermodynamics of DNA–DNA duplex formation are quite different than that of DNA–RNA hybridization. Clearly, thermodynamic parameters will provide better prediction of mispriming than sequence similarity. BLAST also uses a minimum 8 nt "word length," which must be a perfect match; this is used to make the BLAST algorithm fast, but it also means that BLAST will miss structures that have fewer than eight consecutive matches. As GT, GG, and GA mismatches are stable and occur commonly when a primer is scanned against an entire genome, such a short word length can result in BLAST missing thermodynamically important hybridization events. BLAST also does not properly score the gaps that result in bulges in the duplexes. DNA Software, Inc. is developing a new algorithm called ThermoBLAST that retains the computational efficiency of BLAST so that searches genomic can be accomplished rapidly but uses thermodynamic scoring for base pairs, dangling end, single mismatches, bulges, tandem mismatches, and other motifs. **Figure 10** gives some examples of strong hybridization that would be missed by BLAST but detected by ThermoBLAST. The computational efficiency of ThermoBLAST is accomplished using a variant of the bimolecular dynamic programming algorithm that was invented at DNA Software, Inc.

```
GCCCCCAACCTCCGTGGG      GGGCCTGCC-CCCAGG      AGCTCGCAGTGCACCAC
    XX     X                X                X  X    X  X
CGGGGGAGGGAGGCGCCC      CCCGGGCGGAGGGTCC      GCGGGCGGCAGGTGGTG
```

Fig. 10. Three hybridized structures that BLAST misses due to the word length limit of eight. All the structures shown are thermodynamically stable under typical PCR buffer conditions. Note the mismatches (denoted by "x") and bulges (denoted by a gap in the alignment).

3.6. Myth 6: At the End of PCR, Amplification Efficiency Is Not Exponential Because the Primers or NTPs Are Exhausted or the Polymerase Looses Activity

PCR amplification occurs with a characteristic "S" shape. During the early cycles of PCR, the amplification is exponential. During the later stages of PCR, saturation behavior is observed, and the efficiency of PCR decreases with each successive cycle. What is the physical origin of the saturation and why is the explanation important for PCR design? Most practitioners of PCR believe that saturation is observed because either the primers or the NTPs are exhausted or the polymerase looses activity. The idea of lost polymerase activity is historical. In the early days of PCR, polymerase enzymes did loose activity with numerous cycles of PCR. Modern thermostable engineered polymerases, however, are quite robust and exhibit nearly full activity at the end of a typical PCR. The idea of one or more of the NTPs or primers being limiting reagents is perfectly logical and consistent with chemical principles but is not correct for the concentrations that are usually used in PCR. Chemical analysis of the PCR mixture reveals that at the end of PCR there is usually plenty of primers and NTPs so that PCR should continue for further cycles before saturation is observed due to consumption of a limiting reagent. Experimentally, if you double the concentration of the primers, you do not observe twice the PCR product. Thus, the amplicon yield of PCR is usually less than predicted based on the primer concentrations. What is causing the PCR to saturate prematurely? The answer is that double-stranded DNA is an excellent inhibitor of DNA polymerase. This can be demonstrated experimentally by adding a large quantity of non-extensible "decoy" duplex DNA to a PCR and comparing the result to a PCR without the added duplex. The result clearly shows that the reaction with added duplex DNA shows little or no amplification while the control amplifies normally. The reason why duplex DNA inhibits DNA polymerase is that the polymerase binds to the duplex rather than binding to the small quantity of duplex arising from the primers binding to target strands during the early cycles of PCR.

3.6.1. Application of the Inhibition Principle to Multiplex PCR Design

The concept of amplicon inhibition of PCR is particularly important for multiplex PCR design. Consider a multiplex reaction in which there are plenty of NTPs available. It is expected that if one of the amplicons is produced more efficiently than the others, then it will reach saturation and inhibit the polymerase from subsequently amplifying the other amplicons. To achieve uniform amplification of the different targets, the primers must be designed to bind with equal efficiency to their respective targets. Binding equally does not mean "matched T_m's." This requires the use of accurate thermodynamic parameters (i.e., by not using the older methods for T_m prediction) and also accounting for the effects of competing equilibria, which requires the use of the coupled multi-state equilibrium model described in **Subheadings 2.1 and 2.4** as well as the other principles described in this chapter.

3.7. Myth 7: Multiplex PCR Can Succeed by Optimization of Individual PCRs

Not too many people believe this myth, and yet their actions are somewhat irrational as they proceed to immediately use that approach to try to experimentally optimize a multiplex PCR. It is true that well-designed single-target PCRs are useful for developing a multiplex reaction, but for a variety of reasons, this approach alone is too simplistic. The most common experimental approach to optimizing a multiplex PCR design is shown in **Fig. 11**, as suggested by Henegariu et al. ***(30)***. This is a laborious procedure that has a high incidence of failure even after extensive experimentation. The core of this approach is to first optimize the single-target amplifications and then to iteratively combine primer sets to determine which primer sets are incompatible and also to try to adjust the thermocycling or buffer conditions. With such an approach, the optimization of a 10-plex PCR typically takes a PhD level scientist 3–6 months (or more), with a significant chance of failure anyway.

Why does the experimental approach fail? The answer is that there are simply too many variables (i.e., many different candidate primers and targets) in the system and that the variables interact with each other in non-intuitive ways. Anyone who has actually gone through this experimental exercise will attest to the exasperation and disappointment that occurs when 7 of 10 of the amplicons are being made efficiently (after much work) only to have some of them mysteriously fail when an eighth set of primers is added. The approach of trying to adjust the thermocycling or buffer conditions is also doomed to failure because the changes affect all the components of the system in

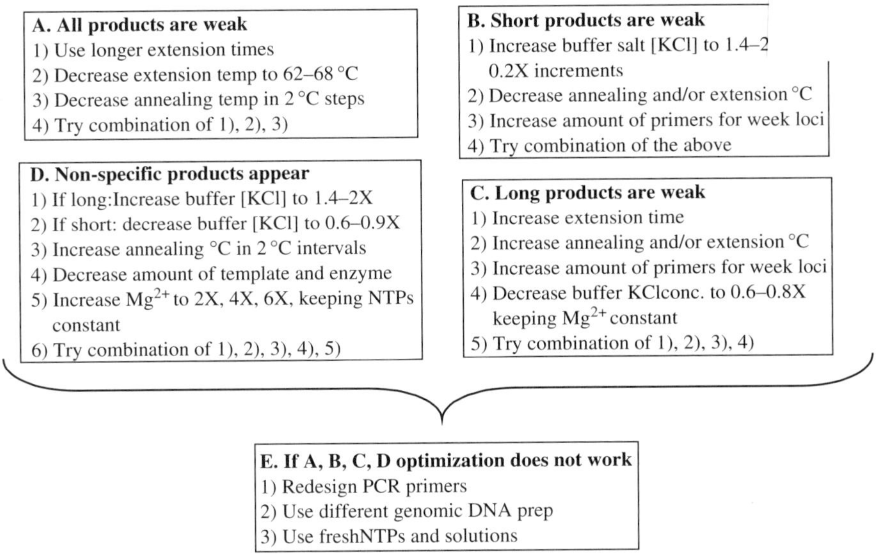

A. All products are weak
1) Use longer extension times
2) Decrease extension temp to 62–68 °C
3) Decrease annealing temp in 2 °C steps
4) Try combination of 1), 2), 3)

B. Short products are weak
1) Increase buffer salt [KCl] to 1.4–2 0.2X increments
2) Decrease annealing and/or extension °C
3) Increase amount of primers for week loci
4) Try combination of the above

D. Non-specific products appear
1) If long:Increase buffer [KCl] to 1.4–2X
2) If short: decrease buffer [KCl] to 0.6–0.9X
3) Increase annealing °C in 2 °C intervals
4) Decrease amount of template and enzyme
5) Increase Mg^{2+} to 2X, 4X, 6X, keeping NTPs constant
6) Try combination of 1), 2), 3), 4), 5)

C. Long products are weak
1) Increase extension time
2) Increase annealing and/or extension °C
3) Increase amount of primers for week loci
4) Decrease buffer KClconc. to 0.6–0.8X keeping Mg^{2+} constant
5) Try combination of 1), 2), 3), 4)

E. If A, B, C, D optimization does not work
1) Redesign PCR primers
2) Use different genomic DNA prep
3) Use freshNTPs and solutions

Fig. 11. Multiplex PCR optimization guidelines suggested by Henegariu et. al. ***(30)***.

different ways. For example, increasing the annealing temperature might be a fine way to minimize primer dimer artifacts (which can be a big problem in multiplex PCR), but then some of the weaker primers start to bind inefficiently. Subsequent redesign of those weak primers might then make them interact with another component of the reaction to form mishybridized products or new primer dimers, or cause that amplicon to take over the multiplex reaction.

The mystery could have been prevented (or at least minimized) with the use of a proper software tool. First, proper design if the single-target PCRs leads to improved success when used in multiplex. Second, software can try millions of combinations with much more complete models of the individual components (as described throughout this chapter) and use more complete modeling of the interactions within the whole system, whereas a human can only try a few variables before getting exhausted.

4. Methods for PCR Design

The flow chart of the primer design protocol used by Visual-OMP is shown in **Fig. 12**. Step 1 is a primer selection algorithm. The quality of each candidate primer pair design is judged by its "combined ranking score," which is the weighted sum of several terms. The scoring method used by Visual-OMP is similar to that implemented in Primer3, wherein each thermodynamic or heuristic

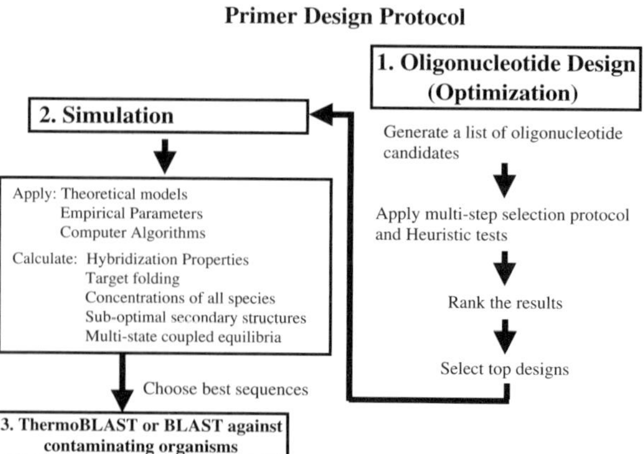

Fig. 12. Primer design protocol used by Visual-OMP. The goal of this protocol is to determine optimized oligonucleotide designs so that problems are identified and solved in silico, thereby reducing trial and error bench work.

properties of the primer is given a numerical score that is compared with user-defined optimal setting, range, weight, and penalty function. Some of the heuristic and thermodynamic properties include cross-hybridization, mismatch specificity filter, $\Delta G°$ and T_m thresholds for hairpins, $\Delta G°$ and T_m thresholds for desired duplexes, oligonucleotide length, %G+C content, polyG filter, 3′-extensibility filter, low complexity filter, and so on. This generates a list of ranked primers with good properties that can be tested further. Step 2 is an advanced simulation that determines the concentration of all species using the multi-state coupled equilibrium methodology. If a candidate primer pair fails to give equal binding to the target strands, then another primer pair from step 1 is automatically tested. Step 3 scans each primer against a genomic database to search for possible mispriming artifacts. If the primer pair fails step 3, then the process is repeated until a primer set is found that satisfies all the tests. Typically, Visual-OMP outputs several of primer pairs, all of which should work effectively; this gives the user a choice of solutions to experimentally test.

5. Future Perspective: Complete PCR Simulation of the Product Distribution During Every Step of PCR

An important goal is the development of algorithms that completely simulate all the physical behavior that occurs in nucleic acid assays and to use these models in algorithms that perform automated optimization of assays. In the case of PCR, the "holy grail" is to develop an algorithm that allows for the

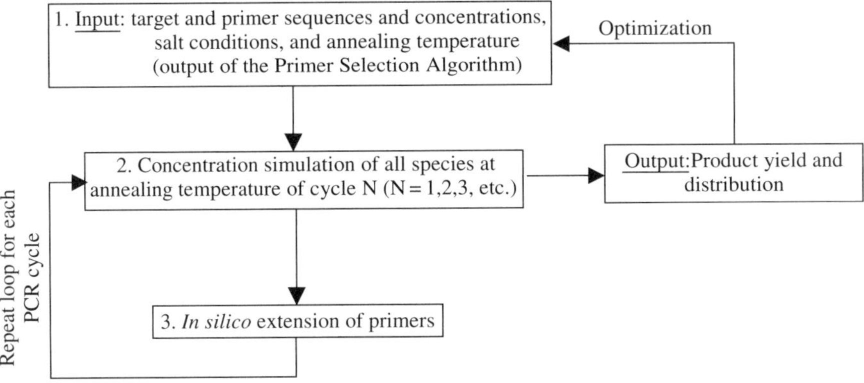

Fig. 13. Scheme for PCR amplification simulation algorithm.

accurate prediction of the product distribution (i.e., concentration of all strands) during every step of each cycle of PCR. Achieving such an algorithm will require not only the methods described in this chapter but also incorporation of the principles of kinetics of polymerase extension, kinetics of DNA folding, unfolding, and hybridization, and simulation of the temperature dependence of the chemical and physical reactions that occur in PCR. Such a model is genuinely within the reach of current scientific methods. To this end, substantial progress has been made at DNA Software, Inc. toward this goal under SBIR funding from the NIH. **Figure 13** shows an overall algorithm for PCR simulation. DNA Software, Inc. has developed a prototype PCR simulator. Description of the details of the prototype simulator is beyond the scope of this chapter.

5.1. Literature Example

Ishii and Fukui ***(31)*** performed an experiment in which two templates (differing only by a single nucleotide) were amplified by the same set of primers. Thus, template 1 is amplified with both primers forming a perfect match, whereas template 2 is amplified with one mismatched primer and the other a match. The experiments showed that with low annealing temperature (<50 °C), both templates are amplified with essentially equal efficiency. As the annealing temperature is raised to 55–60 °C, however, template 2 hybridizes less efficiently to the mismatched primer so that reduced amplification is observed, whereas template 1 continues to be amplified efficiently. Above 60 °C, template 2 amplification is not observed, and template 1 efficiency decreases as the temperature is raised further. These results are consistent with our hypothesis that PCR amplification efficiency depends on the free energy of primer binding to the target.

5.2. OMP PCR Simulation Results

In the OMP PCR simulation, the targets, the primers, and the PCR solution conditions are identical to those used in the study of Ishii and Fukui. The PCR simulator results are shown in **Fig. 14**. The results clearly indicate amplification bias for the matched template over the mismatched template as the annealing temperature is increased, which agrees qualitatively with the exper-

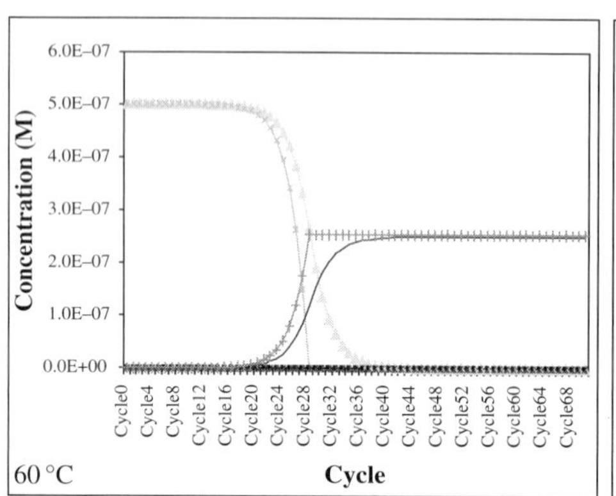

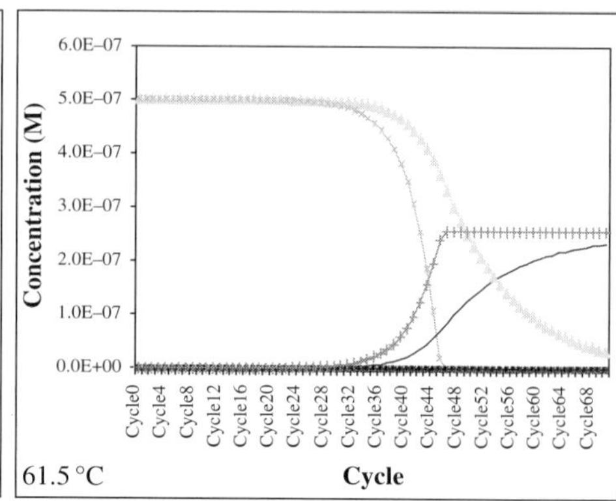

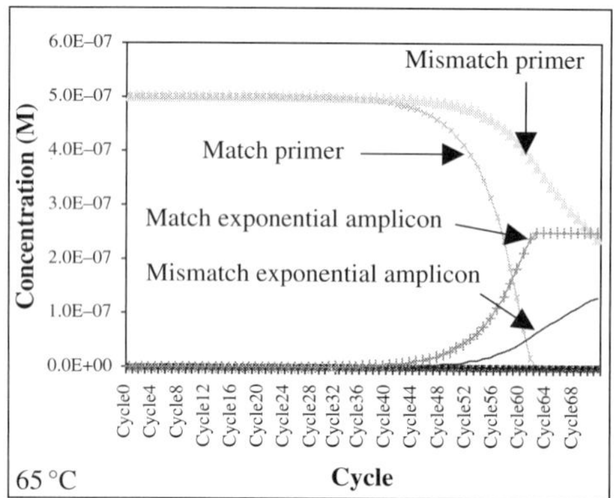

Fig. 14. Output from the prototype PCR simulation from Visual-OMP. The PCR product concentration is plotted versus number of PCR cycles. Annealing temperatures are given at the bottom left of each panel. The panels show the exponentially amplified products (amplicons bracketed by both primers) and the corresponding decrease in primer concentrations. Note that as the annealing temperature is increased, amplicon 2 (mismatch) is amplified less efficiently than amplicon 1 (match). The results clearly indicate amplification bias for the matched template over the mismatched template as the annealing temperature is increased.

imental result. Quantitatively, the OMP simulation shows the amplification bias beginning at approximately 61.5 °C, which is close to the experimentally observed 55–60 °C. Although this is promising, there are two discrepancies that require addressing in the next version of the OMP PCR simulation utility: (1) amplicon sense/anti-sense re-annealing kinetics is neglected, which decreases amplification efficiency, and (2) primer dissociation kinetics are not accounted for. These effects would tend to systematically compete with the desired hybridization and thereby decrease the efficiency of match over mismatch amplification. The availability of a complete PCR simulator will enable nearly perfect PCR design with optimal efficiency and minimal artifacts and provide excellent designs even for the most demanding multiplex and real-time applications.

Acknowledgments

I am thankful to all my previous graduate students and coworkers at DNA Software, Inc. for their hard work and contributions to the fields of DNA thermodynamics and software development. This work was supported by NIH grant HG02020 (to JSL), Michigan Life Sciences Corridor grant LSC1653 (to JSL), and NIH grants HG002555, HG003255, and GM076745 (to DNA Software, Inc.), and NIH SBIR grants HG002555, HG003255, GM076745, and HG003923 (to DNA Software Inc.).

References

1. Royce, R. D., SantaLucia, J., Jr. & Hicks, D. A. (2003). Building an *in silico* laboratory for genomic assay design. *Pharm. Visions* 10–12.
2. SantaLucia, J., Jr. & Hicks, D. (2004). The thermodynamics of DNA structural motifs. *Annu. Rev. Biophys. Biomol. Struct.* 33, 415–440.
3. Puglisi, J. & Tinoco, I., Jr. (1989). Absorbance melting curves of RNA. *Methods Enzymol.* 180, 304–325.
4. SantaLucia, J. J. (2000). The use of spectroscopic techniques in the study of DNA stability. In *Spectrophotometry and Spectrofluorometry. A Practical Approach* (Gore, M. G., ed.), pp. 329–356. Oxford University Press.
5. SantaLucia, J., Jr. & Turner, D. H. (1997). Measuring the thermodynamics of RNA secondary structure formation. *Biopolymers* 44, 309–319.
6. Press, W. H., Flannery, B. P., Teukolsky, S. A. & Vetterling, W. T. (1989). *Numerical Recipes in C*, Cambridge University Press, New York.
7. SantaLucia, J., Jr. (1998). A unified view of polymer, dumbbell, and oligonucleotide DNA nearest-neighbor thermodynamics. *Proc. Natl. Acad. Sci. U. S. A.* 95, 1460–1465.

8. Bommarito, S., Peyret, N. & SantaLucia, J., Jr. (2000). Thermodynamic parameters for DNA sequences with dangling ends. *Nucleic Acids Res.* 28, 1929–1934.
9. Peyret, N., Seneviratne, P. A., Allawi, H. T. & SantaLucia, J., Jr. (1999). Nearest-neighbor thermodynamics and NMR of DNA sequences with internal A-A, C-C, G-G, and T-T mismatches. *Biochemistry* 38, 3468–3477.
10. Allawi, H. T. & SantaLucia, J., Jr. (1997). Thermodynamics and NMR of internal G-T mismatches in DNA. *Biochemistry* 36, 10581–10594.
11. Watkins, N. E., Jr. & SantaLucia, J., Jr. (2005). Nearest-neighbor thermodynamics of deoxyinosine pairs in DNA duplexes. *Nucleic Acids Res.* 33, 6258–6267.
12. Rychlik, W. & Rhoads, E. R. (1989). A computer program for choosing optimal oligonucleotides for filter hybridization, sequencing, and in vitro amplification of DNA. *Nucleic Acids Res.* 17, 8543–8551.
13. Rozen, S. & Skaletsky, H. (2000). Primer3 on the WWW for general users and for biologist programmers. *Methods Mol. Biol.* 132, 365–386.
14. Li, P., Kupfer, K. C., Davies, C. J., Burbee, D., Evans, G. A. & Garner, H. R. (1997). PRIMO: a primer design program that applies base quality statistics for automated large-scale DNA sequencing. *Genomics* 40, 476–485.
15. Haas, S., Vingron, M., Poustka, A. & Wiemann, S. (1998). Primer design for large scale sequencing. *Nucleic Acids Res.* 26, 3006–3012.
16. Hillier, L. & Green, P. (1991). OSP: a computer program for choosing PCR and DNA sequencing primers. *PCR Methods Appl.* 1, 124–128.
17. Proutski, V. & Holmes, E. C. (1996). PrimerMaster: a new program for the design and analysis of PCR primers. *Comput. Appl. Biosci.* 12, 253–255.
18. Hyndman, D., Cooper, A., Pruzinsky, S., Coad, D. & Mitsuhashi, M. (1996). Software to determine optimal oligonucleotide sequences based on hybridization simulation data. *Biotechniques* 20, 1090–1097.
19. Wallace, R. B., Shaffer, J., Murphy, R. F., Bonner, J., Hirose, T. & Itakura, K. (1979). Hybridization of synthetic oligodeoxynucleotides to fX174 DNA: the effect of single base pair mismatch. *Nucleic Acids Res.* 6, 3543–3557.
20. Sambrook, J., Fritsch, E. F. & Maniatis, T. (1989). In *Molecular Cloning: A Laboratory Manual*, 2 edition, Vol. II, pp. 11.46–11.47. Cold Spring Harbor Laboratory Press, New York.
21. Bolton, E. T. & McCarthy, B. J. (1962). A general method for the isolation of RNA complementary to DNA. *Proc. Natl. Acad. Sci. U. S. A.* 48, 1390.
22. Frank-Kamenetskii, M. D. (1971). Simplification of the empirical relationship between melting temperature of DNA, its GC content and concentration of sodium ions in solution. *Biopolymers* 10, 2623–2624.
23. Bonner, T. I., Brenner, D. J., Neufeld, B. R. & Britten, R. J. (1973). Reduction in the rate of DNA reassociation by sequence divergence. *J. Mol. Biol.* 81, 123.
24. Owczarzy, R., Vallone, P. M., Paner, T. M., Lane, M. J. & Benight, A. S. (1997). Predicting sequence-dependent melting stability of short duplex DNA oligomers. *Biopolymers* 44, 217–239.

25. Breslauer, K. J., Frank, R., Blocker, H. & Marky, L. A. (1986). Predicting DNA duplex stability from the base sequence. *Proc. Natl. Acad. Sci. U. S. A.* 83, 3746–3750.
26. Dimitrov, R. A. & Zuker, M. (2004). Prediction of hybridization and melting for double-stranded nucleic acids. *Biophys. J.* 87, 215–226.
27. Mathews, D., Burkard, M., Freier, S., Wyatt, J. & Turner, D. (1999). Predicting oligonucleotide affinity to nucleic acid targets. *RNA* 5, 1458–1469.
28. Mathews, D. H., Sabina, J., Zuker, M. & Turner, D. H. (1999). Expanded sequence dependence of thermodynamic parameters improves prediction of RNA secondary structure. *J. Mol. Biol.* 288, 911–940.
29. Innis, M. & Gelfand, D. H. (1999). Optimization of PCR: conversations between Michael and David. In *PCR Applications: Protocols for Functional Genomics* (Innis, M., Gelfand, D. H. & Sninsky, J. J., eds), pp. 3–22. Academic Press, New York.
30. Henegariu, O., Heerema, N. A., Dlouhy, S. R., Vance, G. H. & Vogt, P. H. (1997). Multiplex PCR: critical parameters and step-by-step protocol. *Biotechniques* 23, 504–511.
31. Ishii, K. & Fukui, M. (2001). Optimization of annealing temperature to reduce bias caused by a primer mismatch in multitemplate PCR. *Appl. Environ. Microbiol.* 67, 3753–3755.

2

OLIGO 7 Primer Analysis Software

Wojciech Rychlik

Summary

OLIGO performs a range of functions for researches in PCR and related technologies such as PCR and sequencing primer selection, hybridization probe design, inverse and real-time PCR, analysis of false priming using a unique priming efficiency (PE) algorithm, design of consensus and multiplex, nested primers and degenerate primers, reverse translation, and restriction enzyme analysis and prediction; based on a protein sequence, oligonucleotide database allows fully automatic multiplexing, primer secondary structure analysis, and more. OLIGO allows for sequence file batch processing that is essential for automation. This chapter describes the major functions of OLIGO version 7 software.

Key Words: OLIGO primer analysis software; probe analysis; TaqMan probes design software; molecular beacons design software; siRNA design software; open reading frames analysis; gene design; PCR primer analysis; real-time PCR primer design; PCR multiplexing; ligase chain reaction; batch processing.

1. Introduction

OLIGO software is the first computer application that appeared on the market. The first freeware version appeared in 1986, and the first commercial release (version 4) was written in DOS environment in 1989. Since then, OLIGO went through many transformations and became a truly comprehensive tool for primer and probe design. The first general description of the software was published in 1993 *(1)*, and much of the theory on primer selection has not changed since then, so the aim of this chapter is not to repeat general considerations on primer and probe selection but rather focus on functionality of our latest software, OLIGO 7, scheduled to be released by the end of 2007.

From: *Methods in Molecular Biology, vol. 402: PCR Primer Design*
Edited by: A. Yuryev © Humana Press, Totowa, NJ

OLIGO software is used for designing and analyzing oligonucleotide primers or probes and synthetic genes. Based on nearest-neighbor thermodynamics *(2–9)*, OLIGO's search algorithms find optimal primers for PCR (including design of multiplex, consensus or degenerate primers, inverse and nested PCR, and site-directed mutagenesis), real-time PCR (TaqMan probes), and sequencing. OLIGO searches also for hybridization, ligase chain reaction (LCR) probes, and molecular beacons and even siRNAs. For each primer or primer/probe set, OLIGO's various analysis windows show a multitude of useful data. The software is also a useful tool for finding restriction enzyme sites (even in reverse-translated proteins) and analyzes all open reading frames (ORFs), including basic information about the translated proteins.

The primer and probe selection may be automated by batch processing several sequence files. OLIGO also selects primers/probes in relatively uniform intervals throughout the entire long DNA sequences.

2. General Information

The source code of OLIGO 7 interface is written in Java, and the search and calculation algorithms are written in C++, compiled for specific machine codes (Windows, Macintosh) to maximize the speed. Presently, OLIGO works only on all newest Windows systems as well as on Mac OS X. There are plans to make it compatible with Solaris and Linux. The minimum free hard disk space is only 5 Mb and the minimum RAM requirement is 25 Mb, or higher, depending on the sequence size.

OLIGO is shipped on a CD and comes with the installer. Once installed and registered, all updates may be easily downloaded using "Check for Update" function in the "File" menu.

2.1. The Interface

If you do not load a nucleic acid sequence file, the main menu of OLIGO displays the menu bar only at the top of the screen, as shown in **Fig. 1**, until you load a sequence file or enter one using the "File/New" command. Once you load or enter a file, the OLIGO "Sequence" window appears.

To call up menu items on any of OLIGO's screens, pull down the menu using the mouse. You may also use the shortcut keys. The shortcut keys are displayed to the right of the command description in the menu.

Oligo File Edit Analyze Search Select Change View Window Help

Fig. 1. The OLIGO 7 menu.

When a sequence file is opened, OLIGO displays one window, called the "Sequence" window (*see* **Fig. 2**), and it includes the elements, functions, and capabilities described below.

Information Boxes: At the top, there are three information boxes. The one on the left shows the sequence length, in nucleotides, current reading frame number (translated protein is on the bottom part of this window—more reading frames are displayed in the "Open Reading Frames" window), Current Oligo length, for which you see the T_m graph (*see* **Fig. 3**), and its 5′-end position number along with its T_m in standarized conditions (100 nM DNA in 1 M salt). In the lower left corner of this box, there is a small Info icon, and clicking on it will open "Analyze—Key Info—Current Oligo" window, containing basic information about the highlighted sequence fragment. The central box displays information on four selected oligos: Forward and Reverse Primers and Upper and Lower Oligos, as well as the length of a PCR product if the primers are selected. The central box has also info and a square icon. By clicking on the square icon, you are selecting the "Primer/Oligo" from the Current Oligo. After the selection, you may also invoke the corresponding "Key Info" windows. Clicking the non-dimmed Info icon in the PCR Product row will open the "Analyze—PCR" window. The box on the right displays all features

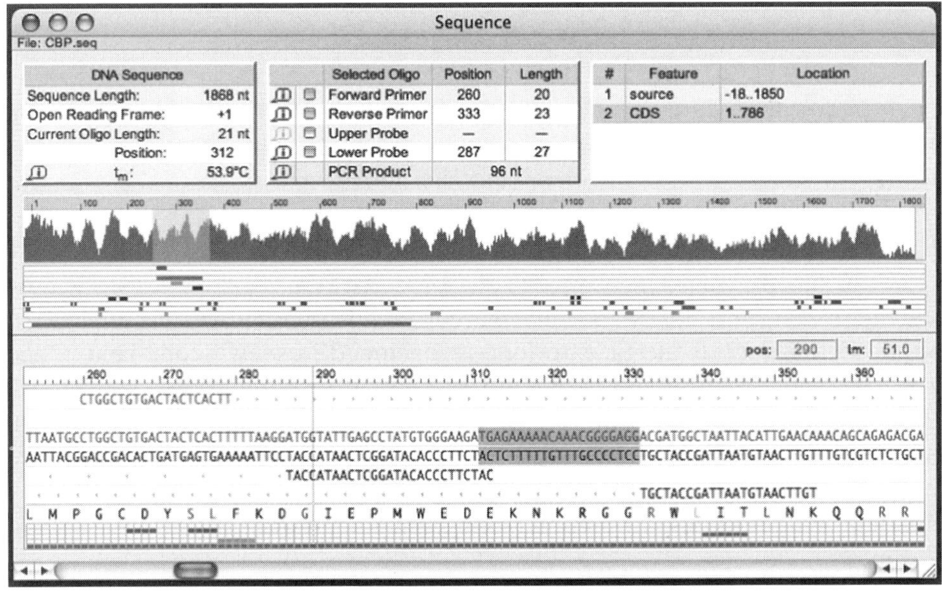

Fig. 2. The OLIGO "Sequence" window.

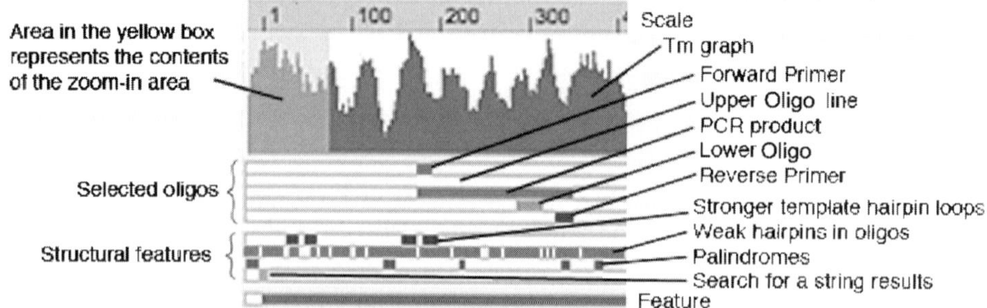

Fig. 3. The OLIGO "Sequence" window: zoom-out area.

(annotations) of the sequence. To browse the list, use the vertical scroll bars, if necessary. By clicking at any feature, the graphical representation of this particular feature is displayed in the "Sequence" window in two locations: zoomed out (center) and zoomed in (at the bottom of the window).

The Zoom-Out Area: Graphical representation of the sequence is displayed below the information boxes in two areas. The top one is the zoom-out area, which shows the scale, the graph representing melting temperatures of the sequence oligos (of the Current Oligo length), positions of the selected oligos, strong hairpin loops in the template, hairpin loops in all oligos of the Current Oligo length, palindromes, results of search for a string, and a selected feature.

Each vertical line of the T_m graph represents the T_m of an oligonucleotide calculated with the nearest-neighbor method ***(2–9)*** or an average of several oligos if the screen has fewer pixels than the number of bases in the sequence. The T_m is calculated for an oligonucleotide length that is set in the "Current Oligo Length" command from the "Change" menu—the default is a 21-mer. The set oligo length is displayed on the window title bar.

The Zoom-In Area: The zoom-in area is located at the bottom of the window and starts from the boxes showing the cursor position and corresponding T_m of an oligonucleotide at that position. The zoomed-in scale is just below, and it is followed by the row for Forward Primer, Upper Oligo, original sequence as saved in the file, its complement, Lower Oligo, and the Reverse Primer. Below these, there is the protein sequence translated from the DNA/RNA active sequence in the selected reading frame, followed by the sequence secondary structure and features in the same order as in the zoomed-out area, as shown in **Fig. 4**.

Sequences of the selected oligos are displayed in lower case letters when mismatched to the active sequence.

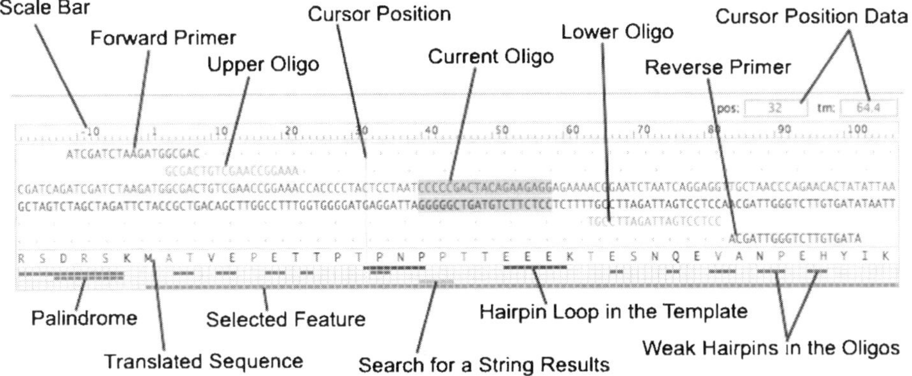

Fig. 4. The OLIGO "Sequence" window: zoom-in area.

As you use the OLIGO program, there are some features that are available at a certain time and others that are not. OLIGO uses the standard system convention of "graying out" items that cannot be accessed on that particular screen or at that particular time. Explaining all menu options goes beyond the scope of this chapter, so only the major functions, particularly those not found in OLIGO 6, will be described here. Besides the "Sequence" window, there are several different "Analyze" menu window types, and if you change something in one window, select another primer for example, the relevant information about the primer will be updated in every opened window.

3. The Analyze Options

There are 18 different options in the "Analyze" menu. They are as follows:

3.1. Key Info

The following information is displayed in this window: the names of the oligonucleotide(s) assigned by OLIGO and its sequence, length, and position number on the sequence file, T_m (dangling ends included, if there is a mismatch between a given oligo and the opened sequence, the melting temperature is displayed with the mismatch), t_m (dangling ends and mismatches excluded), ΔG (free energy), entropy and enthalpy, the 3′-ΔG of the 3′-terminal pentamer, the degeneracy number, the priming efficiency (PE) numbers, and the extinction coefficient properties of the oligo, expressed in nmol/OD at 260 nm and in μg/OD at 260 nm (OD, optical density unit).

3.2. Duplex Formation

The information shown in this window includes the most stable 3′-terminal dimer alignment of the primer or probe, the most stable primer dimer alignment overall, the most stable hairpin structure in the primer or probe, if any, and the stability values of the most stable uninterrupted duplex (*see* **Fig. 5**) in each alignment, and of the hairpin structure, expressed in kcal/mol; T_m of the hairpin is displayed when it is greater than 0°.

3.3. Hairpin Formation

The "Analyze—Hairpin Formation" windows display potential hairpin loop structures in the Forward and Reverse Primers and Upper, Lower, or Current Oligo. The hairpin stems are displayed in descending order of stability, expressed in hairpin loop ΔG and the T_m values.

3.4. Composition and T_m

These windows display the base composition and melting temperatures of the selected oligos, entire sequence, or PCR product at various conditions by various calculation methods: nearest neighbor, %GC, and $2 \times AT + 4 \times GC$. Besides the T_m values, the "Composition" windows display T_m of DNA/RNA hybrids, the A260/A280 absorption ratio (single strand), the molecular weight

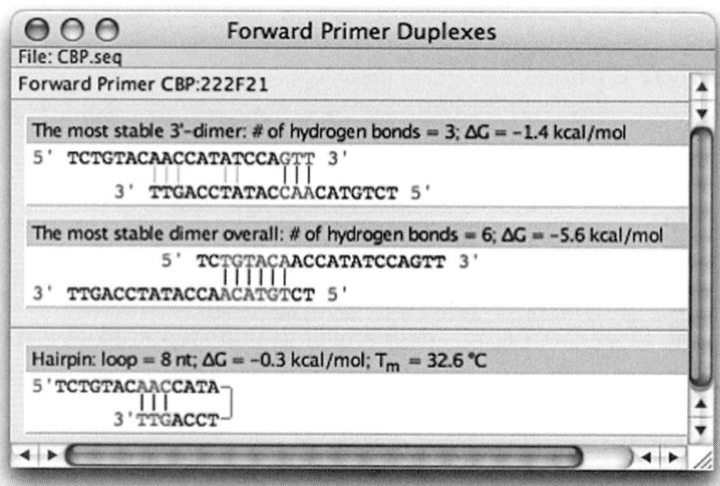

Fig. 5. One of the OLIGO duplex formation windows: Forward Primer Duplexes.

of the given oligo, both single and double stranded, the number and percentage of each base in the oligo plus the number and percentages of A + T and of G + C in this sequence, the melting temperature of the selected nucleic acid molecules with their complements at various salt concentrations (expressed both as mM and as multiples of the SSC buffer) in 0, 10, and 50% formamide, and, for the short oligos only, a calculation of melting temperatures with up to five mismatches.

3.5. False Priming Sites

The "Analyze—False Priming Sites" windows display potential false-priming sites for the selected oligos in the sense and antisense strands of the active sequence (*see* **Fig. 6**). The PE number is a formulation unique to the OLIGO program (a proprietary algorithm) that quantifies the likelihood that a given oligonucleotide will prime at a given site on the template. It has been determined that matching PCR primers by their PE number gives better results than matching the primers by their T_m's. This is perhaps more important in designing multiplex PCR experiments. The PE considers the stability (ΔG) of primer–template duplexes and their distance from the 3′-end, bulge loops, mismatches and their distance from the 3′-end, and the overall primer length.

3.6. Homology

The "Analyze—Homology" windows display the best homology alignments for all the four possible oligos that could be analyzed using OLIGO in both the sense and antisense strands of the active sequence. The homology search is using the Lipman and Pearson's FASTN algorithm ***(10)***.

3.7. Selected Oligonucleotides

The "Analyze—Selected Oligos" window lists all oligonucleotide selections from the most recent search. Choose the kind of oligos to be displayed (Forward Primers, Upper Oligos, etc.) with the button at the upper left corner of the window. The data table shows all the oligos found in the last search, but if the check box "Hide unpaired oligos" is checked, the table shows only the oligos that match with other oligos in the table that create sets. The sets are created during the search for PCR primer pairs, nested primers, TaqMan probes, or molecular beacons. Although PCR primer search data can be displayed using

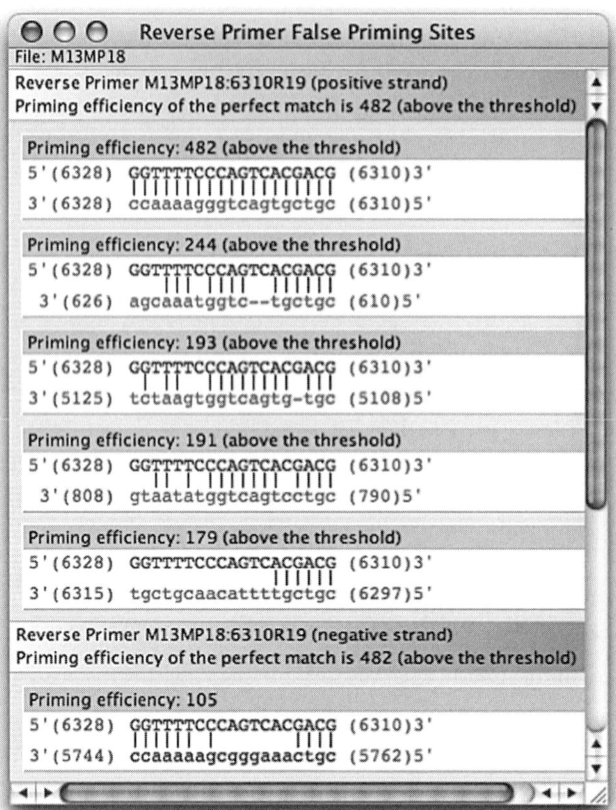

Fig. 6. An example of a "False Priming Sites" window. The site with priming efficiency of 244 points is the real false-priming site as confirmed experimentally ***(11)*** in the m13mp18 DNA.

this option, the window is primarily for displaying sequencing primer and hybridization probe search data. Accordingly, the data table lists T_m of the oligo, the 3′-terminal pentamer ΔG (specificity), and GC clamp ΔG stability (the most stable pentamer in the oligo) associated with each oligonucleotide starting at the displayed position. If you have searched for consensus primers or probes, you may pull down the window separator to reveal all sequence alignments and the consensus sequence, which you may select.

At the default setting, the window sorts oligonucleotides by descending score, ascending position, and descending length. You may also sort the table by oligo's T_m, 3′-ΔG, or GC clamp by clicking on the Sort button above that column. You may implement secondary sorts, tertiary sorts, and so on. Simply hold down the <option> key (Mac) or <Ctrl> key (Win) and click on the name of the row. The primary sort field button will be labeled "1," the secondary "2," the tertiary "3," etc.

As you click on different positions in the table, the Current Oligo position is updated accordingly. Double clicking on a displayed number selects this oligo as a relevant primer or probe.

3.8. Primer Pairs/Probes

This window contains a table of primer pair data from the most recent search for PCR primers, nested primers, TaqMan probes, or molecular beacons. The following information is included in the table: the primer/probe set number, the primers and probe positions on the positive- and negative-strand active sequences, the PCR product score, the length of the PCR product the pair would generate, the calculated optimal annealing temperature recommended for PCR using this primer pair, and the GC content of the PCR product generated by the listed pair of primers.

3.9. PCR

The "Analyze—PCR" window displays various data for PCR based on a user-selected PCR primer pair. The primer pair must be selected before this window can be called up.

Optimal annealing temperature for PCR (Ta^{OPT}) and the maximum annealing temperature (Ta^{max}) are given. For an initial experiment, use Ta^{OPT}, because it usually gives the highest yield. The Ta^{max} should be used mainly for diagnostic purposes because PCR is more specific in these conditions.

The information content of this window is illustrated in **Fig. 7**.

The Comments area cautions when a particular primer or a probe has some design problems. Avoid a primer-product difference of more than 29° because this will produce too much competition between template–template and primer–template annealing. Also avoid designing primers with a high Tm difference whenever possible. When the selected primers have disparate T_m's (greater than 10°), the more stable primer does not work optimally because the annealing temperature has to be reduced to compensate for the less stable primer. Removing nucleotides from the 5′-end of the more stable primer to match T_m's does not decrease PCR efficiency. On the contrary, increasing the length of the less stable primer to match the T_m of the more stable primer will typically improve PCR.

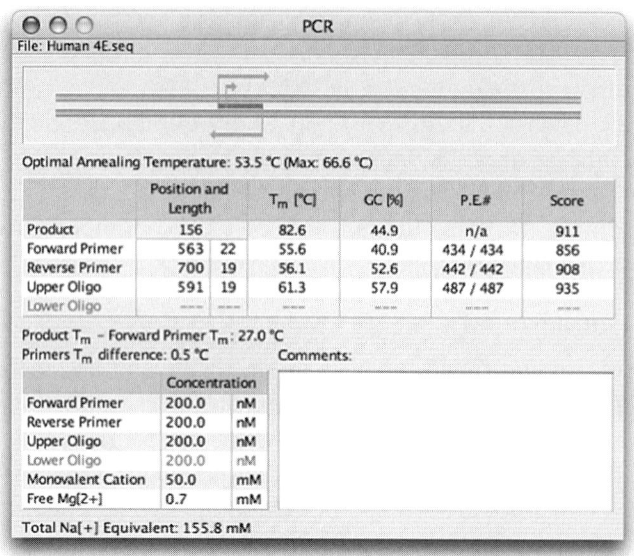

Fig. 7. The "Analyze—PCR" window.

3.10. LCR

LCR is a diagnostic technique designed to confirm the sequence in a target DNA sample. The LCR function designs two pairs of complementary LCR primers for a wild-type DNA sample, and, optionally, an additional pair to detect a point mutation.

3.11. Melting Temperature Graph

The "Melting Temperature" window shows a stability or degeneracy graph displayed on the bottom of the window sequence segment, calculated using the nearest-neighbor method (if stability is displayed) (*see* **Fig. 8**). Each point or bar on the graph represents the T_m or ΔG of an oligonucleotide, the 5′-terminus of which is located at the position listed on the top right of the window (cursor position). The T_m is calculated for oligonucleotide length that is set in the "Current Oligo Length" option from the "Change" menu—the default is a 21-mer. The currently set oligo length is displayed on the window tool bar. The Current Oligo sequence is in red, and a small red circle represents its T_m. This window may display T_m or ΔG profile of oligomers, as well as GC% or degeneracy. There are three modes of displaying T_m and ΔG: without dangling ends, and with dangling ends on either forward or reverse strand.

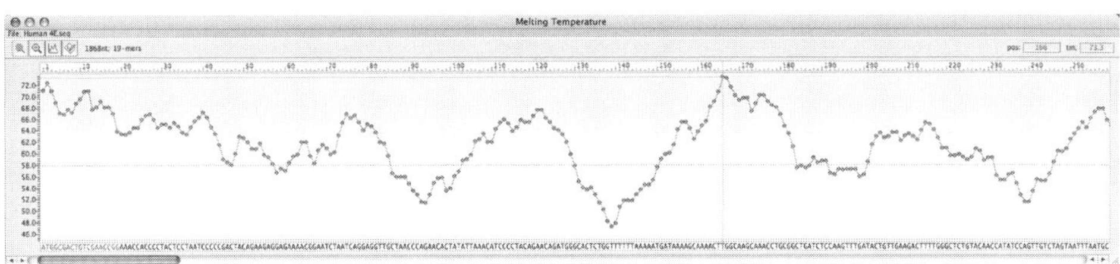

Fig. 8. The "Melting Temperature" window, dot style, using Zoom 1 option. The Current Oligo length was set to 19 nt.

The certain display mode propagates to the "Sequence" window. The default is the melting temperature without the dangling ends, marked as "t_m" throughout entire OLIGO software.

3.12. Internal Stability

These windows plot the ΔG of pentamers (internal stability) of the active sequence (Current Oligo), and all individual primers and probes. There are two display styles available for this window: "bar" and "dot," which may be switched using the Options icon.

Each base of the Current Oligo pentamer is represented by a red dot or a deep-green bar, whereas the rest of the sequence pentamers are represented by green dots or light-green bars. The colors are uniform for the other "Internal Stability" windows that show short oligos. The cursor position number displayed at the top right shows the pentamer number in a given oligo or the position number of the sequence ("Current Oligo" window). Besides this cursor position, you can read also the ΔG value of the pointed-on pentamer (the cursor points at its 5′-end). This window can be used to check whether a given oligo has GC clamps or other features in its ΔG stability. An oligo with a slightly unstable 3′-end, but with higher stability along the remainder of its length, is likely to be an efficient and specific primer, assuming that other design characteristics are optimized.

3.13. Sequence Frequency

This window shows the relative frequency of nucleic acid subsequences (6- or 7-mers) throughout the entire loaded sequence. Oligonucleotides with 3′-ends common in a specific database (subset of GenBank) have a greater likelihood of false priming. The tables of frequencies (located in the Frequencies folder) are user-selected and contain normalized frequencies of GenBank subsequences (6- or 7-mers). By choosing oligos with infrequent 3′-ends, you are decreasing the likelihood of selecting a primer that primes on many sites in a complex substrate, such as genomic DNA. In other words, using oligos with low-frequency numbers decreases the chances of unspecific PCR. A frequency of 1000 is the average hexamer (or heptamer) frequency for a given database. These data tables are also used in the "Eliminate Frequent Oligos" subsearch for primers.

3.14. Open Reading Frames

This window displays all ORFs of the loaded sequence (*see* **Fig. 9**). They are displayed graphically on the top and by the amino acid symbols on the bottom part of the window. ORFs are displayed graphically below the tool bar. The yellow line dividing the positive- and negative-strand ORFs has a green box representing the range of the lower portion of the window with amino acid symbols and the DNA sequence. The extent of the zoomed graph shows also the nucleotide numbers of the most-to-the-right and most-to-the-left nucleotides represented in the graph area.

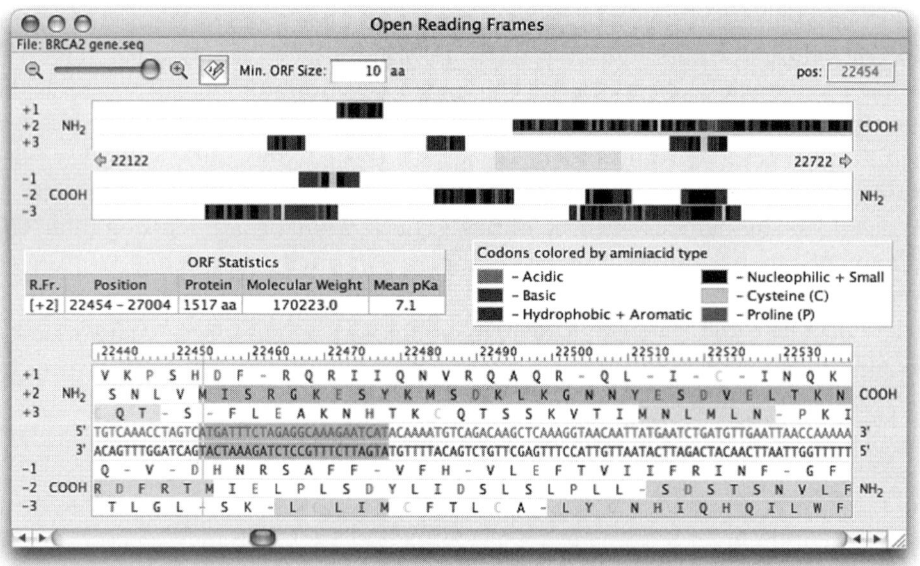

Fig. 9. The "Open Reading Frames" window with the color display showing various amino acid types.

If you click on any of the ORF, displayed graphically or by the letters (the given ORF block becomes underlined in green), the ORF Statistics table will show its reading frame number, positions on the DNA sequence, size of the protein in the number of amino acids and molecular weight, and the mean pK_a value, which is usually similar to its isoelectric point.

To the right of the ORF Statistics table, there is a table explaining the meaning of colors used to present ORFs in the graphic style (above) as well as in the letter style in the zoomed area below. In the zoomed area at the bottom of the window, you can see the DNA sequence in the middle, its positive-strand translation above it, and the translation of its negative strand below it in all reading frames. The amino- and carboxy-termini of the proteins are labeled on the sides. The scale on the top indicates the nucleotide position of the sequence. Clicking on any ORF will set the appropriate reading frame in the "Sequence" window.

3.15. Restriction Enzyme Sites

This "Analyze" window is available after the Search for the Restriction Sites, which is described in the "Search" menu section. There are two parts of this window that display results of the search in two formats, the top is in graphical format and the bottom is the table listing (*see* **Fig. 10**).

The graphical part consists of three columns. The first two are the same as in the table listing below: the number of an enzyme in the restriction enzyme file that was used to perform the search and the enzyme name. The third column shows at the top the nucleotide position scale of the entire sequence file and at the bottom the recognition sites as vertical blue lines spanning from top to bottom of each row representing the space for a given restriction enzyme.

The list, besides the number and the enzyme name, has three more columns, depending on the window width, displaying the actual recognition site, the number of sites in the entire active sequence file, and the recognition sites' positions (in black). The site positions are interrupted by the fragment sizes, displayed in green. Below the table, there is listing of enzymes that do not cut the sequence.

At the window bottom, the two search parameters are displayed: the type of sequence (linear or circular) and the file name of the restriction enzyme database used in the search.

3.16. Restriction Enzyme Sites in Protein

This "Analyze" window is available after the Search for the Restriction Sites in Protein. The window looks exactly the same as in the option described

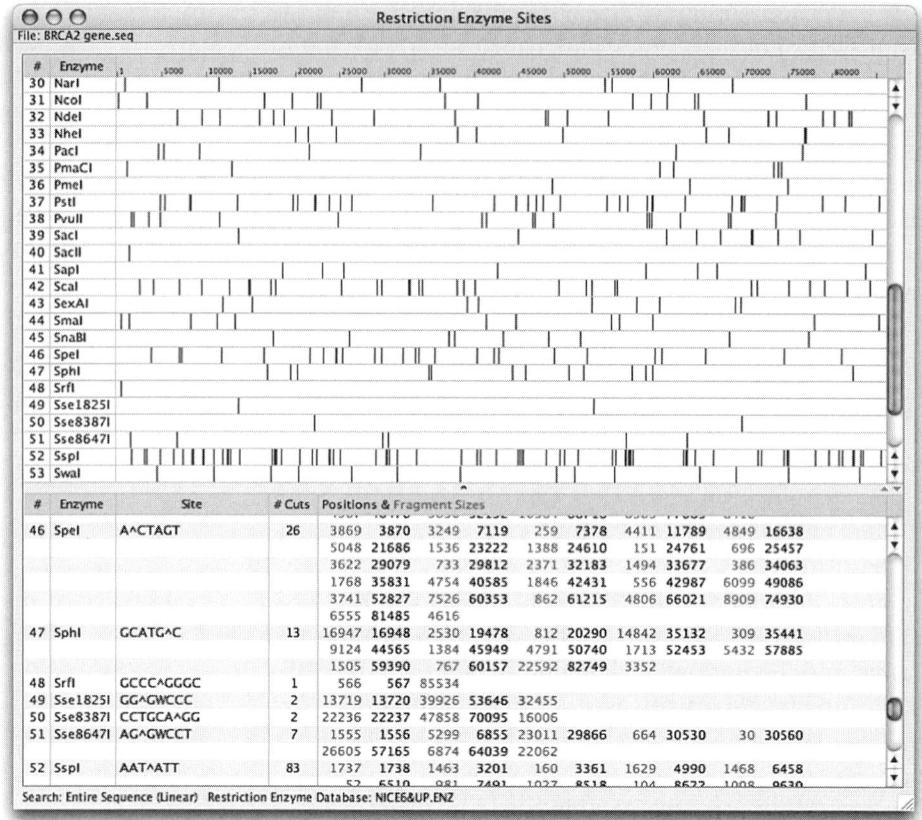

Fig. 10. The "Restriction Enzyme Sites" window.

above, with the same features displayed. The bottom window shows the search range and the type of enzyme cuts that have been considered: blunt ends, 3′- or 5′-overhangs, or other (odd). The layout of this window is essentially the same as shown in **Fig. 10**.

This search can be performed only on the positive-strand sequence in any of the three reading frames. It is easy to use "Open Reading Frames" window, select the ORF of interest from there, and read the exact search range that needs to be performed.

This is the analysis of a protein, so the actual nucleotide sequence has little to do with it. The sites in the potential DNA sequence are calculated only from reverse-translated protein using the degenerate method, and the window displays all possible restriction sites that could be used in the DNA sequence without changing the protein sequence.

3.17. Hybridization Time

The "Analyze—Hybridization Time" command calculates oligonucleotide hybridization time for various user-selected lengths and concentrations. OLIGO uses the following formula for calculation of the time required to reach half of the equilibrium (after this value, half of all possible duplexes should be already formed):

$$T_{1/2} = \ln 2 * \sqrt{L}/3.5 * 10^5 * C_N (\mathbf{12})$$

where N is the total number of base pairs in a non-repeating sequence (molecular complexity), L is the oligonucleotide length (in nt), and C_N is the oligonucleotide concentration (mol nucleotides/liter).

You may enter new DNA length and concentration values, and the hybridization time will be re-calculated.

3.18. Concentrations

This is basically a fancy calculator converting different DNA concentrations units. It also calculates the DNA amounts needed to make certain volumes of a nucleic acid at specific molarity. From this window, you may read: "X micrograms of DNA/RNA or Y optical density (OD) units or Z nanomoles in K microliters makes M micromolar DNA/RNA solution."

4. The Search Options

The "Search" menu lists four types of searches: for Primers and Probes, a Sequence String, Restriction Sites, and Restriction Sites in Protein. The last two options were mentioned in the previous chapter, the search for Sequence String provides for finding short nucleotide subsequences throughout the entire sequence where the degenerate bases are allowed, and the main OLIGO searches are available from the "Primers and Probes" menu. In the previous versions of OLIGO, there were additional search items—for Hairpin Loop Stems in the template and for Palindromes. It is not that OLIGO 7 does not have these search possibilities, quite to the contrary, they are automatically performed each time a new sequence file is opened and results are displayed in the "Sequence" window. Moreover, the search for hairpin loops in the template has been integrated with the primer and probe selection schemes. This chapter will only describe in detail the "Search for Primers and Probes" option.

The Search for Primers and Probes allows you to search for optimized PCR primer pairs, sequencing primers, TaqMan probes, molecular beacons, hybridization probes, or even siRNA probes in the active nucleic acid sequence. Each of these searches consists of several simple procedures, listed in the "Subsearches" window. If you click the Subsearches button, you will see all available subsearches OLIGO may perform. You may manually check/uncheck each button to activate or deactivate a particular subsearch option.

The "Ranges and Parameters" dialogs may be accessed by clicking on the respective buttons in the "Search for Primers and Probes" windows (*see* **Fig. 11**). These options are also available from the "Change" menu. You can change the search parameter values individually or globally using the Parameters button. You may specify the search ranges using the Ranges button and change all parameters to defaults at a given stringency level (except the ranges) using the Defaults button.

4.1. Available Search Modes

Search in Positive or Negative Strand: You can search in the positive strand, negative strand, or both—just check or uncheck the appropriate boxes.

Search Mode: The "Search Mode" is controlled by the Select and Verify buttons. The "Select" search mode is standard for all composite searches and subsearches, wherein all the oligonucleotides in the sequence, within a given search range, are tested. In the "Verify" search mode, all oligos previously found by the search are verified whether they still pass all the tests. This could

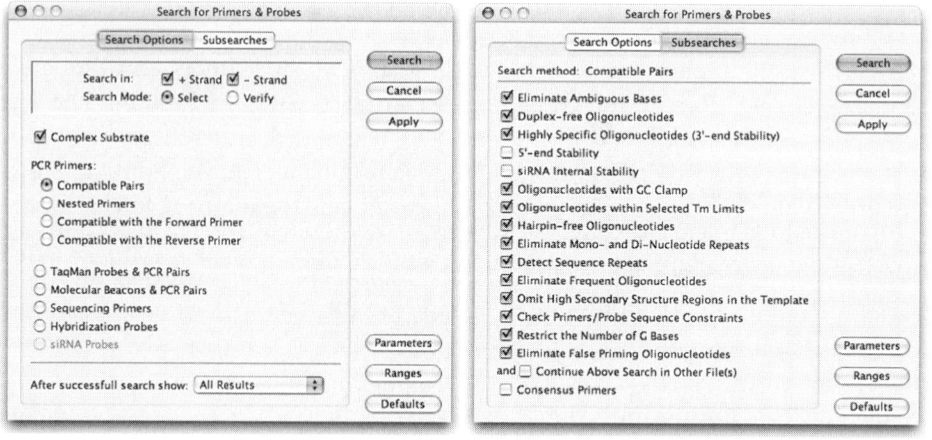

Fig. 11. The "Search for Primers and Probes" windows.

be useful after a search at lower stringency followed by increasing the search stringency parameter(s). Oligos that fail to pass through the various subsearch filters are removed, leaving only oligos that passed those two subsequent searches.

When the "Complex Substrate" box is checked, two additional subsearches are performed: the "Highly Specific Oligos" subsearch, which eliminates primers with excessively stable 3′-ends, and the "Eliminate Frequent Oligos" subsearch, which eliminates primers having 3′-ends which occur more frequently in a given database of sequences. If the substrate for your PCR or sequencing reaction is not complex (i.e., plasmid), checking this box is not typically necessary.

4.2. Composite Search Options

4.2.1. PCR Primers: Compatible Pairs

The "Compatible Pairs" option generates optimal PCR primer pairs from a nucleic acid template. Compatible primer pairs can be selected across the entire active sequence, or they can be selected from specific regions on the file by setting search ranges. Setting search ranges also speeds up this and other OLIGO searches. The stringency of the search can be controlled using the search stringency settings (global) and/or individual parameter settings in the "Parameters" window.

The following subsearches are used to find the PCR primers:

- Eliminate Ambiguous Bases—by default, selects non-degenerate primers
- Duplex-free Oligonucleotides—selects primers free of 3′-end and internal dimers
- Highly Specific Oligonucleotides (3′-end Stability)—selects moderately stable 3′-ends
- Oligonucleotides with GC Clamp—selects oligos with high internal stability regions
- Oligonucleotides within Selected T_m Limits—selects oligos with certain T_m range
- Hairpin-free Oligonucleotides
- Eliminate Mono- and Di-Nucleotide Repeats—selects oligos without internal repeats
- Detect Sequence Repeats—selects oligos without repeats other than those in the above option
- Eliminate Frequent Oligonucleotides—primers with most common 3′-ends, found in GenBank (6- or 7-mers) are eliminated
- Omit High Secondary Structure Regions in the Template—certain sequence areas will not be amplified
- Check Primers/Probe Sequence Constraints—no constraints by default but you may ask OLIGO not to choose primers that start with a specific base, for example

- Restrict the Number of G Bases—good for further multiplexing and overall non-dimerization
- Eliminate False Priming Oligonucleotides

This subsearch can be extended on other sequences by checking the box "Continue Above Search in Other File(s)," thus eliminating primers that could also prime from the other sequences, and the "opposite" search could also be performed by checking the box "Consensus Primers" instead and selecting the sequences on which the primers must prime as well.

The Search for Compatible Primer Pairs proceeds as follows:

- Search for optimized positive-strand primers for PCR
- Search for optimized negative-strand primers for PCR
- Match cross-compatible primers as potential primer pairs by checking the 3′-dimer formation between them and balance their T_m or PE.

4.2.2. PCR Primers: Nested Primers

This option searches PCR primer pairs as described above and additionally selects the nested primer pairs (within the first amplicon). The nested primers are compatible (multiplexed) with the external pair of primers, so there is no dimer formation between any of the four primers. The nested primers are selected and called "oligos," as sometimes these nested oligonucleotides may serve rather as probes than primers.

4.2.3. PCR Primers: Primers Compatible with the Forward Primer

This option searches the negative strand for optimal Reverse Primers that are compatible with a pre-selected Forward Primer. The Forward Primer may be selected from the active sequence or entered with the keyboard. When you choose this option with the default subsearches, OLIGO selects negative-strand primers that are free of dimers with the Forward Primer and have the same characteristics as the primers found in the Search for Compatible Primer Pairs, listed above. When the consensus primer subsearch is selected, OLIGO checks the PE number, not the homology.

At the bottom of the window, there is a pull-down menu allowing to choose which windows should be automatically displayed after a successful search. It could be the selected oligos, primers pairs/probes, or both windows—default "All Results" setting.

4.2.4. PCR Primers: Primers Compatible with the Reverse Primer

This option searches the positive strand for optimal Forward Primers that are compatible with a pre-selected Reverse Primer. The Reverse Primer may be selected from the active sequence or entered with the keyboard in the "Edit" menu.

When OLIGO conducts this search, it looks for potential Forward Primers that are free of dimers with the Reverse Primer and have the same characteristics as the primers found in the Search for Compatible Primer Pairs, listed above.

4.2.5. TaqMan Probes and PCR Pairs

This option searches for and selects optimal PCR primer pairs and TaqMan probes compatible with them. This search proceeds as the "Search for Compatible Primer Pairs" except that not two but three primers are selected. One of them becomes the TaqMan probe, and as a default, its melting temperature is 5° higher than that of the primers. All three oligonucleotides are compatible with each other. By default, OLIGO will choose the probes that lack G-residues at the two last 5′-terminal bases. This could be changed in the "Sequence Constraints" window of the Search Parameters.

4.2.6. Molecular Beacons and PCR Pairs

This option searches for and selects optimal PCR primer pairs and molecular beacon probes. This is essentially the same search as for PCR primer pairs, but to the 3′- and 5′-ends of the probe, two complementary hexamers, not complementary to the probe sequence, are added (default is CGTCAC/GTGACG). The sequence of the complementary tails can be entered in the "Sequence Constraints" window of Search Parameters. OLIGO searches for the probes that the other complementarities within the probes play no or minimal role on the overall hairpin loop stability. Also, the secondary structure of the amplicon is minimized, that means, the search removes DNA regions with stronger hairpin loops from consideration.

4.2.7. Sequencing Primers

The "Sequencing Primers" search option searches for and selects optimal sequencing primers in either or both strands of the active sequence. The OLIGO program automatically adjusts and/or deactivates search parameters to emphasize those characteristics most desirable in a sequencing primer. The following subsearches are used to find the sequencing primers:

- Eliminate Ambiguous Bases—by default, selects non-degenerate primers
- Duplex-free Oligonucleotides—selects primers free of 3′-end and internal dimers
- Highly Specific Oligonucleotides (3′-end Stability)—selects moderately stable 3′-ends

- Oligonucleotides with GC Clamp—selects oligos with high internal stability regions
- Oligonucleotides within Selected T_m Limits—selects oligos with certain T_m range
- Hairpin-free Oligonucleotides
- Eliminate Mono- and Di-Nucleotide Repeats—selects oligos without internal repeats
- Detect Sequence Repeats—selects oligos without repeats other than those in the above option
- Eliminate False Priming Oligonucleotides

4.2.8. Hybridization Probes

The "Hybridization Probes" search option searches for and selects optimal hybridization probes in either or both strands of the active sequence. When OLIGO conducts the hybridization probe search, it uses the following:

- Eliminate Ambiguous Bases—by default, selects non-degenerate primers
- Oligonucleotides with GC Clamp—selects oligos with high internal stability regions
- Oligonucleotides within Selected T_m Limits—selects oligos with certain T_m range
- Hairpin-free Oligonucleotides
- Eliminate Mono- and Di-Nucleotide Repeats—selects oligos without internal repeats
- Detect Sequence Repeats—selects oligos without repeats other than those in the above option
- Eliminate Homologous Probes—those that may anneal to more sites on the DNA

For consensus probes, it checks the homology, not PE.

The default stability T_m setting is high for the hybridization probe search. The stability maximum is not limited and the stability minimum decreases along with the stringency settings.

4.2.9. SiRNA Probes

This search finds RNA oligonucleotides that have certain pentamer stability ratios between 3′- and 5′-ends and the internal 12-nucleotide segment of the probe, essentially as described by Khvorova et al. ***(13)***, and are relatively hairpin-free. This search is available only when the main sequence is RNA. If you are working on a DNA file, use "Change—DNA to RNA" option before attempting this search. The following subsearches are used to find the siRNA 19-mers:

- Eliminate Ambiguous Bases—by default, selects non-degenerate primers
- Duplex-free Oligonucleotides—selects primers free of significant dimers
- Highly Specific Oligonucleotides (3′-end Stability)—selects moderately stable 3′-ends
- 5′-end Stability
- siRNA Internal Stability

- Hairpin-free Oligonucleotides
- Eliminate Mono- and Di-Nucleotide Repeats—selects oligos without internal repeats
- Detect Sequence Repeats—selects oligos without repeats other than those in the above option.

4.3. The Search Parameters

The Parameters settings are essential because the search outcome depends on them. It is therefore very important to fully understand the meaning of each default value before you attempt to improve it. A less "harsh" but still powerful method of changing the search outcome is to change the scoring system. Another simple way to influence the search results is to remove or add non-default subsearches in the "Search for Primers and Probes" windows.

The Search Parameters are laid out in four separate windows. The first one provides access to global stringency settings and to some of the subsearch defaults. **Figure 12** shows the first two windows.

The second window contains the remaining subsearch parameters and the third "Sequence Constraints." From the fourth window, you may change the primer-scoring system, as this OLIGO version assigns a one-number score to each primer or probe that is selected. The score makes the manual primer selection easier, especially if there are many oligos to choose from—instead of looking at several parameters individually, you may sort the results by this

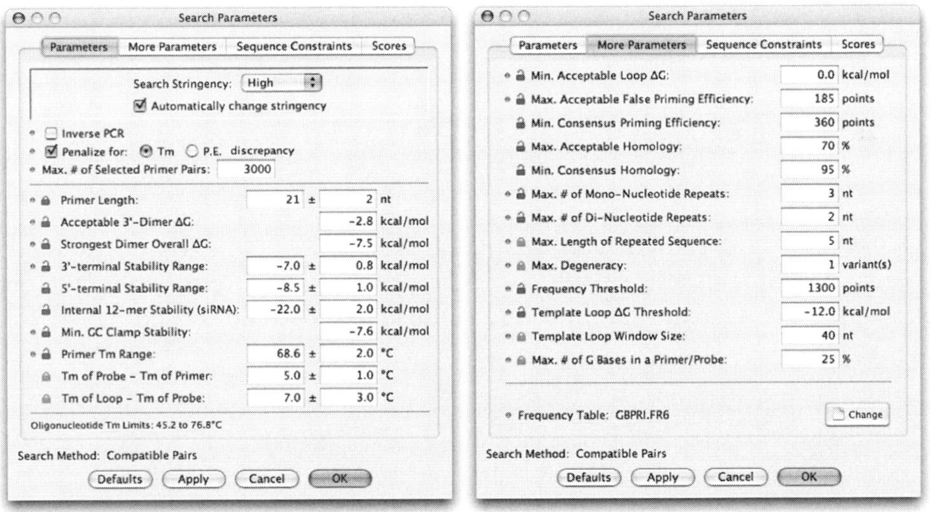

Fig. 12. The "Search Parameters" windows—Parameters and More Parameters.

number and see the top-scoring oligos instantly. Certain applications require different scoring system, as some parameters are more important than others, when compared with standard sequencing, for example. You may rely on this scoring system only if searching for "generic" primers or probes. If your oligos are to be used in specialized assays where the other factors are more important than the standard settings, then you would either do not need to pay attention to the actual scores or re-define the scoring system. This is the only reason why this feature was not implemented in earlier OLIGO versions.

Figure 13 shows the remaining two windows providing access to search outcome control.

In all the windows, you will find common objects that indicate parameters applicability. If a small blue circle is displayed by the side of a given parameter, it means that this parameter is going to be used by the currently selected in the "Search for Primers and Probes" window search type. The opened lock means that a given parameter may be changed during the automatic stringency correction, and the closed lock parameter will not be changed. The status of the dark locks may be changed by clicking at them, whereas the gray locks indicate unchangeable parameters. The check boxes in the "Sequence Constraints" window indicate whether or not the particular constrain will be applied in the search, despite the default setting indicated by the blue circle. In the example in **Fig. 13**, the search for TaqMan probes was selected and

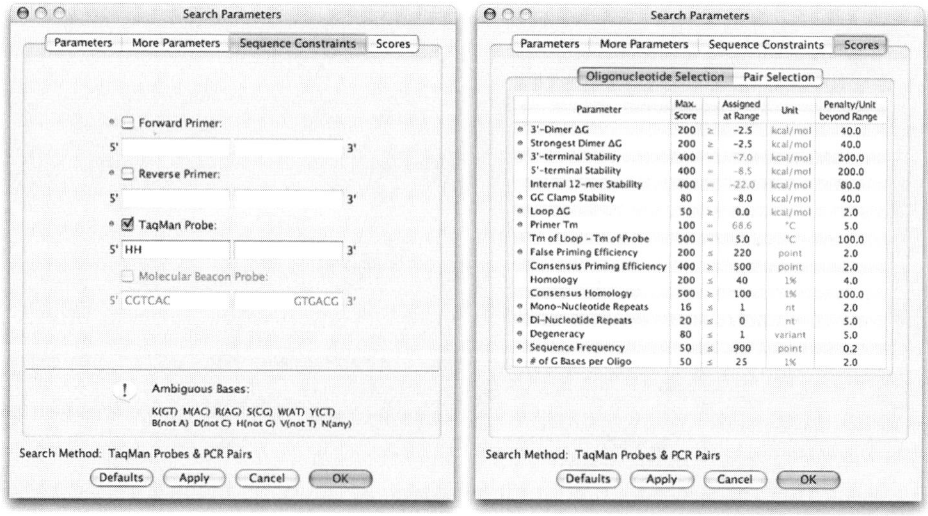

Fig. 13. The "Search Parameters" windows—Sequence Constraints and Scores.

despite the scheduled search for Forward and Reverse primer, the constraints will only be applied to the TaqMan probe itself (lack of GG at its 5′-end) and not to the primers (no matter what would have been typed in the primer constraints boxes).

The "Scores" window (*see* **Fig. 13**, right) is editable. You may change the "Max Score," "Assigned at Range," and "Penalty per Unit beyond Range" values, as long as they are not grayed out. If you want to change the actual "ideal" scoring of the grayed out parameter, you simply need to change this search parameter. There is another "Scores" window, not shown in the figure, that deals with parameters concerning the primer pair and primer set selection. You may assign penalty values for Forward/Reverse 3′-primer dimers, loops in the template, T_m or PE differences between the primers, T_m difference between the TaqMan probe and the primers, template GC content, and a gap between the primer and the probe (small gaps are usually favored).

5. The OLIGO Database

This feature deserves its own chapter as it serves many functions. It provides for not only storing records on oligonucleotides but it is a powerful analysis and multiplex search tool.

There are various format database files now compatible with OLIGO. It could be a list of plain primer sequences, each oligo in a new line, or it could be FASTA format or its modification as used by the AutoDimer (http://yellow.nist.gov:8444/dnaAnalysis/index.do) from National Institute of Standards and Technology. It is also compatible with all versions of OLIGO databases.

Primer sets may be imported to database from various sequence files. The number of records is unlimited. Only some of the database features are described below. Full description of its menus and its functionality can be found only in the OLIGO User Manual.

In the database, oligos and oligo sets may be analyzed for dimer formation and hairpins and displayed in similar windows as described in **Subheadings 3.2** and **3.3**. The window is in a table format. In each row, there is one database record. Each column gives the following information: the sequential number of the record, its ID number, oligonucleotide sequence, 3′-dimer data showing either the ΔG value of self-dimerization or the value between the most incompatible primer (after the manual multiplexing), PE, and the melting temperature of each oligo in four columns (*see* **Fig. 14**). The first PE and T_m columns show the actual values, as if those oligos would have been "annealed" to the

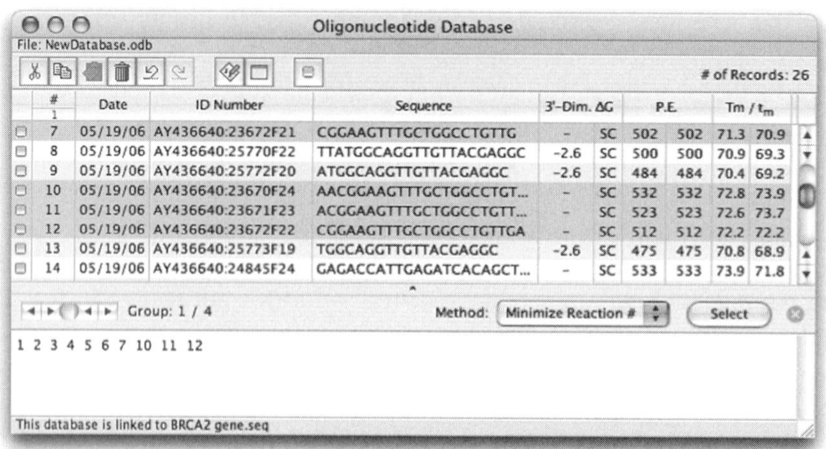

Fig. 14. The "Database" window showing the first multiplexed group.

linked sequence file, with the dangling ends considered. The numbers in the second PE and T_m columns indicate what those values are when annealed to the perfectly matching complementary strand of the same length. Because sometimes dangling ends increase duplex stability, the T_m's in the first column may be higher than the perfect match with its complement (labeled as t_m). There are two more columns, not displayed by default—Reference and Comments, where you may manually enter any relevant information about the oligo. These columns could be revealed through the Database Options from the "Change" menu.

The database provides for fully automatic multiplexing. The simplest results and the fastest method give "Analyze—Multiplex All—Minimize Reaction" option. The multiplexing is performed on all oligos in the database, and OLIGO selects compatible groups. Within each single group, all oligos are compatible with each other (show no 3′-end dimer formation). By clicking on the Select button, all oligos within the given group are selected (highlighted) and become available for export to the linked with the database sequence, saving, or analysis.

The scope of this chapter is to highlight the most important features of OLIGO software and not to explain every single feature available or the theory that stands behind the various algorithms. Therefore, several of the interface features, installation details, explanation of each function, and examples that are not described here can be found in the User Manual, available with the software.

6. Availability

OLIGO software is available from Molecular Biology Insights (MBI) and its affiliates. For more information, please visit our Web site http://www.oligo.net or contact MBI by e-mail (support@oligo.net).

References

1. Rychlik, W. (1993) Selection of primers for polymerase chain reaction, in *Methods in Molecular Biology*, Vol. 15: PCR Protocols: Current Methods and Applications, (B.A. White Ed.), Humana Press Inc., Totowa, NJ. pp. 31–40.
2. SantaLucia, J., Jr. (1998) A unified view of polymer, dumbbell, and oligonucleotide DNA nearest-neighbor thermodynamics. *Proc Natl Acad Sci USA* **95**, 1460–1465.
3. Xia, T., SantaLucia, J., Jr., Burkard, M.E., Kierzek, R., Schroeder, S.J., Jiao, X., Cox, C., and Turner, D.H. (1998) Thermodynamic parameters for an expanded nearest-neighbor model for formation of RNA duplexes with Watson-Crick base pairs. *Biochemistry* **37**, 14719–14735.
4. Allawi, H.T., and SantaLucia, J., Jr. (1997) Thermodynamics and NMR of internal G · T mismatches in DNA. *Biochemistry* **36**, 10581–10594.
5. Allawi, H.T., and SantaLucia, J., Jr. (1998) Nearest neighbor thermodynamic parameters for internal G · A mismatches in DNA. *Biochemistry* **37**, 2170–2179.
6. Allawi, H.T., and SantaLucia, J., Jr. (1998) Nearest-neighbor thermodynamics of internal A · C mismatches in DNA: sequence dependence and pH effects. *Biochemistry* **37**, 9435–9444.
7. Allawi, H.T., and SantaLucia, J., Jr. (1998) Thermodynamics of internal C · T mismatches in DNA. *Nucleic Acids Res* **26**, 2694–2701.
8. Peyret, N., Allawi, H.T., and SantaLucia, J., Jr. (1999) Nearest-neighbor thermodynamics and NMR of DNA sequences with internal A · A C · C G · G T · T mismatches. *Biochemistry* **38**, 3468–3477.
9. Bommarito, S., Peyret, N., and SantaLucia, J., Jr. (2000) Thermodynamic parameters for DNA sequences with dangling ends. *Nucleic Acids Res* **28**, 1929–1934.
10. Lipman, D.J., and Pearson, W.R. (1985) Rapid and sensitive protein similarity searches. *Science* **227**, 1435–1441.
11. Steffens, D.L., Sutter, S.L., and Roemer, S.C. (1993) An alternate universal forward primer for improved automated DNA sequencing of M13. *BioTechniques* **15**, 580–582.
12. Meinkoth, J., and Wahl, G. (1984) Hybridization of nucleic acids immobilized on solid supports. *Anal Biochem* **138**, 267–284.
13. Khvorova, A., Reynolds, A., and Jayasena, S.D. (2003) Functional siRNA and miRNAs exhibit strand bias. *Cell* **115**, 209–216.

3

Selection for 3′-End Triplets for Polymerase Chain Reaction Primers

Kenji Onodera

Summary

Primer extension by thermostable DNA polymerase in PCR starts from the 3′-end of a primer. If the PCR starting process fails, the entire PCR fails. Primer sequences at the 3′-end often interfere with success in PCR experiments. Over 2000 primer sequences from successful PCR experiments used with varieties of templates and conditions were analyzed for finding frequencies of the 3′-end triplets. This chapter discusses a trend in 3′-end triplet frequencies in primers used in successful PCR experiments and proposes requirements for the 3′-end of a primer. Finally, a method to select primers with the best 3′-end triplets is introduced based on the 3′-end analysis result.

Key Words: PCR; primer design; 3′-end; VirOligo, primers; viruses; oligonucleotides.

1. Introduction

The 5′- and 3′-ends of a primer have different meanings for PCR processes. Complementarity of the 5′-end of a primer to the PCR template is not so critical as for the 3′-end, and it is known that longer primers at the 5′-end (such as 30 nt or longer) do not improve specificity of PCR primers. The 5′-end of a primer also allows an addition of a tagging sequence. Complete binding between primers and template is not required at the 5′-end. However, the 3′-end of a primer is different from the 5′-end. Thermostable DNA polymerase starts attaching nucleotides from the 3′-end of a primer during the extension step, and it requires complete annealing of the 3′-end of a primer to a template. Incomplete binding at the 3′-end results in inefficient PCR or sometimes no PCR products. Alternatively, it is possible that too stable annealing of a primer

From: *Methods in Molecular Biology, vol. 402: PCR Primer Design*
Edited by: A. Yuryev © Humana Press, Totowa, NJ

at the 3′-end to a template allows generation of PCR product without complete binding between the rest of primer and template, and tolerance in incomplete binding may amplify unexpected PCR product by primer binding to other templates or different regions in a target template. Thus, the 3′-end of a primer is important in PCR primer design for successful PCR experiments.

1.1. The 3′-End of a Primer Recommended in Literature

Several kinds of recommendations for the 3′-end of a primer can be found in the literature. One recommends one or two S (S stands for C or G) at the 3′-end triplet of a primer for promoting strong annealing at the 3′-end *(1,2)*. Another also recommends C or G at the 3′-end of a primer but no CG or GC due to potential formation of hairpins and primer-dimers *(3)*. Although incorporation of S at the 3′-end is recommended, others recommend addition of W (W stands for A or T) at the 3′-end *(4)*. One recommends low G+C content at the 3′-end *(5,6)*. Another recommends at least one W in the 3′-end triplet but no T at the 3′-end of a primer *(4)*. Recommendations listed here suggest that S and W achieve complete annealing of 3′-end and specificity of a primer, respectively.

Many primer design programs have functions to define the 3′-end of a primer. DS Gene searches for S at the 3′-end in the default configuration. However, most of them, such as Primer3, Primer Premier, and Vector NTI, inactivate such functions in the default configurations. Although primer design programs check dimer or hairpin formations, which partially interfere with the 3′-end sequences of primers, few 3′-end considerations are included by commercial software, currently.

1.2. Needs for Detailed Analysis

Although the importance of the 3′-end of primers is recognized to be a key in the primer design, recommendations listed in the literature were mostly based on theory, and it seemed that the 3′-ends of primers had not been well studied. Further studies of the 3′-end were necessary for primer design. It is difficult to perform comprehensive testing of the effects of all 3′-end triplet types in actual PCR experiments. There are 64 triplet types and the amplifying region of the template is different for every primer for 3′-end triplet testing. The optimized PCR conditions may differ for each primer pairs because primer pairs and amplifying regions of a template differ. In this case, PCR results are confounded with many critical factors: 3′-end triplets, primer and template sequences, and PCR conditions. Thus, another approach was taken to examine the frequencies of the 3′-end triplets in successful PCR experiments with a variety of templates and under a variety of conditions. Experimental conditions and primer sequences in successful PCR experiments have been deposited in and are available through the VirOligo database *(7)*. From the VirOligo database, 2137 PCR primer sequences were retrieved for detailed analysis of the 3′-end triplets of successful PCR primers (*8*; *see* **Note 1**).

1.3. The Properties of the Analyzed Set, 2137 Primers (9)

Primer sequences were obtained for analysis from the VirOligo database on February 6, 2002. On that date, the VirOligo database covered all 1685 articles whose abstracts appeared in PubMed before December 19, 2001 as query results for PCR experiments targeted for alcelaphine herpesvirus, bovine adenovirus, bovine viral diarrhea virus (BVDV), bovine herpesvirus (BHV), bovine respiratory syncytial virus, bovine rotavirus, bovine coronavirus, foot-and-mouth disease virus (FMV), variola (smallpox) virus, cowpox virus, and human adenovirus. No filtration of articles was made from the search results. Simply all PCR conditions and oligonucletides listed in the articles were deposited into the VirOligo database. The articles contained 3985 published virus-specific oligonucleotide sequences and 2300 PCR and hybridization conditions for detection of the viruses. All primer sequences were retrieved from the VirOligo database on that date. Finally, duplicated primer sequences were eliminated from the analyzed set.

Primers in the VirOligo database were mostly between 18 and 22 nt long (65.8%; *see* **Fig. 1**), and 20 nt was the most frequent primer length (30.4%). G + C contents of primers in the VirOligo database were 40–60% mostly

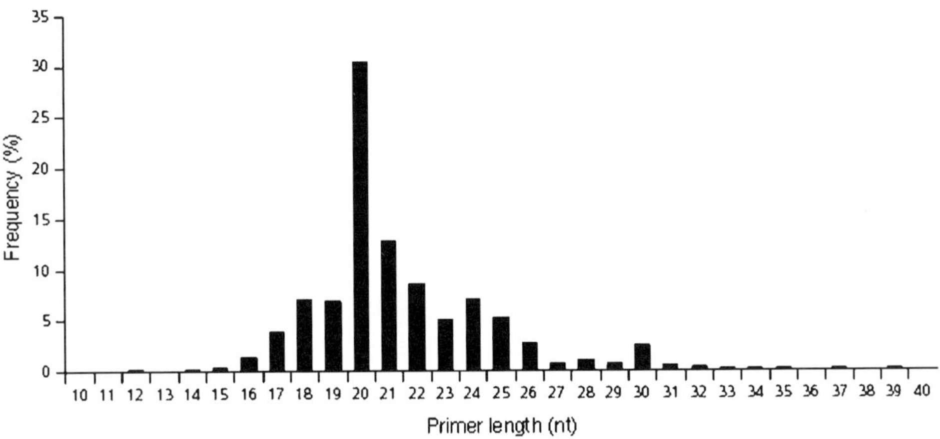

Fig. 1. Distribution of lengths of PCR primers.

(78.0%; *see* **Fig. 2**). Primers were most frequently 50% G+C content (G+C content 47.5–52.4%; 23.7% of all primers). The range of product sizes expected from a distance between primer pairs were most frequently 200–299 bp (18.3%; *see* **Fig. 3**). Less than 100 bp was not used frequently (2.0%). Most of expected PCR product sizes were less than 1 kb (82.3%), and 57.8% of PCR products were less than 500 bp.

In general, the primers in the VirOligo database did not seem to be selected based on T_m of primers. T_m distribution for FMV-specific primers (*see* **Fig. 4**), for instance, was similar to that for any 20 nt fragments from the entire FMV genome sequence (GenBank Accession No. AF377945, 53.1% G+C content, 7813 bp). In this observation, 7794 sequences were obtained from the entire genome by sliding a 20 nt frame from the 5′-end to the 3′-end one nucleotide at a time, and T_m for each 20 nt fragment was calculated by the nearest-neighbor method **Ref. 10**. Only when genome compositions were rich (e.g., BHV-1, GenBank Accession No. AJ004801, 72.4% G+C content, 135,301 bp) and poor (e.g., BVDV, GenBank Accession No. AF220247.1, 45.5% G+C content, 12,294 bp) in G+C contents, were primers with higher than 64 °C and lower than 50 °C, respectively, less frequent than expected from their genome sequences. Frequencies of primers with T_m's between 50 and 64 °C were proportional to those of any random 20 nt fragments in target genome sequences.

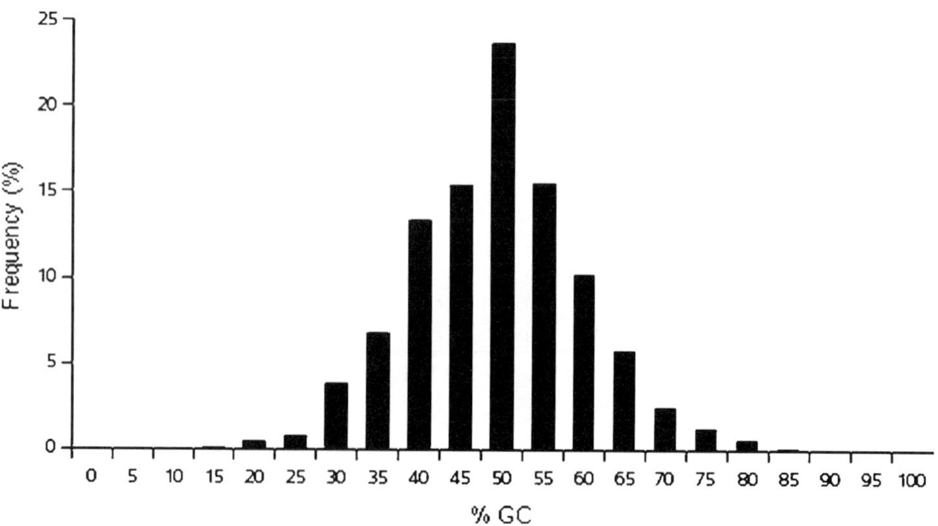

Fig. 2. Distribution of %(G+C) contents of PCR primers.

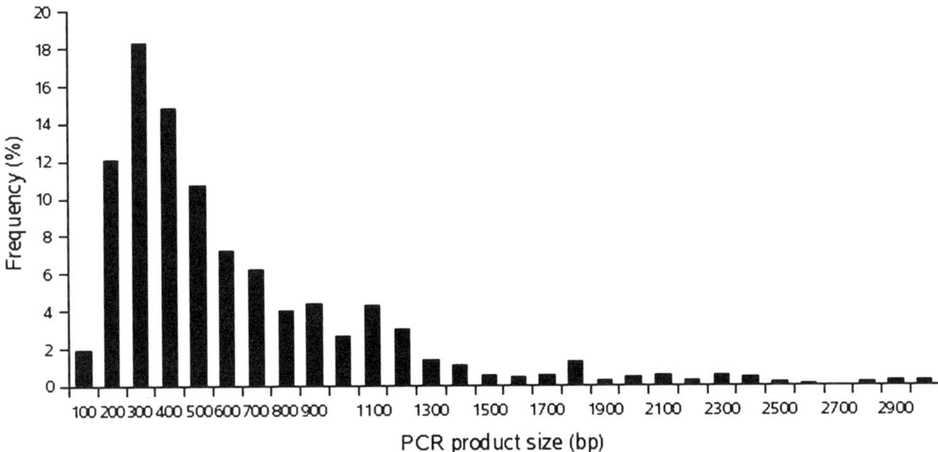

Fig. 3. Distribution of product size for all primer pairs in the VirOligo database.

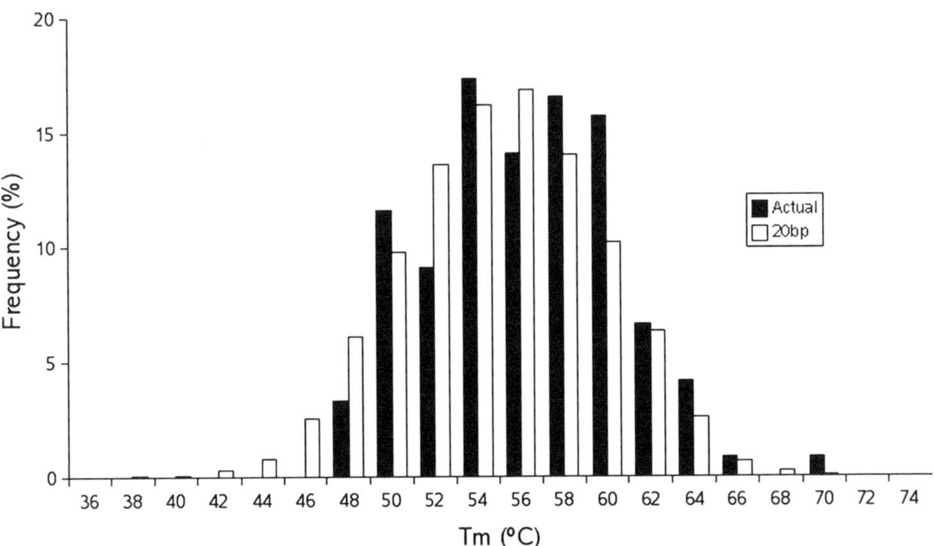

Fig. 4. Distributions of T_m's for foot-and-mouth disease virus (FMV)-specific primers (actual, closed bar) and predicted 20-bp sequences from entire FMV genome (20 bp, open bar).

1.4. Analysis Results for the 3′-End of Primers

The analysis of the 3′-end triplets of primers (*8*; *see* **Fig. 5** and **Table 1**) showed all 64 types were used in successful PCR experiments from the VirOligo database. No triplets completely inhibited a generation of PCR products. However, preferred and not-preferred triplets existed. The possible combinations of three nucleotides at 3′-end triplets give a total of 64 types, as mentioned above. If there are no preferences and selections of target sequences are completely random, the frequencies should be one-64th, which is approximately 1.56%. The mean and standard deviation (SD) of the distribution were 1.56% as expected and 0.63%, respectively. The difference between the most popular triplet, AGG (3.27%), and the least frequent triplets, TTA (0.42%), was high at 2.85% and 7.8 times difference in frequency. Most frequently reported triplets (frequencies greater than mean plus SD) were of eleven types, AGG, TGG, CTG, TCC, ACC, CAG, AGC, TTC, GTG, CAC, and TGC. The least frequently reported triplets (frequencies less than mean minus SD) were of six types, TTA, TAA, CGA, ATT, CGT, and GGG.

Two concerns for the approach used here were addressed. One was that the frequency of a 3′-end triplet may predominantly be determined by its frequency in the target genome sequences. To address this concern, three genome sequences, the most popular templates in the analyzed set (BHV, BVDV, and FMV), were tested for finding whether their genome sequences biased the 3′-end triplet frequencies. Although the details of the genome comparison results were not discussed here, the trends in 3′-end triplet frequencies shown were not due to target genome sequences *(8)*.

Another concern was that the trends in triplet frequencies may be an exact copy of recommendations for primer design in literature. If one design method is used by all primer designers, the trends in triplet frequencies should reflect choices made by the method even though the method may not choose the best 3′-end. Because all recommendations have theoretical support, it was not surprising that the trends in triplet frequencies met some of the recommendations. However, no single set of recommendations was able to predict the most and the least frequent triplets for the analyzed set. The analysis actually revealed triplet frequencies that could not be expected from the recommendations in the literature.

Thus, neither concern was valid for the analyzed set. The recommendation has been derived from the trend in 3′-end triplet frequencies. In the VirOligo database, WSS (19.0%; six of top 11) and SWS (18.0%; four of top 11) were the two most common kinds of 3′-end triplets. Whereas two WSS triplets, the WGC pair, were preferred (top 7th and 11th), another two WSS triplets, the WCG pair, were not favored (worst 14th and 26th). Triplets with “CG” were

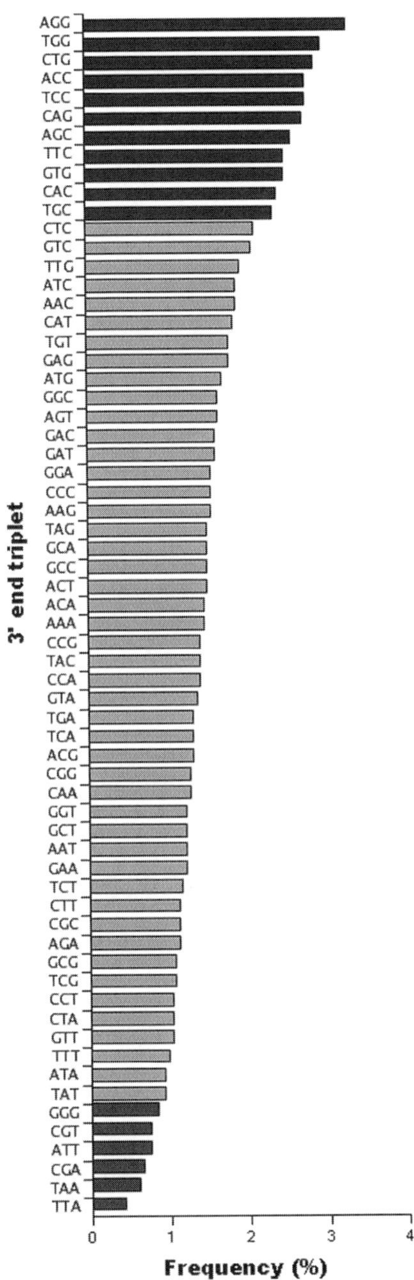

Fig. 5. Distribution of the 3′-end triplets of PCR primers. (Reproduced from **ref.** *8*, Copyright (2004), with permission from Elsevier).

Table 1
3′-End Triplet Frequencies

Triplet	%	Triplet	%	Triplet	%	Triplet	%
AAA	1.45	CAA	1.26	GAA	1.22	TAA	0.61
AAC	1.87	CAC	2.39	GAC	1.59	TAC	1.40
AAG	1.54	CAG	2.71	GAG	1.78	TAG	1.50
AAT	1.22	CAT	1.82	GAT	1.59	TAT	0.94
ACA	1.45	CCA	1.40	GCA	1.50	TCA	1.31
ACC	2.76	CCC	1.54	GCC	1.50	TCC	2.76
ACG	1.31	CCG	1.40	GCG	1.08	TCG	1.08
ACT	1.50	CCT	1.03	GCT	1.22	TCT	1.17
AGA	1.12	CGA	0.66	GGA	1.54	TGA	1.31
AGC	2.57	CGC	1.12	GGC	1.64	TGC	2.34
AGG	3.28	CGG	1.26	GGG	0.84	TGG	2.95
AGT	1.64	CGT	0.75	GGT	1.22	TGT	1.78
ATA	0.94	CTA	1.03	GTA	1.36	TTA	0.42
ATC	1.87	CTC	2.11	GTC	2.06	TTC	2.48
ATG	1.68	CTG	2.85	GTG	2.48	TTG	1.92
ATT	0.75	CTT	1.12	GTT	1.03	TTT	0.98

somehow less frequent in VirOligo, and CGW represented two of the worst six triplets. It should be noted that TTS triplets were also preferred (top 8th and 14th), and all top 16 triplets were NNS at the 3′-end (N stands for any of the four nucleotides). Thus, primers should be designed to have SWS, WSS, or TTS as the 3′-end triplet but no CG in the triplet. If such primers are not available, one should select primers with S at the 3′-end. However, one should never select primers with the worst triplet types: WWW, CGW, or GGG at the 3′-end.

2. Materials

Any primer design programs can be used for the 3′-end selection method described here. However, every primer design program has different searching algorism, and output primer pairs may not be the same. User settings in the programs also change the outputs. If more than one primer design program is available, the investigator should select a program with simple and easy exporting function of candidate primer sequences. For the 3′-end selection, it requires a number of primer sequences, and exporting them for comparison of the 3′-end can be a labored task without helpful exporting function.

Alternatively, a list of primer pairs can be prepared manually. A spreadsheet program (e.g., Excel) is helpful if automated 3′-end selection method (**Subheading 3.2.2.**) is preferred.

3. Methods

Primer design based on the 3′-end triplet recommendation above is simple and straightforward. The investigator needs to filter primer candidates obtained by primer design programs (or manually designed) according to the 3′-end primer recommendations. During selection processes of primer candidates, it is easier to find the one matching the recommendations best with more primer candidates. Thus, more primer pairs should be obtained from the primer design programs than are provided by default.

3.1. Obtaining Primer Pairs

Use a primer design program as usual, but obtain twice or more primer pairs than are given using its default settings (*see* **Notes 2** and **3**).

- Case of Primer3 (by Steve Rozen and Helen J. Skaletsky)
  - "GC clamp" should be set to 0.
  - "Number To Return" should be set to 20 or more.
- Case of Primer Premier 5 (PREMIER Biosoft International)
  - "GC clamp" function does not need to be checked off. GC clamp in Primer Premier checks primer's stability at 5′-end, and it does not interfere with the 3′-end recommendations.
  - Primer Premier returns all primer pairs that passed its requirements. Thus, the number of outputs can not be set. However, if there are too few, conditions such as melting temperature can be relaxed.
- Case of Vector NTI Advance 9.1 (Invitrogen)
  - No modifications of the default settings are required for GC clamp (*see* **Note 4**).
  - "Maximum Number of Output Options" should be 20 or more instead of 3 in the default configuration.
- Case of DS Gene 1.5 (Accelrys)
  - "S" in "3′-dinucleotide" field should be removed (*see* **Note 4**).
  - DS gene returns all primer pairs that passed their requirements and does not need to be set the number of outputs.

After obtaining enough primer pairs, the next step is filtering and ranking the primers. The best suited triplets are TTS, SWS, and WSS (no WCG), and they should be searched from the program's output, a list of primer pairs.

3.2. The 3′-End Triplet Selection Process

3.2.1. Manual 3′-End Selection Method

1. Find TTS, SWS, and WSS but no WCG in the 3′-ends of primers near the top of the list of primer pair candidates (*see* **Note 5**).
   If suitable pairs are found, the selection process is done here. Use them for PCR. If not, the investigator needs to find suitable pairs without selecting the disfavored 3′-end triplets.
2. Find WCG, WWW, CGW, and GGG in the 3′-end triplets of primers and discard them from a list of primer pair candidates.
3. Find primers with S at the 3′-end near the top of the list of primer pair candidates. If no suitable primer pairs are found by **step 3**,
4. Select a primer pair from the top of the list (*see* **Note 6**).

Instead of following **steps 1–4** in **Subheading 3.2.1**, primers can be selected based on the 3′-end triplet frequencies. The recommendation discussed in this chapter may be missing some of important triplets near the top of the frequencies list or avoidable triplets near the bottom of the list. By focusing on the list, the success rate of the PCR experiments can be increased. However, it should be noted that other properties of primers scored by primer design programs may be neglected when the 3′-end triplet frequencies of primer are emphasized although primers with extremely low properties, such as too high or low T_m, have probably been discarded and are not in the list of primer candidates put out by primer design programs.

3.2.2. Automated 3′-End Selection Method (Alternative of **Subheading 3.2.1**.)

Prepare a spreadsheet as per the following steps (*see* **Fig. 6**).

1. Copy and paste, or type in pair ID numbers, forward primer sequences, and reverse primer sequences in columns A, B, and C of a spreadsheet, respectively, from outputs in the primer design program (*see* **Note 7**).
2. Enter the triplet frequencies from **Table 1**, triplet types in column J and percentages in column K.
3. Obtain the right-hand three characters as a substring of the forward primer by entering formula "=Right(B3,3)" in cell D3.
4. Copy the cell D3 and paste it to all cells between D3 and F22.

Microsoft Excel – 3′end_Selection_by_Excel.xls

	A	B	C	D	E	F	G	H	I	J	K
1		Primer seqence		3′ end triplet		Triplet Frequency				DATA	
2		Forward Primer	Reverse Primer	Forward	Reverse	Forward	Reverse	F*R		Triplet	%
3	0	ACGGGTACTTCTGCCAAATG	GGGTGCTTTTGACCTGTTGT	ATG	TGT	1.68	1.78	2.99		AAA	1.45
4	1	ATCACCGACCACATGACAGA	TACGGTACCCCTGTGCTTTC	AGA	TTC	1.12	2.48	2.78		AAC	1.87
5	2	TCACCTACTTTGCGCTGATG	GGTTGCCTACCATGAGCATT	ATG	ATT	1.68	0.75	1.26		AAG	1.54
6	3	GCTGATGGATGGAAAGGTGT	GGTTGCCTACCATGAGCATT	TGT	ATT	1.78	0.75	1.33		AAT	1.22
7	4	TGGGCCTCAAAATCACCTAC	GGTTGCCTACCATGAGCATT	TAC	ATT	1.40	0.75	1.05		ACA	1.45
8	5	TTGCCCCTTATCAAAGCAAC	TGAGTGCTGCCACTAACCTG	AAC	CTG	1.87	2.85	5.33		ACC	2.76
9	6	GGTGCGAAGATGGGAAGTTA	GGGCAGAGTCCAAATTCAAA	TTA	AAA	0.42	1.45	0.61		ACG	1.31
10	7	CTGCCTGCTTACCCAAGAAC	CAACGTGACTTTGTGGGATG	AAC	ATG	1.87	1.68	3.15		ACT	1.50
11	8	CATGAGTGGAACAAGCATGG	CTGTTAAGGGTGTGGGGCTA	TGG	CTA	2.95	1.03	3.03		AGA	1.12
12	9	ACACGAAAGAAGAGGGAGCA	GCGTGATTCCTGAGGTTTGT	GCA	TGT	1.50	1.78	2.66		AGC	2.57
13	10	GATCAGGCTGCAACAGATGA	CCCTTGTGAGTCCATCCCTA	TGA	CTA	1.31	1.03	1.35		AGG	3.27
14	11	TCACAGAGCACGAAAAATGC	GTGTAAGCCCAGGCAGTCTC	TGC	CTC	2.34	2.10	4.92		AGT	1.64
15	12	ACACGAAAGAAGAGGGAGCA	ATGGTACAAGCCGTCCTGAG	GCA	GAG	1.50	1.78	2.66		ATA	0.94
16	13	GAGCGTGGTGATACTGAGCA	TGGACTCAATAGCGTTGTGC	GCA	TGC	1.50	2.34	3.50		ATC	1.87
17	14	GATCAGGCTGCAACAGATGA	TTTCCCTTGTGAGTCCATCC	TGA	TCC	1.31	2.76	3.61		ATG	1.68
18	15	CTCAGGACGGCTTGTACCAT	ATCCCTTCCGTCCCATTATC	CAT	ATC	1.82	1.87	3.41		ATT	0.75
19	16	CATGAGTGGAACAAGCATGG	AAGGGTGTGGGGCTATCTCT	TGG	TCT	2.95	1.17	3.44		CAA	1.26
20	17	GGTCACCACATGGCTAGCTT	TGGCGTCAGGCTCTACTTTT	CTT	TTT	1.12	0.98	1.10		CAC	2.38
21	18	CACCGACCACATGACAGAAG	TACGGTACCCCTGTGCTTTC	AAG	TTC	1.54	2.48	3.82		CAG	2.71
22	19	TAGCCCCACACCCTTAACAG	TTCCTTGTCTGGCACCTTCT	CAG	TCT	2.71	1.17	3.17		CAT	1.82
23										CCA	1.40
24		Primer3 output in the defualt configuration (Number To Return set at 20)								CCC	1.54
25			GI: 9626649 (Bovine viral diarrhea virus 1)							CCG	1.40
26										CCT	1.03
27		D3	=RIGHT(B3,3)							CGA	0.65
28		F3	=VLOOKUP(D3,$J$3:$K$66,2)							CGC	1.12
29		H3	=F3*G3							CGG	1.26

Sheet1 / Sheet2 / Sheet3

Fig. 6. Screen image of automated 3′-end selection method.

5. Look up its frequency value from column J and K by entering formula "=vlookup(D3, j3:k66,2)" in cell F3.
6. Copy the cell F3 and paste it to all cells between F3 and G22.
7. Multiply triplet frequencies of forward and reverse primers by entering formula "= F3*G3" in cell H3.
8. Copy the cell H3 and paste it to all cells between H3 and H22.
9. Sort the table based on the frequencies of products in column H.

Then, report the top-scoring pairs in the spreadsheet. Because it can be automated, this enhancement could allow entry of large sets of primer pairs.

4. Conclusion

In this chapter, a primer selection method based on 3′-end triplet sequence was introduced. The disfavored triplets do not mean that primers with such triplets do not work at all in PCR experiments and only mean that such primers have more chances of failing to amplify the target templates. Such outputs should not be discarded from the primer design program even though all primer candidates obtained from the program did not meet the 3′-end recommendations described in this chapter.

If a pair that matches the 3′-end recommendations is found using the recommendations, the success rate of PCR experiments is increased and that will reduce costs and one's stress from unsuccessful PCR experiments and low amplification efficiencies. Such effects are especially significant in large-scale PCR experiments such as the amplification of genes or gene fragments for preparation of microarrays for gene expression analysis.

5. Notes

1. The VirOligo database: For the analysis of the 3′-end triplets here, all primer sequences data were retrieved from the VirOligo database. The VirOligo database has been developed by the author of this chapter and Dr. Melcher at Biochemistry and Molecular Biology Department, Oklahoma State University *(7)*. The database's entries have been significantly increased after the retrieval made for the analysis in this chapter, and the database is currently maintained by Ms. Song and Dr. Melcher. It is a publicly available free Web-based database accessible at any time at http://VirOligo.okstat.edu/. It is not only collections of primer sequences. It is actually collections of successful PCR experiments in published articles from refereed literature including target virus names, target region specific to in the target, thermal cycles, and concentrations and names of reagents and polymerase used in the experiments. As of December 2005, the database consists of 8151 primer sequences specific for 525 virus species. If PCR experiments on viral genomes are planed, it is recommended to check any primers and conditions applicable before designing new primers. The successful PCR primers and knowledge of experimental conditions may save elaborate optimizations, and it can be faster than designing new PCR primers.
2. Many primer design programs allow the user to define the number of primer pairs as an output. The favored triplets are 11 of 64 types (17%). Because there are two primers in a pair for PCR, the chance of getting both forward and reverse primers with the recommended triplets is 17% multiplied by itself, or as small as 3%. Fifty output pairs are expected to have at least one pair that matches the requirements if nothing interferes with 3′-end sequence. A larger primer output is better, but more lower ranked pairs appear in the output in such case, and low ranked pairs have undesirable properties in terms of G + C contents or T_m of primers. In addition, it simply is not easy to handle very large numbers of primers. So, try to obtain about twice the default number.
3. When primers are obtained by a primer design program, the user needs to note that some primer search settings such as 3′-end "GC clamp" interfere with the 3′-end triplet selection. Such settings should be turned off. Otherwise, all 3′-ends in output primer set will have uniform triplets as a result of the search configuration.

   GC clamp has been used for point mutation detection by adding GC-rich sequences to 5′-end of PCR primer ***(11)***. When PCR products with G + C-rich 5′-end primers were differentiated mutated sequences from wide type, GC clamp helps avoiding complete denaturation of DNA strands during a denaturing gradient gel electrophoresis process. GC clamp for point mutation detection is simply by adding it to the 5′-end of a primer, and it is essentially not a complementary sequence of the gene sequence.

In primer design programs, GC clamp means different from that for point mutation detection. G + C-rich sequence creates strong annealing of primers to the template. Thus, some primer design programs search for G + C-rich or stable region in a target template. Some of them look for GC clamp at the 5′-ends of primers (e.g., Primer Premier) as that of point mutation detection does at the 5′-end. On the contrary, some programs (e.g., Primer3) have a function to search G + C-rich region at the 3′-ends of primers. Although such function can be used for ensuring complete annealing of the 3′-end of a primer to the template, it interferes with the recommendations in this chapter. Please consult a manual for primer design program using and find out a type of "GC clamp" that the primer design program searches for. In case the program has the searching function for GC clamp at the 3′-end, such function should be turned off or the number of GC clamp required should be set to zero.

4. Both Vector NTI Advance 9.1 and DS Gene 1.5 let the user decide on a 3′-end sequence to search for primer candidates. Because the recommendations described here consist of three types, SWS, WSS, and TTS, it is not possible to search all three types using this function in one primer search session. The user may try one triplet type at a time using such 3′-end restriction function, but the user should note that a better primer pair might be a forward primer with SWS type and a reverse primer with WSS type. Thus, such functions are not easy to apply for the present 3′-end recommendations, but it may be worth trying to find "S" at the 3′-end as all top 16 triplets contain "S" at the 3′-end.
   The following are the steps for finding "S" at the 3′-end:

   - Case of Vector NTI Advance 9.1
     In "3′-end" tab of "find primer" window of Vector NTI, click off A and T for first nucleotide and leave the rest for both sense and anti-sense primer 3′.
   - Case of DS Gene 1.5
     Use the default settings ("S" should be appeared in "3′-dinucleotide" field).

5. In most programs, e.g., in Primer3, the list of primer candidates is sorted from the best to the worst. So, try to find primers with TTS, SWS, and WSS at the 3′-ends near the top of the list to find the best scored primer pairs with the best triplet selection.
6. Primer pairs that should be avoided have been discarded by **step 3**. Rest of primer pairs are neither extremely preferred nor not-preferred ones. Thus, an average success rate in the PCR experiment can be expected from the primer pairs.
7. Recording primer pair ID number from the primer design program will help tracking primer pairs after sorting the spreadsheet in later processes.

Acknowledgment

The author expresses appreciation to Dr. Ulrich Melcher of Department Biochemistry and Molecular Biology, Oklahoma State University for advice on and review of the manuscript.

References

1. Mitsuhashi, M. (1996) Technical report: Part 2. Basic requirements for designing optimal PCR primers. *J. Clin. Lab. Anal.* 10, 285–293.
2. Sharrocks, A.D. (1994) The design of primers for PCR, in *PCR Technology: Current Innovations* (Griffin, H.G., and Griffin, A.M., eds.), CRC Press, Boca Raton, FL, pp. 5–12.
3. Sambrook, J., and Russell, D.W. (2001) In vitro amplification of DNA by the polymerase chain reaction, in *Molecular Cloning: A Laboratory Manual* (Sambrook, J., and Russell, D.W., eds.), Cold Spring Harbor Laboratory, New York, pp. 8.1–8.17.
4. Kidd, K.K., and Ruano, G. (1995) Optimizing PCR, in *PCR 2: A Practical Approach* (McPherson, M.J., Hames, B.D., and Taylor, G.R., eds.), IRL Press, Oxford, pp. 1–21.
5. Hyndman, D., Cooper, A., Pruzinsky, S., Coad, D., and Mitsuhashi, M. (1996) Software to determine optimal oligonucleotide sequences based on hybridization simulation data. *BioTechniques* 20, 1090–1094,1096,1097.
6. O'Connell, J. (2002) The basics of RT-PCR, in *RT PCR Protocols* (O'Connell, J., ed.) vol. 193, Humana Press, Totowa, NJ, pp. 19–25.
7. Onodera, K., and Melcher, U. (2002) VirOligo: a database of virus-specific oligonucleotides. *Nucleic Acids Res.* 30, 203–204.
8. Onodera, K., and Melcher, U. (2004) Selection for 3′ end triplets for polymerase chain reaction primers. *Mol. Cell. Probes* 18(6), 369–372.
9. Onodera, K. (2002) Construction, application and analysis of the oligonucleotide database, VirOligo, PhD thesis, Oklahoma State University, Stillwater, OK.
10. SantaLucia, J. (1998) A unified view of polymer, dumbbell, and oligonucleotide DNA nearest-neighbor thermodynamics. *Proc. Natl. Acad. Sci. USA.* 95(4), 1460–1465.
11. Traystman, M.D., Higuchi, M., Kasper, C.K., Antonarakis, S.E., and Kazazian, H.H. Jr. (1990) Use of denaturing gradient gel electrophoresis to detect point mutations in the factor VIII gene. *Genomics* 6(2), 293–301.

4

The Reference Point Method in Primer Design

Thomas Kämpke

Summary

The conflicts between several design objectives for PCR primers are computationally resolved by specifying a set of ideal parameters and searching for primer pairs whose parameters approximate this ideal point as close as possible. It thus becomes feasible to identify an "optimum" and to efficiently compute it. User-specified target regions, primer conditions, and procedural conditions are obeyed.

Key Words: Computational search; ideal point; molecular design; multicriteria decision making; nearest neighbor thermodynamics; scoring.

1. Introduction

PCR primers must meet multiple criteria that are usually set by the experimenter. In most cases, there are multiple primer pairs that conform to the input parameters. The goal of the reference point method is to choose the best primer pair. The method ranks all primer pairs by some score that is indicative of how close a primer pair is to a user-specified ideal pair. Below we describe how to the set the ideal pair, called reference point, and how to aggregate the scoring functions.

The design of PCR primers adheres to a multitude of objectives in the same way as many technological products and processes do. While it appears to be familiar that the "input" to a problem consists of multiple variables, multiple objectives account for the fact that the "outcome" may also consist of multiple variables. These are synonymously called objectives, scores, criteria, or performance measures. Examples are the melting temperature, the GC content, etc.

From: *Methods in Molecular Biology, vol. 402: PCR Primer Design*
Edited by: A. Yuryev © Humana Press, Totowa, NJ

The set of objectives is specified by the developer of the design method and remains fixed from the experimenter's perspective, but the experimenter sets ideal values in each instance of a PCR design. Objectives may be reached more or less by the oligonucleotide strings of a pair of forward and reverse primers.

The most common notation for situations with multiple outcome variables is that of multicriteria problems or multicriteria decision problems *(1)*. The entities under consideration in PCR are the primers, which are understood as alternatives indicating that there always are alternative choices.

It is typical for multicriteria problems that no single choice of design parameters simultaneously optimizes all the objectives though this may happen in rare exceptions. Normally, after some initial improvement, the decision situation gets close in the sense that a further improvement with respect to one objective can only be achieved by a sacrifice in another objective. We focus on these trade-offs in the sequel. They are made explicit and a solution method—the reference point method—is adapted to primer design.

Most primer design programs use some sort of reference point method even if the underlying method is not named so or named otherwise. Details are exemplified below. But instead of stipulating that the given details are the only ones possible, their interplay should become clear below. Most users of primer design programs have a good sense of few or all objectives but not of their aggregation. Some light will be cast onto this matter. We will deal with primer design software in a generic sense rather than with one particular program.

The process of primer design by the reference point method, which is also named as utopia point method and ideal point method *(2)*, is split into two stages. The first stage consists of the computational generation of primer pairs whereas the second stage consists of their evaluation. In principle, the stages are unrelated but a loose coupling provides for computational efficiency.

The second stage, the evaluation stage, is pivotal and its intuition is as follows. An ideal value is set for every objective such as an ideal melting temperature, an ideal primer length, etc. The vector of these values forms a point, the reference point, in multidimensional space. Actually, the reference point is a vector but it is called "point" for traditional reasons. The corresponding vector of objective values is now supposed to come as close as possible to the reference point. The vector of objective values varies with the primers generated in the first stage. The primer pair whose objective value has smallest distance to the reference point is considered as the best pair, the pair with second smallest distance to the reference point is considered as second best, etc. It is intrinsic to the reference point method that it yields a ranking of all alternatives. The user will only see the high scorers of that ranking similarly to result lists of Internet search engines.

Setting an ideal value for each objective is by no means unique. These settings often reflect subjective experience or judgments by users or by programmers—whether recognizable as subjective or not. As a consequence, solving a multicriteria decision problem by the reference point method may involve the use of several reference points. This will allow one to inspect the sensitivity of the proposed primer rankings. However, certain settings for the reference point may turn out to be reasonable, and variation of the reference point is not a mandatory element of the reference point method.

Actually, the ranking itself can be considered as a mild form of sensitivity analysis. This saves the experimenter's time and effort by not requiring him or her to play with input parameters while optimizing the primer design. Reasonable values for reference points are given below in section 3.1.

The term "reference point" must not be confused with a point of interest of a real-valued function of several variables, for example, of a molecular potential function *(3)*. Noteworthy, PCR *(4)* today attracts attention from many fields including even type-of-work analysis in psychology *(5)*.

2. The Situation

Primer assessment extends beyond mere string matching by involving such criteria as the proximity between primer-melting temperatures, minimization of hybridization effects between forward and reverse primers, and the avoidance of hybridization of primers with themselves. The latter two criteria are dealt with by annealing scores. This takes into account that PCR primers can freely float in an assay. When primers are used as probes on microarrays on which they are immobilized, annealing of a primer with itself is less important and annealing with another primer is even more so.

All criteria for the design of one pair of PCR primers yield real numbers. The reference point method as set out here is based on 12 criteria for which an implementation and evaluations were performed *(6)*. The adaptation of the method to another number of criteria and to other sets of criteria, as indicated further below, is straightforward. To be very clear, the criteria assess one pair of primers rather than a single primer. In our case, the reference point method requires specifying 12 ideal values, one for each criterion. The differences between the reference point values and the actual objective values are then aggregated by a weighted sum for each primer pair.

Although the computational generation of primers precedes the computational evaluation of primers, the evaluation step is stated first. The reason is that the particular form of the evaluation makes understandable certain computational speed-ups in the generation step.

Although computationally forecasting the outcome of PCR hybridizations is all but certain, we assume exactly that. For the sake of simplicity, no formal uncertainty method is invoked. It is left to the fast growing experience of a user of PCR software to adapt to the varying reliability of outcomes. The proposed method supports this important aspect by suggesting a ranked list of primer pairs rather than a single pair. The final decision if and which primer pair to select must be made by a user.

A pair of primers (p, q) with forward primer p and reverse primer q is specified by nucleotide strings.

2.1. Primer Length

The length of a primer is nothing but the number of its oligonucleotides.

2.2. GC content

The GC content measures the fraction of G's and C's of the primer $GC(p) = (\#G \text{ in } p + \#C \text{ in } p)/|p|_* 100$. The reason for considering this quantity is the presence of three hydrogen bonds in GC pairs as compared with only two bonds for AT pairs in any bonding of the primer.

2.3. Melting Temperature

A universally valid computation of the primer-melting temperature does not seem to exist. A prominent approximation for the melting point is a formula that is based on so-called nearest neighbor thermodynamics ***(7,8)***. Several different parameter settings exist for these nearest neighbors ***(9)***, and the primer length for which these are valid is difficult to assess. The nearest neighbor thermodynamics is quite versatile because it considers the type and concentration of the assay. A simple approximation ***(10)*** of the melting temperature in degrees Celsius, called Wallace formula, is the linear function $T_{\mathrm{m}}(p) = 2*\ (2\ \#G \text{ in } p + 2\#C \text{ in } p + \#A \text{ in } p + \#T \text{ in } p)$.

2.4. Self-Annealing

Each primer is tested and scored for unintended hybridization with itself by testing for self-annealing and for self-end-annealing ***(11)***. The test for self-annealing accounts for the primer-dimer effect, which is hybridization of one part of a primer with another part.

The self-annealing score of a primer is computed by aligning the primer and its reversal in all theoretically possible overlap positions. The reversal of a primer is not to be confused with the reverse primer of a primer pair. For each overlap position, a match of an A with a T contributes two units, each match of a G with a C contributes four units, and all other matches contribute zero units to the self-annealing score. The contributions are summed for each fixed overlap position. The maximum over all overlap positions is retained as self-annealing score.

2.5. Self-End-Annealing

The test for self-end-annealing considers only those alignments for which the 3′-end belongs to the overlapping region. More important, only the subsequences of uninterrupted bonds beginning at the 3′-end of either the primer or its reversal are considered for self-end-annealing. The self-end-annealing of a primer never exceeds its self-annealing value, $sea(p) \leq sa(p)$.

2.6. Pair Annealing

The two primers of a pair must also be tested for unintended hybridization with each other. This test again consists of two modes that are called pair annealing and pair end-annealing. For pair annealing, both primers are arranged in all overlapping antithetic alignments and the maximum score is retained as pair annealing value. Note that, now, the forward primer is aligned with the reverse primer and no reversal of any primer is considered.

2.7. Pair End-Annealing

The test for end-annealing involves only alignments with at least one of the 3′-end belonging to the overlapping region, and only subsequences of uninterrupted bonds beginning at one of the 3′-ends are considered for the pair end-annealing value of p and q.

Example: The primers of a typical pair (p, q) with their individual assessments are given in Table 1 and the common assessment of the primer pair is given in Table 2.

Table 1
Primers with Individual Assessments

	Forward primer		Reverse primer
p	GGATTGATAATGTAATAGG	q	CATTATGGGTGGTATGTTGG
$\|p\|$	19	$\|q\|$	20
$GC(p)$ [in %]	32	$GC(q)$ [in %]	45
$T_m(p)$ [in °C]	50	$T_m(q)$ [in °C]	58
sa(p)	12	sa(q)	20
sea(p)	0	sea(q)	4

Table 2
Primer Pair with Common Assessments

(p, q)	(GGATTGATAATGTAATAGG,CATTATGGGTGGTATGTTGG)
$pa(p,q)$	16
$pea(p,q)$	4

3. Method

3.1. Aggregation of PCR Objectives by Reference Points

Any primer pair (p, q) is assigned a scoring vector with 12 coordinates:

$$\mathrm{sv}(\mathrm{p}, \mathrm{q}) = (|\mathrm{p}|, |\mathrm{q}|, \mathrm{GC}(\mathrm{p}), \mathrm{GC}(\mathrm{q}), \mathrm{T_m}(\mathrm{p}), \mathrm{T_m}(\mathrm{q}), \mathrm{sa}(\mathrm{p}), \mathrm{sa}(\mathrm{q}), \mathrm{sea}(\mathrm{p}), \mathrm{sea}(\mathrm{q}), \mathrm{pa}(\mathrm{p}, \mathrm{q}), \mathrm{pea}(\mathrm{p}, \mathrm{q})).$$

All primers are designed to have ideal values of length, GC content, melting temperature, and annealing scores. The six ideal values for length, GC content, and melting temperature—three for the forward primer and three for reverse primer—are specified by the designer of the hybridization experiment, and they are part of the input of a primer design program based on such a reference point. The ideal settings for the annealing scores, which are the remaining six coordinates, are all zero. The reason therefore is that zero annealing values are ideal for all design instances. Thus, they cannot be edited by the user.

The ideal score vector or reference point for primer pairs is

$$sv_{\mathrm{ref}} = (length_{\mathrm{f}}, length_{\mathrm{r}}, GC_{\mathrm{f}}, GC_{\mathrm{r}}, T_{\mathrm{m,f}}, T_{\mathrm{m,r}}, 0, 0, 0, 0, 0, 0).$$

Though not necessary, the typical settings will require that the GC content and, in particular, the melting temperatures of the primers should be the same. Thus, $GC_{\mathrm{f}} = GC_{\mathrm{r}}$ and $T_{\mathrm{m,f}} = T_{\mathrm{m,r}}$. The six input coordinates of the reference point are thus reduced to four values: ideal lengths of forward and reverse primer, which may be different, ideal GC content, and ideal melting temperature, with the latter two being valid for both primers.

The evaluation of any primer pair (p, q) can be performed by taking the sum of absolute distances between the two vectors:

$$||sv(p, q) - sv_{\mathrm{ref}}|| = \sum_{i=1}^{12} |sv(p, q)_i - sv_{\mathrm{ref}\ i}|.$$

The reason to not apply the celebrated sum of squared distances is that the units of different coordinates may lead to strongly differing contributions of the sum. The trade-offs between coordinates can be controlled better by linear instead of quadratic differences. Therefore, we even propose to use a weighted distance of the form

$$||sv(p, q) - sv_{\text{ref}}||_w = \sum_{i=1}^{12} w_i{}^* |sv(p, q)_i - sv_{\text{ref}\ i}|.$$

The weights do not have to fulfill any other condition than being positive. They can be considered as reflecting the relative importance of the objectives at given units.

The weights can always be divided by the weight of the melting temperature if this weight is unique; deviations from the common ideal temperature of the forward and the reverse primer are supposed to weigh equally. Then, normalization by the temperature weight trivially gives a new, unit temperature weight. The weighted difference between a coordinate of the score vector and the reference point can thus be considered as a degree Celsius equivalent. One set of reasonable weights with the foregoing property is given in Table 3.

The following reference point is considered to allow a sample computation of a weighted distance:

$$\begin{aligned} sv_{\text{ref}} &= (length_{\text{f}}, length_{\text{r}}, GC_{\text{f}}, GC_{\text{r}}, T_{\text{m,f}}, T_{\text{m,r}}, 0, 0, 0, 0, 0, 0) \\ &= (21, 22, 50\%, 50\%, 57\,^{\circ}\text{C}, 57\,^{\circ}\text{C}, 0, 0, 0, 0, 0, 0). \end{aligned}$$

Table 3
Weight Set for Deviations

Deviation from ideal lengths $length_{\text{f}}$, $length_{\text{r}}$	$w_1 = w_2 = .5$
Deviation from ideal GC content $GC_{\text{f}} = GC_{\text{r}}$	$w_3 = w_4 = 1$
Deviation from ideal melting temperature $T_{\text{m,f}} = T_{\text{m,r}}$	$w_5 = w_6 = 1$
Deviation from ideal self-annealing value 0	$w_7 = w_8 = .1$
Deviation from ideal self-end-annealing value 0	$w_9 = w_{10} = .2$
Deviation from ideal pair annealing value 0	$w_{11} = .1$
Deviation from ideal pair end-annealing value 0	$w_{12} = .2$

All coordinates of the reference point are dimensionless; the % and °C units are only specified for better intuition. The weighted distance between this reference point and the primer pair from **Subheading 2** is

$$\begin{aligned} &||sv(GGATTGATAATGTAATAGG, CATTATGGGTGGTATGTTGG) - sv_{\text{ref}}||_w \\ =&.5_*|19-21|+.5_*|20-22| \\ &+1_*|32-50|+1_*|45-50| \\ &+1_*|50-57|+1_*|58-57| \\ &+.1_*|12-0|+.1_*|20-0| \\ &+.2_*|0-0|+.2_*|4-0| \\ &+.1_*|16-0| \\ &+.2_*|4-0| \\ =&39.4. \end{aligned}$$

The foregoing pair competes with the value 39.4 in any comparison with other primer pairs. The smaller the value, the better the primer pair, and a pair with smallest weighted distance to the reference point is considered as optimal.

Not all objectives bear a general direction of monotonicity with respect to the reference point. Although this applies to the last six coordinates where a smaller annealing score is always preferable to a larger score with all other criteria being equal, there is no such general monotonicity in primer lengths, GC content, and melting temperature. In the non-monotone objectives, deviations of the same size from the reference point are considered as equally unwanted whether they overshoot or fall short of the ideal value. But, for example, overshooting of the ideal melting temperature might be considered worse than falling short of it. The distance term for the forward primer temperature—the fifth term—can therefore be modified as follows to an unsymmetric penalty function:

$$w_{5*}|T_{\text{m}}(p) - T_{\text{m,f}}| \rightarrow w_{5*}D(T_{\text{m}}(p), T_{\text{m,f}})$$

with

$$\begin{aligned} D(T_{\text{m}}(p), T_{\text{m,f}}) &= |T_{\text{m}}(p) - T_{\text{m,f}}|, \text{ if } T_{\text{m}}(p) \geq T_{\text{m,f}} \\ D(T_{\text{m}}(p), T_{\text{m,f}}) &= 1/2_*|T_{\text{m}}(p) - T_{\text{m,f}}|, \text{ if } T_{\text{m}}(p) \leq T_{\text{m,f}}. \end{aligned}$$

The modification for the reverse primer—the sixth term—would be the same, but we continue to discuss the unmodified weighted distance.

It may turn out that one primer pair is as least as good as another pair in all objectives and it is strictly better in at least one objective. Then, from the viewpoint of merely computing an optimal pair, the other pair, which is called dominated, can be neglected. However, the second best pair may already be dominated. Because this and other primer pairs may be of interest for comparison with the optimal pair, the dominated primer pairs are not excluded from further consideration. This is a technical but important feature of the present method, and it makes this method different from some other approaches in multicriteria decision making, even some in molecular biology ***(12)***.

The computational load for calculating the objective values may exceed that for calculating the weighted distance to the reference point. This is true, in particular, for the annealing scores.

3.2. The Geometry of Assessment by Reference Points

The set of isosurfaces that contain all equally evaluated primer pairs forms a diamond. The diamond is centered about the reference point, and the ratio of any two axes equals the inverse ratio of the corresponding weights. A hypothetical situation with only two objectives is sketched in **Fig. 1**. The objectives are not further specified.

The hypothetical situation allows to return to the PCR situation by considering projections into two dimensions. As an example, the self-annealing score and the self-end-annealing score are considered for the forward primer. Their contribution to the weighted distance from the reference point is

$$||sv(p, q) - sv_{ref}||_w = \ldots + .1_*|sa(p) - 0| + .2_*|sea(p) - 0| + \ldots$$

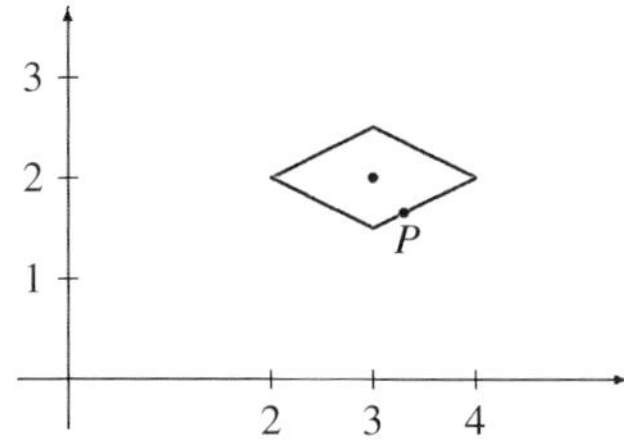

Fig. 1. Reference point (3, 2) and diamond of unit distance with coordinate weights 1 and 2. The points $P = (x_P,\ y_P)$ on the isosurface are thus given by the equation $||P - (3,\ 2)||_w = 1 \cdot |x_P - 3| + 2 \cdot |y_P - 2| = 1$.

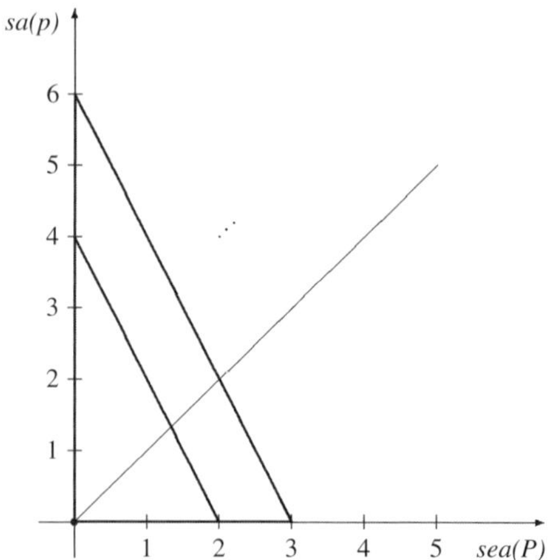

Fig. 2. Projected reference point at the origin and triangles whose slanted lines denote the isosurfaces for the levels 0.4 and 0.6 of the self-annealing and self-end-annealing scores. Actual scoring points of primer pairs lie above or on the diagonal because the self-anneal score is always greater or equal to the self-end-anneal score.

The isosurfaces reduce to the upper right quadrant of a diamond as sketched in **Fig. 2**. All forward primers, which lie on the same line, contribute equally to the overall score.

3.3. Computation of the Primer Selection Set

The first stage of the reference point method amounts to generating a set of primer pairs that compete for the shortest weighted distance to the reference point. The pairs may overlap in the sense of one primer possibly being the one primer of several pairs. Also, a string that appears as forward primer in one pair may appear as reverse primer in another pair.

Candidates for forward and reverse primers are generated from two user-specified windows of the DNA double string that enclose the section of interest or target (*see* **Fig. 3**). The section of interest may have a length of several thousand base pairs.

Ideally, the section of interest should be contained in the amplified area, but the amplified area should exceed the section of interest only by little. Because the hybridization product extends between the primers, the windows should be placed as close as possible to the section of interest. There is no law for

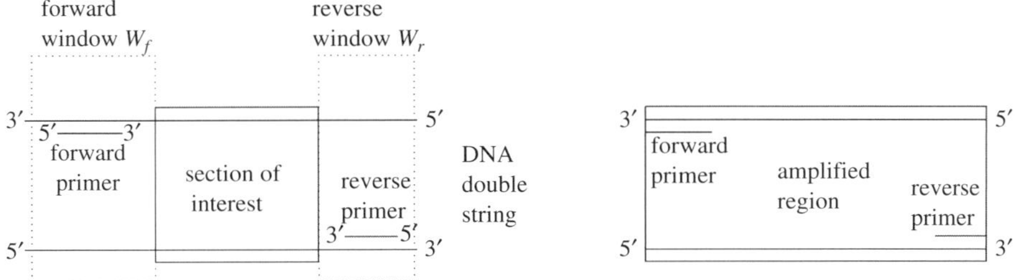

Fig. 3. DNA section of interest enclosed by a pair of windows from which primers are selected (left) and ideal terminating situation (right). The figures are not in scale.

choosing the window size. As a rule of thumb, the size of a window should be at least of triple ideal primer length. A reasonable choice for an ideal primer length of 20 thus is a window size of 60 or more.

The candidates for forward and reverse primers are chosen independently as (complementary) strings from the windows as indicated in the left part of **Fig. 3**. The reason for this kind of candidate selection is that such strings ideally match a section within the respective window.

In principle, every candidate for the forward primer is combined with every candidate for the reverse primer to form pairs. Because mere pairing tends to yield an unnecessary large set of primer pairs, each primer that enters the pair formation can be constrained by user-specific thresholds. These thresholds must be compatible with the reference point. For example, constraining primers to have lengths between 14 and 20 is totally unreasonable for ideal primer length parameters $length_{\mathrm{f}} = 22$ and $length_{\mathrm{r}} = 23$.

The ideal primer length is difficult to choose. The value 16 seems to be a reasonable lower bound for human nucleic DNA. This DNA has a total length of about three billion. Assuming that all four nucleotides appear randomly and independently with probability 0.25—which is not the case over some non-coding sections of the DNA—length λ yields a potentially unique string if $4^{\lambda} > 3 \cdot 10^{9}$. This requires $\lambda > 15.741$. Thermal considerations favor primers that are slightly longer than 16 base pairs ***(13)***.

Constraints are imposed on other objectives as well. These constraints must be set so that the reference point satisfies all of them. This is understood as loose coupling of the first and second stages of the reference point method. Besides a reduction in size of the primer candidate set, these constraints give additional control on the design. A sample list of lower and upper thresholds for the design objectives is as follows:

$$\lambda_l \leq |p|, |q| \leq \lambda_u$$

$$GC_l \leq GC(p),\ GC(q) \leq GC_u$$

$$T_{m,l} \leq T_m(p),\ T_m(q) \leq T_{m,u}$$

$$0 \leq sa(p),\ sa(q) \leq sa_u$$

$$0 \leq sea(p), sea(q) \leq sea_u$$

$$0 \leq pa(p,q) \leq pa_u$$

$$0 \leq pea(p,q) \leq pea_u.$$

A set of feasible primer pairs is generated by listing all strings of suitable length from the forward window and all strings of suitable length from the reverse window and then successively thinning out the lists until all constraints are satisfied.

Feas:

1. Input windows W_f, W_r and thresholds λ_l, λ_u, GC_l, GC_u, $T_{m,\,l}$, $T_{m,\,u}$, sa_u, sea_u, pa_u, pea_u.
   Initialization $F = \{\text{string } p | p \in W_f \text{ and } \lambda_l \leq |p| \leq \lambda_u\}$
   $R = \{\text{string } q | q \in W_r \text{ and } \lambda_l \leq |q| \leq \lambda_u\}$.
2. Thinning
   (a) For all $p \in F$
   if ($GC_l \leq GC(p) \leq GC_u$ is false) or ($T_{m,\,l} \leq T_m(p) \leq T_{m,\,u}$ is false) or ($sa(p) > sa_u$) or ($sea(p) > sae_u$)
   then $F = F - \{p\}$.
   For all $q \in R$
   if ($GC_l \leq GC(q) \leq GC_u$ is false) or ($T_{m,\,l} \leq T_m(q) \leq T_{m,\,u}$ is false) or ($sa(q) > sa_u$) or ($sea(q) > sae_u$)
   then $R = R - \{q\}$.
   (b) $FR_{feas} = F \times R$.
   (c) For all $(p, q) \in FR_{feas}$
   if ($pa(p, q) > pa_u$) or ($pea(p, q) > pea_u$) then $FR_{feas} = FR_{feas} - \{(p, q)\}$.
3. Output set of feasible primer pairs FR_{feas}.

Pairs of primers are considered from **step 2(b)** onwards. Thus, if a particular primer is eliminated when paired with another particular primer, it may be retained when paired with still another primer. The forward window contains $w_f - \lambda + 1$ strings of length λ when the window size or window length is

w_f. Thus, the window contains strings whose length ranges between the given thresholds.

$$\sum_{\lambda=\lambda_l}^{\lambda_u} w_f - \lambda + 1 = (w_f - \lambda_l + 1 + w_f - \lambda_u + 1)_* (\lambda_u - \lambda_l + 1)/2$$

This also is the number of strings in the list F at initialization. The number of feasible strings in the reverse window, which equals the number of strings in the initial list R, is computed similarly.

In case one of the individual feasibility sets F or R becomes empty in **step 2(a)**, the bounds on objectives involving single primers are to be relaxed. In case only the feasibility set FR_{feas} is empty, the upper bound on pair annealing or on pair end-annealing must be increased.

For performance reasons, the computation of the primer selection set by procedure Feas can be enhanced. This is achieved by exploiting the monotonicity of the melting temperature formula, whenever a string exceeds the upper bound of the melting temperature, every superstring does so and, hence, can be eliminated from the primer selection set without further test. More important, because some strings are superstrings of others, the computations of annealing values can be interleaved by dynamic programming *(6)*.

4. Notes

4.1. Note 1: Overall Algorithm

The reference point method can now readily be stated in complete form in generic terms. It requires three kinds of input parameters. The first kind describes the DNA section of interest and thresholds for the objectives. The second kind of input consists of all coordinates of the reference point, and the third kind refers to the weights for aggregating distances from the reference point. Practically, the reference point and the weight set will be changed significantly less often than the DNA section of interest. Thus, the reference point coordinates and the weight set are specified by data files, which are named rp.dat and w.dat. From the implementation viewpoint, some or all of the remaining input can also be read from files.

RP:

1. Input windows W_f, W_r and thresholds λ_l, λ_u, GC_l, GC_u, $T_{m,l}$, $T_{m,u}$, sa_u, sea_u, pa_u, pea_u.
   Initialization. Read reference point sv_{ref} from rp.dat and weights from w.dat.

2. Computations

   (a) Computation of FR_{feas} by Feas. If $FR_{feas} = \emptyset$ then request new input, else continue.
   (b) For all $(p, q) \in FR_{feas}$ compute $Dist_{ref}(p, q) = ||sv(p, q) - sv_{ref}||_w$.
   (c) Sort R_{feas} by increasing values of $Dist_{ref}(p, q)$.

3. Output sorted list FR_{feas}.

More conveniently, the input to a reference point algorithm may not require interactively setting upper bounds on the annealing objectives for each problem instance. But, in all other objectives, more control on the ideal and tolerable values can be given to the user by requiring that these are specified interactively. This requires slightly modifying the entries in the reference point file rp.dat making it slightly less systematic, but the procedure essentially remains the same.

RP-alt:

1. Input windows W_f, W_r and values $\lambda_l \leq length_f$, $length_r \leq \lambda_u$, $GC_l \leq GC_f$, $GC_r \leq GC_u$,
   $T_{m,\,l} < T_{m,\,f}$, $T_{m,\,r} < T_{m,\,u}$.
   Initialization. Read remaining coordinates (all zero) of reference point and upper annealing bounds from rp.dat and weights from w.dat.
2. Computations

   (a) Computation of FR_{feas} by Feas. If $FR_{feas} = \emptyset$ then request new input, else continue.
   (b) For all $(p, q) \in FR_{feas}$ compute $Dist_{ref}(p, q) = ||sv(p, q) - sv_{ref}||_w$.
   (c) Sort R_{feas} by increasing values of $Dist_{ref}(p, q)$.

3. Output sorted list FR_{feas}.

Schematically, the parameter input to the latter procedure may appear as in **Fig. 4**. This is only part of the whole input which also requires specifying the forward and the reverse window.

A typical output from generic primer design procedures based on reference points looks as follows:

1. p=GGATTGATAATGTAATAGG q=CATTATGGGTGGTATGTTGG
2. p=GGATTGATAATGTAATAG q=CATTATGGGTGGTATGTTGG
3. ...

4.2. Note 2: Variations

The reference point method allows one to consider a variety of other scoring criteria. These include testing for unintended hybridization within the DNA

```
                              Parameter input
                              ===============

                       -----            -----           -----
Primer length    min : |     |  ideal : |     |  max : |     |
                       -----            -----           -----

                       -----            -----           -----
Melting temp.    min : |     |  ideal : |     |  max : |     |
                       -----            -----           -----

                       -----            -----           -----
GC content       min : |     |  ideal : |     |  max : |     |
                       -----            -----           -----
```

Fig. 4. Schematic parameter input to primer design software.

section of interest such that an objective for the alignment of the primer with the section of interest is as low as possible. Another criterion is the number of "GC" substrings in one primer, which indicates stable bonding. A third criterion could consist of testing the primer for ending with "GC," which may be considered as favorable. More subtle criteria and some refinements of the foregoing scoring criteria are given in **ref.** ***14***. These include a dimer score that increases with proximity to the 3′-end of the primer.

The reference point method can also be adapted to significant changes of the primer design protocol. These changes admit the PCR of chemically modified rather than original DNA ***(15)***.

4.3. Note 3: Outlook

Multicriteria methods such as the reference point method are likely to gain importance for more complex versions of PCR and for other design tasks in molecular biology. One is so-called multiplex PCR, another is absorption distribution metabolism and excretion (ADME) cycle, and still another is the design of compound libraries.

The design complexity for primers increases for multiplex PCR, which amounts to performing multiple PCR reactions simultaneously in a single tube or in as few tubes as possible. This requires designing several primer pairs so that physical parameters such as cycle number, cycle duration, and annealing temperature are identical for all the PCR reactions. Moreover, the analysis of unintended primer–primer interactions becomes more intricate than for ordinary PCR. Designing multiplex primers with the intent of minimizing the number of diverse primers to use them in as many pairs as possible is investigated in **ref.** ***16*** by a combinatorial approach.

Multicriteria methods are also a critical ingredient to the conjoint evaluation of absorption, distribution, metabolism, and excretion in the so-called ADME approach for drug discovery and design. The retrieval function in chemical compound libraries, which may exceed one million entries, follows various objectives describing similarity between molecules and fitting behavior. There is an increasing use of multicriteria methods for the aggregation of these objectives ***(17)***.

References

1. Keeney, R.L. and Raiffa, H. (1976) *Decisions with Multiple Objectives*. Wiley, New York.
2. Yu, P.L. (1989) Multiple criteria decision making: five basic concepts, in *Handbooks in OR and Management Science I* (Nemhauser, G. et al. eds.), North-Holland, Amsterdam, 663–699.
3. Kim, K.H. and Kim, Y. (2004) Variational transition state theory calculations for the rate constants of the hydrogen scrambling and the dissociation of BH_5 using the multiconfiguration molecular mechanics algorithm. *The Journal of Chemical Physics* 120, 623–630.
4. Mullis, K.B. and Faloona, F.A. (1987) Specific synthesis of DNA in vitro via a polymerase-catalyzed chain reaction. *Methods in Enzymology* 155, 335–350.
5. Sahdra, B. and Thagard, P. (2003) Procedural knowledge in molecular biology. *Philosophical Psychology* 16, 477–498.
6. Kämpke, T., Kieninger, M. and Mecklenburg, M. (2001) Efficient primer design algorithms. *Bioinformatics* 17, 214–225.
7. Breslauer, K.J., Frank, R., Blocker, H. and Marky, L.A. (1986) Predicting DNA duplex stability from the base sequence. *Proceedings of the National Academy of Sciences of the United States of America* 83, 3746–3750.
8. Rychlik, W., Spencer, W.J. and Rhoads, R.E. (1990) Optimization of the annealing temperature for DNA amplification in vitro. *Nucleic Acids Research* 18, 6409–6412.
9. Owczarzy, R., Vallone, P.M., Gallo, F.J., Paner, T.M., Lane, M.J. and Benight, A.S. (1997) Predicting sequence-dependent melting stability of short duplex DNA oligomers. *Biopolymers* 44, 217–239.
10. Thein, S.L. and Wallace, R.B. (1986) The use of synthetic oligonucleotides as specific hybridization probes in the diagnosis of genetic disorder, in *Human Genetic Diseases: A Practical Approach* (Davis, K.E. ed.), IRL Pres, Herndon.
11. Hillier, L. and Green, P. (1991) OSP: A computer program for predictions of DNA duplex stability. *PCR Methods and Applications* 1, 124–128.

12. Fleury, G., Hero, A., Yoshida, S., Carter, T., Barlow, C. and Swaroop, A. (2002) Pareto analysis for gene filtering in microarray experiments. Proceedings of the 11th European Signal Processing Conference, Toulouse.
13. Prezioso, V.R. (2005) General notes on primer design in PCR. Eppendorf North America, Westbury, http://www.eppendorfna.com.
14. Yuryev, A., Huang, J., Pohl, M., Patch, R., Watson, F., Bell, P., Donaldson, M., Phillips, M.S. and Boyce-Jacino, M.T. (2002) Predicting the success of primer extension genotyping assays using statistical modeling. *Nucleic Acids Research* 30, e131.
15. Tusnady, G.E., Simon, I., Varadi, A. and Aranyi, T. (2005) BiSearch: primer-design and search tool for PCR on bisulfite-treated genomes. *Nucleic Acids Research* 33, e9.
16. Huang, Y.-C., Chang, C.F., Chan, C.H., Yeh, T.J., Chang, Y.C., Chen, C.C. and Kao, C.Y. (2005) Integrated minimum-set primers and unique probe design algorithms for differential detection on symptom-related pathogens. *Bioinformatics* 21, 4330–4337.
17. Agrafiotis, D.K. (2001) Multiobjective optimization of combinatorial libraries. *IBM Journal of Research and Development* 45, 545–566.

5

PCR Primer Design Using Statistical Modeling

Anton Yuryev

Summary

I describe the approaches for choosing primer parameters and calculating primer properties to build a statistical model for PCR primer design. Statistical modeling allows you to fine-tune the PCR primer design for your standard PCR conditions. It is most appropriate for the large organizations routinely performing PCR on the large scale or for the instruments that utilize PCR. This chapter shows how to use the statistical model to optimize the PCR primer design and to cluster primers for multiplex PCR. These methods have been developed to optimize single-nucleotide polymorphism–identification technology (SNP-IT) reaction for SNP genotyping and implemented in the Autoprimer program (http://www.autoprimer.com). The approaches for combining the individual primer scores into statistical model are described in the next chapter.

Key Words: Statistical modeling; multiplex PCR; primer scoring and selection; PCR primer design optimization and customization.

1. Introduction

When PCR is performed routinely for thousands of experiments using the same technique and preferably the same instruments, the success rate of amplification can be further improved by tailoring the primer design for your experimental setup. You can optimize primers for your PCR buffer, for standard amplification conditions, for DNA polymerase, for the thermo-conductive properties of your PCR machine, or for adhesive properties of your

From: *Methods in Molecular Biology, vol. 402: PCR Primer Design*
Edited by: A. Yuryev © Humana Press, Totowa, NJ

tubeware. In addition, you can optimize for the manufacture of your oligonucleotide by excluding primer sequences that tend to have more errors due to synthesis mistakes. The optimization can also be done for primers synthesized using non-standard nucleotides with poorly investigated thermodynamic properties. In this chapter, we describe PCR primer design optimization using statistical modeling. In our experience, such optimization improved the number of successful reactions by 10–40%, achieving an overall success rate as high as 96%.

As you will see in the following sections of this chapter, the development of a statistical model takes some investment of resources. Because it involves some trial-and-error approaches for finding optimal primary scores and combining them into the model, we estimate that the effort should take about 2–4 months with the participation of a statistician and a molecular biologist or physical chemist. As pointed out in Chapter 1 of this volume, PCR succeeds without optimization in about 75% cases. This level of success can be acceptable for many individual laboratories. Therefore, further optimization is only necessary when PCR is used on the large industrial scale of tens of thousands of reactions per day using dedicated instrumentation and personnel. In this case, statistical modeling can save millions of dollars for your company in the long term. Optimization is also useful for the development of an instrument utilizing PCR. In this case, the software for primer calculation complements the instrument and can be supplied with it. Another application for a statistical model is the selection of optimal markers for a genome-wide screen. Very often there is an abundance of genetic markers exceeding the marker density necessary to achieve statistical association or linkage power. You can select the markers that are more likely to be successfully amplified by using a statistical model.

The primer design optimization using a statistical model requires a database of PCR primers with a thoroughly recorded success/failure rate for every primer pair. The bigger the database size, the more reliable are the predictions made by a statistical model about the success rate of a new primer pair. We used three databases collected by scientists at Orchid Biosciences, Inc. during the development of the single-nucleotide polymorphism–identification technology (SNP-IT). The SNP-IT reaction has two steps: the first step includes PCR amplification of the small genomic region containing an SNP; the second step is the primer extension step to detect the polymorphism ***(1)***. The Orchid Biosciences primer databases contained the results for 20,000–30,000 SNP-IT reactions along with primers and amplicon sequences for three different types of SNP-IT reactions: single-plex, 12-plex, and 24-plex. Three separate statistical models predicting the primer success rate were developed for each of three databases corresponding to three SNP-IT reactions.

Upon development, the statistical model must be implemented into the PCR primer design program. The model calculates the success probability for every primer set, thus allowing the selection of the best primer set that is most likely to succeed for every amplicon. We also developed two models for a multiplex PCR to evaluate the compatibility of primers in a multiplex reaction. These models were used for clustering individual primer sets to achieve the highest average success probability for all primer sets used in clustering.

2. Building the Statistical Model

The statistical model is a set of multiple individual scores that are put together with different weights into a formula calculating the overall success probability of a primer set. The statistical approach for combining individual scores into a formula is called logistic regression and is thoroughly described in Chapter 6 of this volume, as well as in **ref.** ***1***. Here, we focus on how to develop the individual scores for logistic regression.

2.1. Developing Individual Scores for the Statistical Model of the Single-Plex Reaction

Each individual score measures one characteristic of a primer, a primer pair, the amplicon, or a combination of primers and amplicon. As the first step, the standard set of PCR primer parameters described in many chapters of this volume can be used as individual scores. It includes primer-melting temperature (T_m); primer and amplicon GC-content; primer self-complementarity; primer pair complementarity; number, size, and type of nucleotide repeats in primer sequence; and amplicon-melting temperature. A score is usually calculated using a formula that quantifies a certain sequence characteristic. From a program architecture point of view, the score is a result of a C-function that takes primers and/or amplicon sequence as string parameters and returns a number—the score.

In Table 1 of the next chapter and in **ref.** ***1***, we describe an additional set of characteristics of a PCR measuring stability of amplicon structure and 3′-end of PCR primers. Unlike standard PCR parameters, such as T_m and GC-content, the scores used for measuring the oligonucleotide and amplicon structure can be calculated in a variety of ways. For example, a score can be used to calculate the free energy of the most stable oligonucleotide dimer or the free energy of the most stable amplicon secondary structure. Alternatively, it can be used to calculate an average of free energies of all possible oligonucleotide dimers or a sum of the 10 most stable dimers or the number of possible dimers with free energy above a certain threshold. The ultimate choice of this function to calculate an individual score is determined by the score's statistical power. The statistical power can be evaluated by plotting the receiver operating characteristic (ROC) curve ***(1,2)*** for every score and calculating the c-statistics (the area under the curve). The sensitivity axis in an ROC curve corresponds to the success rate, that is, the percentage of successful experiments for a given score value. The ROC specificity axis corresponds to the failure rate for the same score value.

The score for categorical properties such as the nucleotide at a certain position in the primer sequence can be represented as binary variable with value 1, indicating that the nucleotide is present in the position and value 0 is not (*see* **Note 1**). We and other authors ***(1,3)*** found that the last three 3′-end nucleotides in the primer sequence significantly influence the success rate of the PCR and therefore has to be included in the model. We also found that first two nucleotides in the amplicon sequence immediately following the primer-annealing site must be considered together with the sequence at primer 3′-end ***(1)***. Because no ROC curve is possible for a binary score, its statistical significance can only be estimated by comparing the performance of the statistical model with and without the score (*see* **Note 1**). If the score failure rate curve appears to deviate significantly from the linear approximation, we suggest including the square and cube score value in addition or instead of the primary score value in the model. The complete set of individual scores developed for the SNP-IT genotyping reaction can be found in Table 1 in the next chapter and in **ref.** ***1***. Over the course of constructing the model, we have rejected several scores after they failed to significantly improve the predictive power of the model. The failed scores are not shown in Table 1 of the next chapter, but we mention them here to emphasize the amount of effort needed for model construction.

Table 1 of the next chapter contains several scores, such as template-dependent and template-independent noise or amplicon structure around PCR primer-annealing sites that are not calculated by the standard primer design programs. These scores have been developed from the careful consideration of possible bimolecular interactions that may occur between primers and amplicon molecules during the PCR. The importance of these interactions depends on the nature of the reaction and PCR conditions; therefore, the design of such scores has to be tailored for an individual assay. We used the standard methods ***(1,4,5)*** for the free energy prediction of the oligonucleotide structures to estimate the stability of every possible structure. In theory, the more accurate the prediction of the free energy is, the higher predictive power of the score. We recommend using multi-state coupled equilibrium equations described in Chapter 1 of this volume and parameters described in **ref.** ***4*** for the most precise calculations of free energies and T_m. The multi-state coupled equilibrium approach was not published at the time of construction of the SNP-IT reaction, and therefore, we have not used it in our work.

Keep in mind that the methods developed for calculating individual scores for the model will be also used later in the primer design program. During primer design, the model has to evaluate thousands of primer combinations and the calculation speed becomes the critical issue. Therefore, the accuracy of the individual score can be sacrificed in favor of faster calculation times during the primary score design. In addition, the advantages of more accurate free energy calculations can be devalued by the interactions with other scores in the model and thus become unnecessary.

2.2. Developing Individual Scores for a Multiplex Statistical Model

Scores for the statistical model of a multiplex PCR are used in addition to the scores developed for a single-plex reaction. They are developed from the careful consideration of bimolecular interactions that can occur when multiple primer sets are combined in one reaction, interfering with one another during PCR amplification. Again, we recommend using multi-state coupled equilibrium equations described by John SantaLucia in this volume to evaluate the interference from non-specific intra- and inter-molecular interactions. Another approach for developing multiplex scores is measuring the homogeneity of the primer sets in a multiplex cluster. If primers in a cluster have very different thermodynamic and sequence properties, the "strongest" primers will out-compete the "weaker" primers, causing the failure of the correspondent amplicons (*see* **Note 2**). The examples of primary scores for a multiplex reaction are given in Table 1 of the next chapter. The multiplex score is calculated for every individual primer set but also depends on the sequences of all other primers in the cluster. Therefore, the C-function in the primer design program that calculates a multiplex score for one primer set has to take sequences of all primers in a cluster and the index for the primer set under evaluation as parameters. The data structure representing a multiplex cluster that is used for passing the primer sequences to the C-function can be also used for marker clustering algorithm that uses the statistical model to optimize PCR primer clusters.

2.3. Combining Individual Scores into Statistical Model

As evident from the previous sections, the selection of individual scores and the model construction are interdependent. The set of primary scores can be described as a parametric vector containing all possible scores. For example, the model for a 12-plex SNP-IT reaction had 211 components, including 186 binary scores for representing a single nucleotide, a dinucleotide, or a nucleotide triplet in a certain position of a primer. The ultimate goal of model construction is to combine scores in such a way that the final score can be used to separate best the failed PCRs from the successful ones in your training set. Different statistical methods for calculating weights for primary scores and combining them into one global score are possible. The logistic regression is probably the most advanced and sensitive method. Detailed approaches for the construction of the logistic regression model can be found in the next chapter.

It is extremely important to understand the molecular interactions contributing to the success and failure rate of the PCR amplification to develop individual scores. However, we found that the statistical model can tolerate a certain level of theoretical misconceptions or approximations as long as its primary scores correlate well with the failure rate. This robustness is also important for the optimization of the computational speed of primer design using a statistical model. The accuracy of the individual score can be sacrificed in favor of faster computation.

The development of the statistical model for PCR can help you to better understand the molecular mechanisms important for reaction success. For example, if the primary score has significant predictive power by itself but does not improve the model, it usually indicates that the molecular mechanism evaluated by the score is already reflected by other primary scores. For example, we found that the primer length score was significant for the success of the SNP-IT reaction but had no effect on the model. The most likely explanation is that primer length effects are already included in the T_m score, which is proportional to the primer length. Some scores actually decrease the model quality while having significant predictive power by themselves. In this case, the primary score interfering with another score can be identified and a hypothesis can be generated explaining the score interference on the molecular level. The scores that have little effect on the model should be omitted to minimize the computational time for primer design.

3. Using Statistical Model for Primer Design

Once the statistical model is created, it can be used to optimize the primer design. In theory, the model must be used to score all possible primer pairs so that the pair with the best score can be selected. In practice, however, owing to the large number of possible primer pairs, the evaluation of every pair will take an impossibly long time using a standard personal computer. The large number of primary scores used for calculating the free energy of the oligonucleotide structures can significantly affect the performance of the program, sometimes making primer calculations prohibitively slow. Although the program acceleration is possible by using parallel or multithreading computing, in most cases it is sufficient to use some cutoff values for primer parameters to minimize the number of primer pairs. For example, we used a basic cutoff based on the primer length and T_m *(6)*. Additional filtering can be done using minimum allowed free energy for primer dimer and primer self-complementarity.

From the software architecture perspective, the scoring is the last step of the primer design. The subroutine for statistical scoring takes sequences of the primer pair and the amplicon as an input and returns its success probability. The primary score weights pre-calculated by logistic regression are written into the file and used as parameters during primer design. Therefore, the output from any other primer design algorithm described in this volume can be used as the input for scoring subroutine. Most of these algorithms output the collection of possible primers all satisfying a user's constraints. The selection of the best primer pair is still necessary to fully optimize the PCR.

A lot of parameters used for pre-filtering also serve as primary scores in the statistical model; therefore, they can be recycled during the program run to avoid repetition of identical calculations. Because PCR primers are part of the amplicon sequence, the result of the analysis of their secondary structures can be also re-used for amplicon structure analysis.

4. Using Statistical Modeling for Marker Clustering

Once the best primers are selected for all amplicons, their combinations in multiplex clusters can be also optimized using the statistical model for multiplex reaction. Most primary scores for the SNP-IT multiplex reaction recycle the primary scores for the single-plex, and consequently, clustering was not the computationally intensive in our program. We have suggested three possible algorithms for marker clustering *(6)*. All of them are based on the idea that multiplex clusters are built in such a way that the clusters having the best score are selected for the next iteration. The clustering algorithms either search for the best cluster for every primer pair gradually depleting the pool of non-clustered primers or randomly reshuffle primers between clusters accepting only swaps that increase the overall success probability of two clusters (*see* **Figs. 1** and **2**).

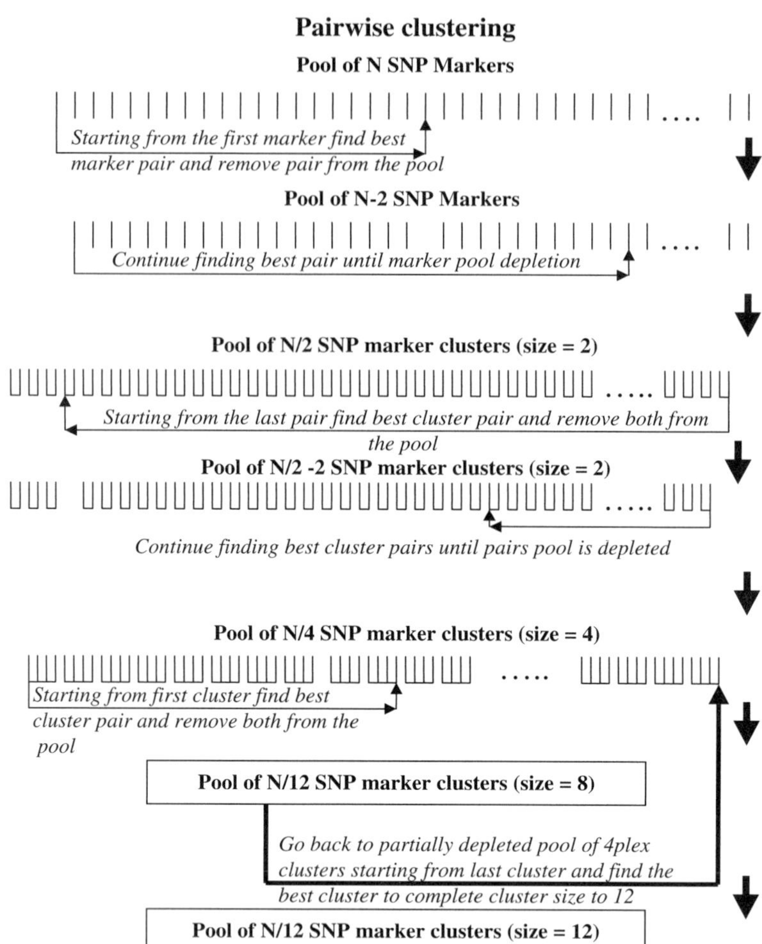

Fig. 1. Pairwise clustering algorithm. This algorithm is reminiscent of a classic hierarchical clustering algorithm. It starts with a pool of clusters obtained from the previous cycle or from the algorithm's original input. At the initialization step, all individual primer sets serve as a pool of clusters with a size equal to one. The cluster size is defined as the number of markers in the cluster. Next, the algorithm finds the most compatible cluster for every cluster in the pool. Both clusters are removed from the pool immediately after the best cluster pair is found. The cycle continues until all cluster pairs are found. Because every step depletes the cluster pool, the clusters that are processed at the end of the cycle have fewer choices of pairs and, thus, may have poorer probability of success. To compensate for this disadvantage, the clusters

The random clustering must be stopped by limiting the number of swaps or by calculating the local minimum. We found that random clustering provided better results than hierarchical clustering for multiplex SNP-IT reactions while running for the same amount of computational time as non-random algorithms. For all approaches, we found that selecting the best primer pairs for every amplicon and sorting them by the probability of success prior to clustering significantly improved clustering results and decreased computational time. These results are most likely because the success rate of a multiplex cluster depends on the homogeneity of the individual primer pairs. The primer pairs with similar individual success probability tend to have similar sequence characteristics and thus form better multiplex clusters.

The simulation annealing method can be applied for random clustering to avoid trapping in the local minimum during the program run. Although we have not investigated the efficiency of simulation annealing, its implementation is straightforward: certain swaps between two multiplex clusters that actually decrease the success probability by no more than a certain threshold are allowed. The threshold can be set as a parameter for the clustering program as the maximum allowed percentage of success probability decrease. The threshold can be set as a parameter for the clustering program as the limit for the relative decrease of success probability.

Fig. 1. obtained at the end of previous clustering cycle are considered first at the next clustering cycle. To obtain the final clusters with a size equal to the desired multiplex fold 12 or 24, two and three cycles of pairwise clustering are performed to first produce clusters of size four and eight, respectively. During the last step, one pairwise clustering is performed to obtain the clusters of size eight (for 12-plex reaction) and of size 16 (for 24-plex reaction). The number of these clusters is equal to the number of multiplex clusters in the output of the program. Thus, at this last step the cluster pool from the previous cycle is not depleted completely yet. The "incomplete" clusters are merged with unused clusters of size four or eight, respectively, from the previous cycle pool to form the final clusters. Each vertical bar represents a single-nucleotide polymorphism (SNP) marker: a set of two PCR primers and one SNP-IT extension primer. The bars connected with a horizontal line represent a cluster. The cluster size is equal to the number of vertical bars in it. The arrows connecting two clusters represent the best pair chosen to combine into the cluster of a bigger size. The boldface arrows represent the sequence of steps in the algorithm flow.

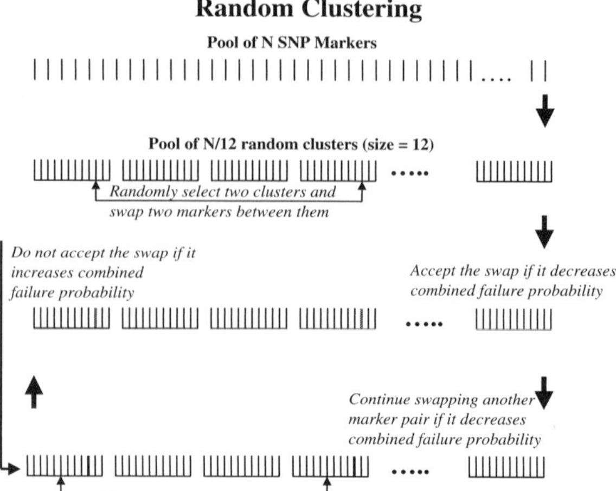

Fig. 2. Depiction of the random clustering algorithm. The symbols are the same as in **Fig. 1**. The vertical bars shown in red depict the markers randomly selected for a swap between two clusters. At the next step, the same markers are shown in bold lines representing that the swap was accepted. Random clustering is reminiscent of classical non-hierarchical clustering algorithm. It begins with generating a pool of random clusters, containing a multiplex fold of randomly selected markers. Each minimization step consists of randomly selecting two clusters from the pool and switching two randomly selected markers between clusters. The new failure probabilities are calculated for all markers in both clusters. If the marker swap leads to a decrease in the combined failure probability of the two clusters, it is accepted and the next random swap is performed. If the swap leads to an increase of the failure probability, it is rejected and markers are returned to the original clusters. During clustering, the number of consecutive successful cluster swaps is monitored. If no minimization of the clustering function occurs during pre-defined number of consecutive cluster swaps, clustering is stopped. In essence, this approach finds the local minimum of a pre-defined "depth" in the overall minimization rate to break the process earlier. The random clustering algorithm induces higher randomization of markers compared with other methods. This can explain its superior performance.

5. Notes

1. To incorporate the last three 3′-end nucleotides of the primer sequence into the model, start by including all 64 binary scores corresponding to every possible triplet combination of A, G, C, and T nucleotides. Parametric score vector for every PCR must have at most two scores equal to one corresponding to 3′-end of two PCR primers. All other scores must be equal to zero. After an evaluation of how every binary score influences the predictive power of the entire model,

you can omit some triplet combinations as being non-significant. In the statistical model for SNP-IT genotyping reaction, we found that only about 10–15% of all possible 3′-end nucleotide triplets significantly influence success probability ***(1)***.

2. We used the following scores to measure the homogeneity of the individual PCR in the multiplex cluster: dispersion of amplicon complexity and amplicon-melting temperature ***(1)***.

References

1. Yuryev A, Huang J, Pohl M, Patch R, Watson F, Bell P, Donaldson M, Phillips MS, and Boyce-Jacino MT (2002) Predicting the success of primer extension genotyping assays using statistical modeling. *Nucleic Acids Res* **30**(23), e131.
2. Egan JP (1975) *Signal Detection Theory and ROC Analysis*, Academic Press, New York.
3. Onodera K and Melcher U (2004) Selection for 3′ end triplets for polymerase chain reaction primers. *Mol Cell Probes* **18**(6), 369–372.
4. SantaLucia J Jr (1996) A unified view of polymer, dumbbell, and oligonucleotide DNA nearest-neighbor thermodynamics. *Proc Natl Acad Sci USA* **95**, 1460–1465.
5. Mathews DH, Sabina J, Zuker M, and Turner DH (1999) Expanded sequence dependence of thermodynamic parameters improves prediction of RNA secondary structure. *J Mol Biol* **288**, 911–940.
6. Yuryev A, Huang J, Scott KE, Kuebler J, Donaldson M, Phillips MS, Pohl M, and Boyce-Jacino MT (2004) Primer design and marker clustering for multiplex SNP-IT primer extension genotyping assay using statistical modeling. *Bioinformatics* **20**(18), 3526–3532.

6

Developing a Statistical Model for Primer Design

Jianping Huang and Anton Yuryev

Summary

This chapter describes the statistical method that can be used to predict the success and failure of a designed primer based on properties of genomic sequence surrounding the primer extension, using user's own existing genotyping database. After scores that measure properties of genomic sequence surrounding primer extension are developed as described in previous chapters, this chapter first shows how to use simple statistics to evaluate the correlation between the score and the likelihood of primer success and failure based on user's own empirical data. All scores that show significant correlations with the primer success are kept for further analysis. Next, logistic regression method is described in detail to estimate the contribution of each primer score to the overall primer success/failure rate when all significant scores are weighted simultaneously to produce the logistic regression model. Statistics that evaluate model fit and model discrimination are provided as well. Last, all significant scores are combined into one measure that can predict overall success/failure rate of a given primer design. The estimated logistic regression score allows prioritization of primers, selection of the best possible primer pair, and combining primers into best clusters for multiplex PCR. Software and hardware requirements and sample SAS programs are also included.

Key Words: Primer design; statistical modeling; logistic regression; logistic model; primer success; primer failure; genotyping success; genotyping failure; primer score; PCR clustering.

1. What Can Statistical Model Do for Primer Design?

In the past, primer design has mostly based on theoretical estimation of the molecular mechanisms that lead to assay success or failure. Until recently, there have been very few empirical methods in literature that can be used to

From: *Methods in Molecular Biology, vol. 402: PCR Primer Design*
Edited by: A. Yuryev

estimate how big the effects various factors contribute to primer assay failure. Even when empirical estimates from other experiments exist, they can often be limited and biased by particular experimental conditions under which data were generated. There is virtually no developed method for scientists to utilize their own experiment data to produce more accurate parameters for primer assay design.

When working with molecular biologists in single-nucleotide polymorphism (SNP) genotyping [SNP-identification technology (SNP-IT)], the authors found that statistical modeling is a useful tool that had yet to be introduced in the field of primer design. Working with molecular biologists, we demonstrated that statistical modeling, specifically logistic regression, is a good method that can be used to improve prediction of assay success ***(1)***.

From previous chapters, molecular mechanisms that lead to assay success or failure of any designed assay can be identified and calculated. Furthermore, these scores can be stored, along with the actual success or failure of the assay, in primer design databases. Through statistical modeling, each molecular factor score can then be assigned a weight that accurately predicts its contribution to the probability of assay failure. Specifically, the logistic regression method allows the weights of all the score to be estimated simultaneously from existing primer databases. Moreover, an estimated overall fail rate can be calculated by adding the weights of all factors that are present in a primer assay and then finding the log transformation of the sum. As genotyping databases grow overtime, the weights and predictions will become more and more accurate.

The estimated logistic regression score allows prioritization of primers, selection of best possible primer pair, and combining primers into best clusters for multiplex PCR, something that cannot be done without the scoring.

This chapter intends to show readers, step by step, how to develop logistic regression models to estimate the weights of various molecular mechanisms that affect primer assay success and how to calculate the probability of failure for any assay design, based on the empirical weights of the factors that are present in the primer assay.

2. Materials

2.1. Hardware

Any PC or server that has at least 10 GB of storage space and 1 GHz processor with more than 500 MB RAM should be sufficient for doing the analysis on 30,000 PCR primer reactions in the database.

2.2. Primer Database and Primer Primary Scores

One of the most important components of developing a sound statistical model is a well-designed database. A well-designed primer database should capture all important factors that would affect primer assay outcome. In the case of SNP genotyping, the tables that are most relevant to our model developing are the sequence table, the SNP marker table, the primer table, the PCR product table, and the GBA plate table GBA (genetic bit analysis) primer is the primer used for extension to determine the SNP variation. Variables such as SNP source, DNA sequence, and SNP position can be stored in the sequence table. Extension mix, PCR upper primer, PCR lower primer, and GBA primer synthetic temperature can be stored in the SNP marker table. The primer table should have DNA sequence of the primer. Anneal temperature, melting temperature, delta temperature, and percent GC information can be stored in the PCR product table. SNP outcomes can be stored in the GBA plate table. A complete list of all important factors affecting SNP genotyping success can be found in **Subheading 3.3**.

In addition, batch variation, operators, quality of DNA assay, quality of sample, quality of oligo, genotyping equipments, and the order of plates can all affect the success or failure of genotyping. Such information should be captured as much as possible when designing database so that they can later be used for quality control and improvement.

2.3. Statistical Software

There are many statistical software that can be used to develop basic statistical models. All programming examples in this chapter are written in SAS *(2)*. However, it is possible to use other statistical software such as SPSS *(3)*, Stata *(4)*, JMP *(5)*, S-plus *(6)*, and Systat *(7)*. In addition, software that are designed for genotyping database and statistical analysis such as Spotfire *(8)* may also be used. Be aware that some software only offer limited number of variables that can be included in the data set as well as in model development. If the statistical software of choice is something other than SAS, make sure the software offers reasonable flexibility in data management and preparation, and few limitations on number of variables to be included in working data file and in model development.

The SAS modules needed for the analysis include SAS Connect *(9)*, SAS SQL *(10)*, SAS Base *(11,12)*, SAS Stat *(13)*, and SAS Graph *(14)*. SAS Connect enables the connection between the user's PC and the databases residing on any PCs or servers in the same network, thus allowing users to manage and access data *(9)*. SAS SQL allows users to retrieve, extract, subset, and process data in database tables and to create new tables while running in SAS *(10)*. SAS Base is the basic module that is required for data input and output and data management *(11,12)*. SAS Stat provides users statistical tool for statistical data analysis and developing statistical models *(13)*. SAS Graph module lets users to turn data into varieties of graphs and charts in SAS *(14)*.

2.4. Connection to the Databases and Servers

In order for user's computer to access the database on different servers, connections between the client computer and the databases and servers need to be established. The technical details of how to establish network connections between the client computer and the network servers are beyond the scope of this chapter. Reader should consult their network or database administrators on how to establish network connections.

After the server connections are established, we can refer to the data on the server by using a SIGNON statement and a LIBNAME statement that specifies the REMOTE engine. Different types of servers require different connections. One way that links the SAS on PC to a database on Oracle server is as follows:

```
SIGNON server-ID;
LIBNAME libref REMOTE 'datalib' SERVER= server-ID userid=
" myuserid" password="mypassword";
```

where *server-ID* is the ID of the server which we can get from our network administrator and *libref* is the library name we choose for referencing the data library. *Datalib* is the name of the database. *Myuserid* and *mypassword* are user ID and password for logging into the database (*see* **Note 1**).

3. Methods

3.1. Data Preparation

3.1.1. Prepare Analytical Data File

The data that will be used for primer design often resides in several tables from one or more databases. One way is to create a table or view that extracts variables from different tables by directly joining tables in the database. Another way is to select relevant data elements from various tables and store them into several SAS data sets on PC. Then we can merge them into a single analytical SAS data file using SAS merge statement. Here is an example of merging a SNPMarkers data set with a PCRProducts data set to get melting temperature and delta temperature from the PCRproducts data using markerid as the key for merge. The newly merged data are stored in the data set gba.analysis (*see* **Note 2**).

3.1.2. Prepare Categorical Data for Analysis

After analytical SAS data set is created, the next step will be to clean and recode data to be used for further analysis to develop the logistic regression model. All categorical data need to be recoded into binary variables of 1 or 0:

Where, 1=if condition is true;
0=if condition is false.

Categorical data include variables such as extension mix and most of the last 1–3 bases at the 3′-end of the PCR primer. Sometimes the binary variable can be a threshold, for example,

1=when number of consecutive Gs in SNP-IT primer is 6 and more;
0=otherwise

The array statement in SAS is an easier way to code binary variables (*see* **Note 3**).

3.1.3. Prepare Continuous Data for Analysis

The first step in preparing continuous data for analysis is to learn about the distributions of the score variables. The basic statistics that describe score distributions are range, mean, standard deviation, median, and quartiles. Box plot and histogram help users learning about the distribution visually. User should use the PROC UNIVARITE procedure in SAS to get all the basic statistics and plots for all primer scores in one simple step before any further analysis (*see* **Note 4** for the SAS program on PROC UNIVARIATE procedure).

It is usually a good idea to create scores that are within a limited range and spread somewhat evenly along the range, which makes it simpler to understand and easier to interpret. Most continuous scores can be transformed into 1–100 scale using the following formula:

$$\text{Transformed score} = \frac{\text{original score} - \text{lowest score}}{\text{highest score} - \text{lowest score}} \times 100 \qquad (1)$$

Sometimes the original primer scores are skewed toward the larger values, that is, the smaller values are squeezed closely together, whereas large values are spread out. Other time scores may have outliers or have unequal variances.

This is when log transformation becomes a useful tool. Log transformation tends to stretch out the small values and squeeze together the large values; thus the transformed scores are more evenly distributed. The log transformation is also useful when the scores have a multiplicative effect such as percentages or ratios. In general, log transformation should be considered when data are bound below by zero, data are defined as a ratio, and/or when the largest value in our data is at least three times as big as the smallest value ***(15)***.

$$\text{Transformed score} = \log(\text{original score}) \tag{2}$$

To make sure the log transformed scores yield equal or better correlation with primer fail/pass, graphs and chi-square (χ^2) tests should be examined between the fail/pass variable and the transformed scores (*see* **Subheadings 3.4.1** and **3.4.2** for details).

3.2. Logistic Regression Method

Logistic regression is a commonly used regression method when the dependent variable is a categorical variable ***(16–19)***, e.g., pass or fail. Logistic regression first transforms the dependent variable into natural log of odds ratio,

$$\text{Logit}\,(p) = \log_e \frac{p}{1-p} \tag{3}$$

where p is the probability of the event (e.g., fail) ***(18)***. The effects of independent variables, called logit coefficients or weights, can then be calculated with maximum likelihood estimation from the specified model:

$$\ln \frac{p}{1-p} = \alpha + \beta_1 X_1 + \beta_2 X_2 + \cdots + \beta_k X_k + \varepsilon, \quad i = 1 \text{ to } K \tag{4}$$

where α is a constant referred to as intercept, β_is are the regression coefficients, X_is are the independent variables, and K is the number of independent variables included in the model ***(16–19)***. In this chapter, p represents the probability of PCR failure, β_is are the weights of primary scores to be calculated from the model using the training sets, X_is are the primary scores, and ε is random error.

After weights are calculated from the model, estimated probability of fail, p, for any new PCRs can then be calculated as

$$p = \frac{e^{\alpha+\beta_1 X_1+\beta_2 X_2+\cdots+\beta_k X_k}}{1+e^{\alpha+\beta_1 X_1+\beta_2 X_2+\cdots+\beta_k X_k}}, \quad i = 1 \text{ to } K \tag{5}$$

Separate models should be developed for different PCR settings, such as single-plex setting versus multiplex setting, to account for the different effects primary scores have on PCR success in different settings. All data in a given setting should be utilized to develop the model for that setting to get most variations in molecular property and most accuracy in estimated weights.

3.3. What Variables Have Been Shown to Be Important for Primer Success?

From our work with SNP-IT genotyping, we found the following variables to be significant factors affecting SNP-IT success or failure (*see* **Table 1**). For detailed methods on how to compute primary scores, please refer to chapter 5.

3.4. Which Variables from User's Database Are Likely to Be Important for Primer Success?

3.4.1. Statistical Graphs

The first step in data analysis to determine which scores are important is to examine the graph of each score against percentage of failed reactions. An easy way of graphing is to divide all observations into 10 groups based on the transformed scores of 1–100. The SAS codes in **Note 5** enable users to divide a transformed score of 1–100 into 10 groups.

Second, calculate the percentage of failed reactions in each group. The SAS code for this task is

```
PROC FREQ DATA= mydataset;
TABLE score10*fail/NOPERCENT CHISQ OUT= test;
RUN;
```

where *mydataset* is the analysis data set, *score10* is the transformed and grouped score, $fail = 1$ for failed genotyping and $fail = 2$ for successful genotyping, and *test* is the output SAS data to be used for graphs. The row percentage shown in the SAS output is the percentage of fail for each score group.

Percentage of fail data can then be exported into spreadsheet to make line graphs (*see* **Note 6** and **Fig. 1**).

3.4.2. χ^2 Test

The next step is to calculate χ^2 to test for the association between the grouped score and genotyping failure. χ^2 test answers the question of how the PCR fail rate changes in accordance with change in primer score. In statistical terms, the

Table 1
Complete List of Primary Scores, Wald Chi-Square (χ^2), Degrees of Freedom (*df*), and *p*-Values from Logistic Models for Single-Plex and 12-Plex Multiplex SNP-IT Reaction

Score name	Score description	Single-plex model (χ^2, df, p-value)	Multiplex model (χ^2, df, p-value)
TIN	The free energy of the most stable structure of extension primer with itself	2.4, 1, <0.0001	
Most stable 3′-end dimer	The free energy of the most stable dimer formed by annealing SNP-IT primer with itself. Only the dimers formed by last nine 3′-end bases are considered for this score. In addition, a binary becoming 1 when no dimers are formed at the 3′-end is used for single-plex model	65.2, 1, <0.0001	
TDN	The free energy of the most stable structure between extension primer and PCR product	28.1, 1, <0.0001	12.3, 1, 0.0005
Cumulative TDN	Free energies' sum of all possible dimers (no loops) between extension primer and PCR product	36.6, 1, <0.0001	7.1, 1, 0.0077
Number of consecutive Gs in SNP-IT primer	Binary score becomes 1 when number of consecutive Gs in SNP-IT primer is six and greater	7.1, 1, 0.0078	
Stability of last nine bases of extension primer	The free energy of the SNP-IT primer 3′-end. In addition to this score, single-plex model uses its cube approximation and multiplex model uses its square approximation		
	Linear term of stability of last nine bases of extension primer	60.1, 1, <0.0001	284.1, 1, <0.0001

	Square term of stability of last nine bases of extension primer		172.0, 1, <0.0001
	Cubic term of stability of last nine bases of extension primer	32.1, 1, <0.0001	
Number of C3 linkers	Number of C3 linkers in the extension primer		149.3, 1, <0.0001
	Binary score becomes 1 when number of C3 linkers is equal to one	16.4, 1, <0.0001	
	Binary score becomes 1 when number of C3 linkers is greater than three	8.9, 1, 0.0029	
GC percent of PCR primers		29.7, 1, <0.0001	43.4, 1, <0.0001
Amplicon melting temperature			106.0, 1, <0.0001
	Binary score becomes 1 when amplicon T_m is less than 73 °C		106.0, 1, <0.0001
Number of ambiguous bases	Number of ambiguous bases in amplicon	27.6, 3, <0.0001	
Number of repeats	Number of repeats in amplicon		109.1, 1, <0.0001
	Binary score becomes 1 when number of repeats in amplicon eight or nine	22.0, 1, <0.0001	

(continued)

Table 1 (*Continued*)

Score name	Score description	Single-plex model (χ^2, df, p-value)	Multiplex model (χ^2, df, p-value)
Amplicon structure around SNP site	The free energy of the most stable amplicon structure (with loops) containing five bases of extension primer-annealing site and two bases upstream of primer-annealing site		231.5, 1, <0.0001
Amplicon structure around PCR primer-annealing sites	The free energy of the most stable amplicon structure (with loops) containing five bases of PCR primer-annealing site and two bases upstream of primer-annealing site. Only one of the most stable structures between two PCR primers was considered		80.2, 1, <0.0001
Extension mix change date and 13 interactions with combinations of extension mix and last 3′-end of SNP-IT	This binary score was introduced for single-plex model to reflect historical assay modification. New extension nucleotides were introduced at Orchid Bioscience into SNP-IT assay	491.6, 14, $<0.0001^*$	

Extension mix+last two bases at 3′-end of SNP-IT primer	24 different combinations were used for single-plex model and 67 combinations for multiplex model out of 96 possible. Only combinations, which show significant correlations with the failure/success rate, were considered. In addition, three interaction scores with extension mix change date score are used for single-plex model	187.7, 24, <0.0001*	543.0, 67, <0.0001*
Extension mix+last three bases at 3′-end of SNP-IT primer	42 different combinations were used for single-plex model and 56 combinations for multiplex model out of 384 possible. Only combinations, which show significant correlations with the failure/success rate, were considered. In addition, seven interaction scores with extension mix change date score are used for single-plex model	208.2, 42, <0.0001*	255.5, 56, <0.0001*
Extension mix+last four bases at 3′-end of SNP-IT primer	34 different combinations were used for single-plex model and 60 combinations for multiplex model out of 1536 possible. Only combinations, which show significant correlations with the failure/success rate, were considered	101.8, <0.0001*	155.1, <0.0001*
Extension mix+last five bases at 3′-end of SNP-IT primer	3 different combinations were used for single-plex model and 18 combinations for multiplex model out of 6144 possible. Only combinations, which show significant correlations with the failure/success rate, were considered	9.3, 0.03*	53.6, <0.0001*

(*continued*)

Table 1 (*Continued*)

Score name	Score description	Single-plex model (χ^2, df, p-value)	Multiplex model (χ^2, df, p-value)
Last two bases at the 3′-end of the PCR primer+one amplicon base next to PCR primer-annealing site	Two different combinations were used for single-plex model and seven combinations for multiplex model out of 64 possible. Only combinations, which show significant correlations with the failure/success rate, were considered	5.9, 0.05*	36.0, <0.0001*
Last three bases at the 3′-end of the PCR primer+one base next to PCR primer-annealing site	22 different combinations were used for single-plex model and 24 combinations for multiplex model out of 256 possible. Only combinations, which show significant correlations with the failure/success rate, were considered	91.1, <0.0001*	70.5, <0.0001*

Last two bases at the 3′-end of the PCR primer+two amplicon bases next to PCR primer-annealing site	19 different combinations were used for single-plex model and 18 combinations for multiplex model out of 256 possible. Only combinations, which show significant correlations with the failure/success rate, were considered	67.4, <0.0001*	48.5, 0.0001*
Last three bases at the 3′-end of the PCR primer+two amplicon bases next to PCR primer-annealing site	40 different combinations were used for single-plex model and 49 combinations for multiplex model out of 1024 possible. Only combinations, which show significant correlations with the failure/success rate, were considered	135.4, <0.0001*	128.8, <0.0001*
External TIN	The free energy of the most stable structure formed between marker extension primer and other extension primers in multiplex cluster		15.0, 1, 0.0001

(continued)

Table 1 (*Continued*)

Score name	Score description	Single-plex model (χ^2, df, p-value)	Multiplex model (χ^2, df, p-value)
External TDN	The free energy of the most stable structure between marker extension primer and other PCR products in multiplex cluster		39.5, <0.0001
External PCR TDN	The free energy of the most stable structure formed between any of two marker PCR primers and other PCR products in the multiplex cluster		55.4, <0.0001
Dispersion of amplicon melting temperature	The score is calculated as the sum of absolute differences between T_m of marker amplicon and T_m of all other amplicons in multiplex cluster		16.3, <0.0001
Dispersion of amplicon complexity	The score is calculated as the sum of absolute differences between marker amplicon complexity and complexity of all other amplicons in multiplex cluster		22.6, <0.0001

SNP-IT, single-nucleotide polymorphism–identification technology; TDN, template-dependent noise; TIN, template-independent noise. An empty cell for the score indicates that it was not included into the corresponding model.

* The difference in Wald χ^2s between the final model and the model without the set of binary variables, *df*, and *p*-values.

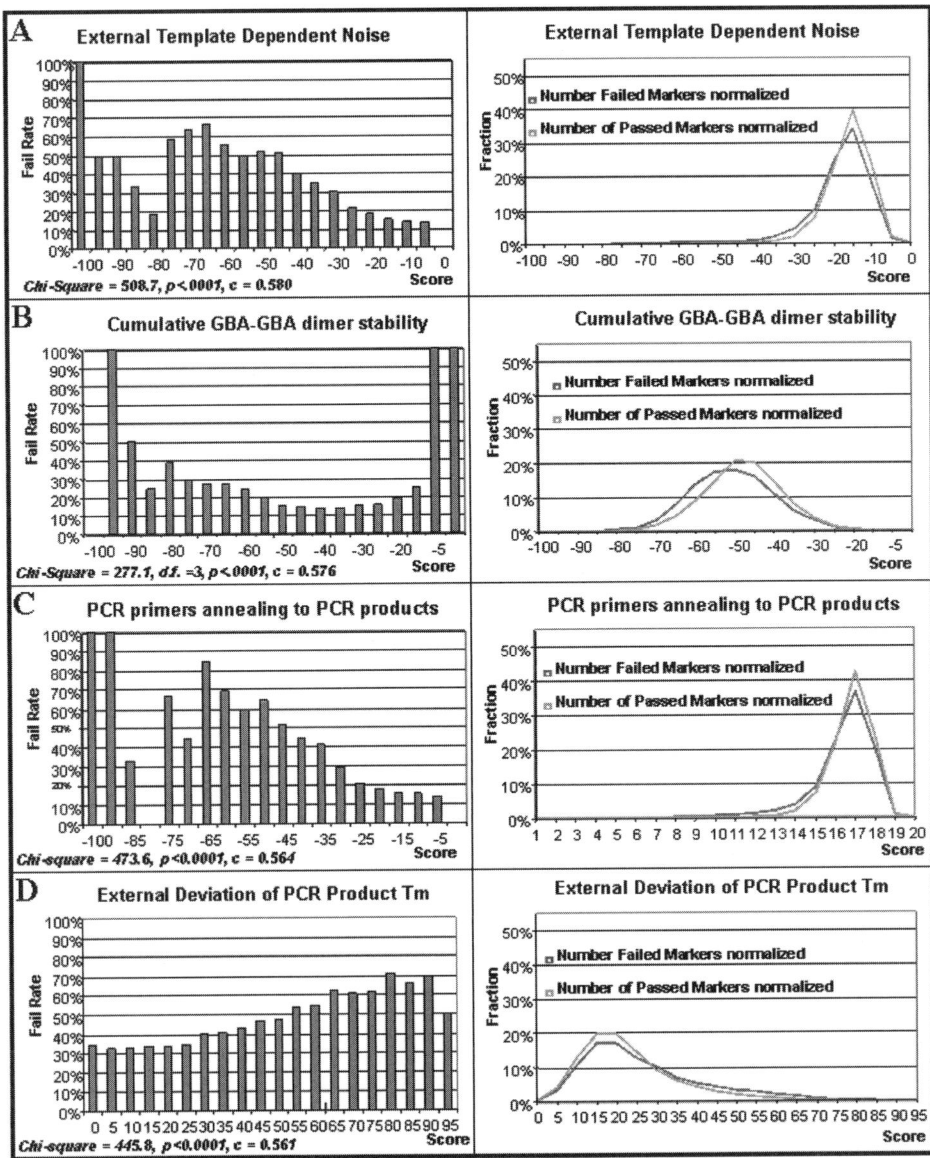

Fig. 1. The correlation curve of the failure rate of 12-plexed single-nucleotide polymorphism–identification technology (SNP-IT) reactions with several primary scores that measure the "external" noise caused by the interference of 11 SNP markers with the one, which is considered for calculation (left panels). The distribution of failed and passed SNP markers against the same scores (right panels). The score values in

null hypothesis (H0) is that there is no difference in the fail rate of genotyping among groups of primers with different primer scores.

$$\chi^2 = \sum_{r=1}^{r}\sum_{c=1}^{c}\frac{(f_o - f_e)^2}{f_e} \quad (6)$$

where r is the number of rows, c is the number of columns of the table, f_o is the observed frequency, and f_e is the expected frequency of each cell ***(16–18)***. The sampling distribution for the χ^2 test statistic is the χ^2 distribution with $(r-1)(c-1)$ degrees of freedom (*df*). Usually a significance level of 0.05 is used in this type of statistical testing, which simply means that 95% of the time we are correct about the statistical inference we make based on the χ^2 test. If the probability, p, is 0.05, we are 95% confident that genotyping failure rates are different for primers with different ranges of primer scores.

To obtain χ^2 and its related statistics in SAS, simply add the CHISQ option in the table statement (*see* example in **Note 7**). Primer scores that have χ^2 p values that are less than 0.05 are the variables most likely to be included by the logistic regression model, or the so-called candidate variables. In order for the model to reflect the true effect of the underlying primer factors instead of random effects, it would be a good idea to keep only the candidate factors that have more than 30 occurrences in the database. In the next section, we will discuss in more detail how to develop logistic regression model using the candidate variables.

3.4.3. How to Develop Logistic Regression

The ultimate goal in developing a logistic regression model is to select an "optimal" group of scores that best predicts primer success/fail, meanwhile

Fig. 1. the correlation curves (horizontal *X*-axis) and number of markers in the distribution curves (vertical *Y*-axis) are normalized for comparison. The normalization of scores is done by dividing all score values by (maximal) score. The distribution curves are normalized by plotting fraction of passed or failed markers instead of total number of markers. Distributions of passed markers are shown in light gray, and distributions of failed markers are shown in dark gray. **(A)** External template-dependent noise (TDN) measured as the free energies' sum of most stable structures formed between SNP-IT primer and 11 "external" PCR products. **(B)** External TDN measured as a sum of all possible dimers formed between SNP-IT primer and 11 "external" PCR products. **(C)** Correlation and distribution curves of the deviation of the SNP-IT primer melting temperature from other 11 SNP-IT primers. **(D)** Correlation and distribution curves of the deviation of the GC content of PCR primers pair from other 11 PCR primer pairs.

keeping the number of scores included in the model to a minimum, so called a parsimonious model. To put it in another way, a parsimonious model is the model that utilizes the least number of scores but still achieves a satisfactory level of goodness of fit. To achieve parsimony, one tests to see whether a restricted model does not significantly differ from the model that includes all scores and all possible interactions among scores. If there is no significant difference, then the user can be confident that the scores dropped from the model were not needed to explain the observed variation of data. The user explores until the most parsimonious model that still has acceptable fit is found.

When comparing different models for the same data, the −2 Log Likelihood statistics provide good information on model fitting. After each logistic regression model is estimated, the −2 Log Likelihood statistics along with two adjusted statistics, the Akaike Information Criterion and the Schwartz Criterion, are printed in the SAS output. The −2 Log Likelihood statistics have χ^2 distribution under the null hypothesis that all the score effects in the model are zero, and the *p*-values for the statistics are also produced. Lower values of the statistics indicate a more desirable model ***(20)***.

There are several different SAS procedures that allow one to perform multivariate logistic regressions and select the "optimal" model automatically. The basic logistic regression procedure is PROC LOGISTIC. PROC LOGISTIC allows user to estimate multiple regression coefficients, model fit statistics and tests, and measures that assess the predictive ability of the model ***(20)***.

PROC LOGISTIC provides four variable selection methods: forward selection, backward elimination, stepwise selection, and best subset selection. When estimating a model with a lot of variables, backward selection is often preferred to other methods. When SAS options SELECTION = BACKWARD coupled with SLSTAY = level are used, all scores specified in the model statement are estimated first. Results of the Wald test for individual scores are examined automatically. The least significant score that does not meet the SLSTAY = level for staying in the model is removed. Once a score is removed from the model, it remains excluded. The process is repeated until no other score in the model meets the specified level for removal ***(20)***.

Here is the general SAS statement for using PROC LOGISTIC with backward selection method:

PROC LOGISTIC DATA=*data-set-name*DESCENDING;
MODEL *fail*=$score_1$+$score_2$+...+$score_k$/
SELECTION=BACKWARD SLSTAY=*0.10* LACKFIT RISKLIMITS;

OUTPUT OUT=*output-data-set-name*
PREDITED=predicted-probability-name;
RUN;

where *K* is the number of candidate primary scores, that is, scores that are significant in χ^2 test, that will be included in the model. The DESCENDING option in the PROC LOGISTIC statement reverses the order of the fail variable so that PROC LOGISTIC models the probability of fail = 1. The SELECTION = BACKWARD option specifies the model to use backward method to select the variables in the model. The SLSTAY = 0.10 specifies the model to use 90% confidence level for the Wald χ^2 for an effect to stay in the model in a backward selection step. SLSTAY is usually set at 0.10 when cross-validating models, whereas for the final model SLSTAY is usually set at 0.05. The LACKFIT option requests the Hosmer–Lemeshow goodness-of-fit test. The RISKLIMITS option requests confidence intervals (CIs) of the odds ratios to be printed for all the scores selected by the model.

3.4.4. Model Development

Once the candidate variables were identified by examining the graph and χ^2 test as illustrated in **Subheadings 3.4.1** and **3.4.2**, models can then be developed and validated in the following order: model selection, cross-validation, and calculation of model adequacy measures.

3.4.4.1. Model Selection

All data to be used are first randomly split into two samples: the development data and the validation data. To include similar number of passed and failed primers in each data set, all failed and passed primer data should be randomly split into development and validation samples, respectively. The first half of failed data and the first half of the passed data are then combined to form the development sample. The second half of the failed sample and the second half of the passed sample are combined to form the validation sample (*see* **Note 8** for SAS codes).

Model is first developed on the development data with all candidate variables included by running the SAS codes from **Subheading 3.4.3**. The selected variables from the development model will be listed in the SAS output and can then be re-entered for model selection using the validation data with the same SAS codes. Finally, variables selected from the validation sample are then re-estimated with all data combined to produce the final model (*see* **Note 9**).

When building the development model, all candidate variables, i.e., those significant from χ^2 tests and have 30 or more occurrences, are entered into the logistic regression model. The backwards selection method will remove variables that do not meet the SLSTAY criterion (details on PROC LOGISTIC and SAS codes have been discussed in **Subheading 3.4.3.**). Usually in cross-validation stage, the criterion for removing variables from the model is 0.10. Backward option in logistic regression procedure with a cutoff p-value <0.1 is used to find the optimal set of scores.

However, because of large number of scores and their inter-correlation among each other, the individual effects cannot be estimated reliably when all scores are entered into the model together *(21)*. Careful attention should be paid to multicolinearity each time a set of new scores is introduced to the model. When multicolinearity occurred, the scores that contribute most to the model predictability should be chosen for the final model (*see* **Subheading 3.4.4.4** for details on multicolinearity).

3.4.4.2. Model Cross-Validation

After development model is built, the model is cross-validated using the validation sample (*see* **Note 9**). To determine which factors remained significant when the development model was applied to the cross-validation sample, the model built in the model selection process should be re-estimated using the cases in the validation sample. Variables with a p-value ≥ 0.10 will be removed from the validation model.

3.4.4.3. Final Model Estimation

Final model is re-estimated using all data from both development and validation data sets. All variables that are cross-validated from the validation model are entered into the backward logistic regression procedure. The SLSTAY criterion should be set to 0.05 when developing the final model. The logistic model coefficients from the final model are then used to calculate expected probability of fail for each primer in design (*see* **Subheading 3.5.2** for details on how to calculate predicted fail rate).

3.4.4.4. How to Deal with Multicolinearity?

For multiple regression to produce unbiased estimates, none of the independent variables can be highly correlated with another independent variable or linear combination of other independent variables in the model. Multicolinearity becomes a problem when the correlations between some of the variables in the model are high, usually over 0.8. For example, template-dependent noise (TDN) and cumulative TDN are correlated. Consequently, parameter estimates become unreliable. The regression coefficients may differ considerably from sample to sample or often fail to achieve statistical significance even when the independent variable *X* is actually associated with the dependent variable *Y* ***(21)***.

There are several common symptoms of high multicolinearity that may alert the researcher to the problem. One of the symptoms is high *c*-statistic for the model but statistically insignificant regression coefficients for the scores. A second symptom is that the values of regression coefficients change greatly when adding or dropping certain scores to the model. A third symptom involves unexpectedly large (small) regression coefficients for scores. A fourth symptom is a coefficient of a score with a wrong sign.

It is always a good idea to calculate Pearson correlations among all primer score variables before developing the logistic regression model. When any of the correlation coefficient is 0.8 or greater, there is high multicolinearity. In SAS, PROC CORR procedure will generate the correlation matrix in one simple step. A better way to detect multicolinearity is to regress each primer score on all the other primer scores using ordinary least square regression. When any of the R^2 of these equations is close to 1.0, there is high multicolinearity. The highest R^2 can serve as an indicator of the amount of multicolinearity (*see* **Note 10**) ***(21)***.

In the presence of high multicolinearity, there are several ways to get around the problem. The first solution is to increase sample size. However, sometimes the researcher is unable to increase the sample. Other times even the increased sample is not able to eliminate high multicolinearity. The second strategy is to combine scores that are highly correlated into a single score if it makes conceptual sense. Nevertheless, some scores may be different in conceptual sense like "apples and oranges" making it impractical to combine them. The third strategy is to leave the highly correlated scores in the model as is if the purpose of the model is to predict the success/fail only, as is our case here most of the time, but not to interpret the independent effect on success/fail of a change in the value of a single score. If the interest also is to accurately measure the effect of each score on prime success, then the last strategy is to estimate several models by discarding one offending score at a time. After that, one can then evaluate all models together comparing with the original model to assess the damage done by specification error ***(21)***. The chosen model should have the set of scores that contribute most to the model predictability, yet not cause too much increase in −2 Log Likelihood statistics when compared with the original model.

3.4.4.5. Measures of Model Adequacy

The *c*-statistic is often used as a measure of discriminatory power of the logistic regression ***(18,19)***. Specifically, *c*-statistic measures how many times the model assigns higher failure probability to "failed" SNP markers compared with the failure probability of "passed" markers. The *c*-statistic ranges from around 0.5, equivalent to no discrimination power, to 1, equivalent to perfect discrimination power. Any model which has a *c*-statistic higher than 0.75 is considered as having good discriminatory power.

The Hosmer–Lemeshow goodness of fit is another statistic that can be used to measure model goodness of fit for binary response models. First, data are sorted into approximately 10 groups according to increasing order of the predicted probability of fail. The Hosmer–Lemeshow goodness-of-fit statistic is then obtained by calculating the Pearson χ^2 statistic from the 2×10 table of observed and expected frequencies with $df = 8$. Large values of the χ^2 (and small *p*-values) indicate a lack of fit of the model ***(19)***. Both *c*-statistic and Hosmer–Lemeshow goodness-of-fit statistic are printed in the SAS output.

3.4.5. How to Interpret SAS PROC LOGISTIC Output

In the 12-plex example given here, **Output 1** shows that the PROC LOGISTIC procedure models on the probability of genotyping failure. There are 10,536 failures and 17,908 successes. The χ^2 test for the null hypothesis that all coefficients equal to zero is significant: Wald $\chi^2 = 3640.9$, $df = 74$, $p<0.0001$.

Output 2 shows all the significant primer scores in the 12-plex primer model. For example, the maximum TDN on the GBA primer in a 12-plex cluster, TDN_GBA_M score, would increase the log of odds ratio of primer failure by 0.1349 for each increase of 1 unit in the score, all other scores being the same. The χ^2 test for the null hypothesis that the effect of TDN_GBA_M=0 is 16.6, $df = 1$, $p<0.0001$, which confirms that the effect of TDN_GBA_M on primer failure is statistically significant. The sum of absolute differences between temperature of marker amplicon and temperature of all other amplicons in multiplex cluster, EXT_AMPTMDIFSUM, would increase the log of odds ratio of primer failure by 0.00512, all other scores being the same. The χ^2 test for the null hypothesis that the effect of EXT_AMPTMDIFSUM=0 is 30.7, $df = 1$, $p<0.0001$, which confirms that the effect of EXT_AMPTMDIFSUM on primer failure is also statistically significant.

Outputs 3 and **4** show the odds ratio estimates and their 95% Wald CI. In terms of odds ratio, each unit increase in the TDN_GBA_M score would increase the odds ratio of fail by 14.4%, with the 95% CI between 1.072 and 1.221. Each unit increase in the EXT_AMPTMDIFSUM score would increase

Output 1

```
                         The SAS System
       12-plex Primer Scores --- External and Internal Scores

                     The LOGISTIC Procedure

                       Model Information

            Data Set                        J.PLEX12C
            Response Variable               fail
            Number of Response Levels       2
            Number of Observations          28444
            Model                           binary logit
            Optimization Technique          Fisher's scoring


                        Response Profile

                  Ordered                      Total
                    Value          fail    Frequency

                        1             1        10536
                        2             0        17908

                  Probability modeled is fail=1.

NOTE: 1544 observations were deleted due to missing values for the response or explanatory
      variables.


                    Model Convergence Status

          Convergence criterion (GCONV=1E-8) satisfied.


                      Model Fit Statistics

                                                Intercept
                                Intercept            and
              Criterion              Only     Covariates

              AIC               37501.125      33058.049
              SC                37509.381      33677.226
              -2 Log L          37499.125      32908.049


              Testing Global Null Hypothesis: BETA=0

       Test                 Chi-Square       DF     Pr > ChiSq

       Likelihood Ratio      4591.0764       74         <.0001
       Score                 4352.0596       74         <.0001
       Wald                  3640.8719       74         <.0001
```

Output 2

The SAS System

12-plex Primer Scores --- External and Internal Scores

The LOGISTIC Procedure

Analysis of Maximum Likelihood Estimates

Parameter	DF	Estimate	Standard Error	Wald Chi-Square	Pr > ChiSq	Standardized Estimate
Intercept	1	34.1869	10.7287	10.1538	0.0014	
tdn_gba_m	1	0.1349	0.0331	16.5852	<.0001	0.0435
cumTDN_GBA	1	-0.00049	0.000147	11.0359	0.0009	-0.0326
TDN_PCR_min	1	-0.0546	0.00646	71.6438	<.0001	-0.0684
max_gba3end9	1	0.9412	0.0606	241.2249	<.0001	0.9534
max_gba3end9_sq	1	0.0412	0.00311	175.8468	<.0001	0.7857
SNPsite_struct	1	-0.0332	0.00453	53.8367	<.0001	-0.0545
Ampl_End_struct	1	-0.0310	0.00405	58.7512	<.0001	-0.0780
x_count	1	0.2880	0.0177	265.2656	<.0001	0.1243
GC_Max	1	1.1410	0.1880	36.8524	<.0001	0.0631
Melttm_New	1	0.1016	0.00910	124.8042	<.0001	0.2176
melttm_new66	1	15.7777	2.7651	32.5582	<.0001	2.5151
Melttm_Ne*melttm_new	1	-0.2366	0.0425	30.9218	<.0001	.
RealTm_Dif	1	0.0140	0.00430	10.5968	0.0011	0.0432
Ampl_RepNum	1	0.0828	0.00781	112.2599	<.0001	0.0907
Ext_TDN_min	1	-0.0197	0.00434	20.6236	<.0001	-0.0379
GBAGBA_cumSum	1	0.0589	0.0162	13.2683	0.0003	4.7925
gbagba_cumsum_sq	1	0.000028	8.124E-6	11.8054	0.0006	8.9734
gbagba_cumsum_cb	1	4.396E-9	1.355E-9	10.5288	0.0012	4.2286
Ext_PCRTDN_min	1	-0.0485	0.00792	37.5264	<.0001	-0.0473
Ext_AmpTmdifSum	1	0.00512	0.000924	30.7255	<.0001	0.0509
Ext_AmpComplexitydif	1	0.000017	3.41E-6	23.8849	<.0001	0.0412
AT_AA	1	1.2480	0.3067	16.5612	<.0001	0.0305
AT_AC	1	1.3864	0.3821	13.1661	0.0003	0.0276
AT_AT	1	1.5663	0.4001	15.3271	<.0001	0.0307
AT_CA	1	1.1015	0.3359	10.7535	0.0010	0.0228
AT_CC	1	1.2825	0.3729	11.8290	0.0006	0.0262
AT_CG	1	2.8284	1.0915	6.7142	0.0096	0.0277
AT_CT	1	1.6858	0.5675	8.8252	0.0030	0.0227
AT_GA	1	0.9587	0.3352	8.1805	0.0042	0.0198
AT_GC	1	1.3612	0.4186	10.5737	0.0011	0.0244
AT_GG	1	2.4260	0.4953	23.9905	<.0001	0.0469
AT_GT	1	1.2819	0.4573	7.8593	0.0051	0.0201
AT_TA	1	0.9208	0.4105	5.0312	0.0249	0.0165
AT_TC	1	1.9782	0.4337	20.8094	<.0001	0.0398
AT_TG	1	0.8324	0.3795	4.8115	0.0283	0.0154
AT_TT	1	1.4083	0.3724	14.2967	0.0002	0.0315
CA_AA	1	-0.8490	0.2007	17.8857	<.0001	-0.0342
CA_AT	1	-0.6706	0.2754	5.9295	0.0149	-0.0198
CA_GA	1	-0.8214	0.2524	10.5931	0.0011	-0.0271
CA_GC	1	0.5850	0.2089	7.8417	0.0051	0.0197
CA_GG	1	0.5737	0.1982	8.3793	0.0038	0.0202
CA_TA	1	-0.9822	0.2478	15.7109	<.0001	-0.0321
CG_AC	1	0.5025	0.2057	5.9696	0.0146	0.0171
CG_AG	1	0.6917	0.1645	17.6882	<.0001	0.0298

Output 3

The SAS System
12-plex Primer Scores --- External and Internal Scores

The LOGISTIC Procedure

Analysis of Maximum Likelihood Estimates

Parameter	DF	Estimate	Standard Error	Wald Chi-Square	Pr > ChiSq	Standardized Estimate
CG_CC	1	1.1231	0.1886	35.4700	<.0001	0.0429
CG_GC	1	0.7245	0.2596	7.7913	0.0052	0.0198
CG_GG	1	0.8189	0.1946	17.7012	<.0001	0.0297
CG_GT	1	0.4644	0.2322	3.9995	0.0455	0.0140
CG_TC	1	1.2037	0.2740	19.3037	<.0001	0.0322
CG_TT	1	0.5181	0.1641	9.9722	0.0016	0.0220
CT_AC	1	0.5526	0.1048	27.7880	<.0001	0.0370
CT_AG	1	0.8195	0.1040	62.0651	<.0001	0.0554
CT_CC	1	0.5171	0.0980	27.8268	<.0001	0.0375
CT_CG	1	0.7368	0.2544	8.3859	0.0038	0.0206
CT_GC	1	0.4329	0.1201	13.0023	0.0003	0.0255
CT_GG	1	0.6258	0.1124	31.0206	<.0001	0.0392
CT_GT	1	0.2854	0.1459	3.8260	0.0505	0.0139
CT_TA	1	-0.3091	0.0985	9.8392	0.0017	-0.0239
CT_TC	1	0.8521	0.0952	80.1127	<.0001	0.0632
CT_TG	1	0.6880	0.0984	48.9038	<.0001	0.0493
CT_TT	1	0.5921	0.0932	40.3896	<.0001	0.0452
GA_AA	1	-0.2607	0.0690	14.2693	0.0002	-0.0295
GA_AT	1	-0.6374	0.0802	63.1615	<.0001	-0.0659
GA_CA	1	-0.1506	0.0652	5.3369	0.0209	-0.0177
GA_CT	1	-0.4789	0.0759	39.7780	<.0001	-0.0508
GA_GA	1	-0.2892	0.0777	13.8559	0.0002	-0.0288
GA_GT	1	-0.4741	0.0874	29.4554	<.0001	-0.0437
GA_TA	1	-0.5935	0.0917	41.8727	<.0001	-0.0525
GA_TT	1	-0.6347	0.0933	46.2694	<.0001	-0.0567
GT_AG	1	1.1958	0.3285	13.2540	0.0003	0.0262
GT_GA	1	0.9036	0.4221	4.5822	0.0323	0.0151
GT_GG	1	0.6910	0.2840	5.9182	0.0150	0.0170
GT_TG	1	1.0737	0.3720	8.3291	0.0039	0.0208
GT_TT	1	1.3326	0.3137	18.0437	<.0001	0.0314

Odds Ratio Estimates

Effect	Point Estimate	95% Wald Confidence Limits	
tdn_gba_m	1.144	1.072	1.221
cumTDN_GBA	1.000	0.999	1.000
TDN_PCR_min	0.947	0.935	0.959
max_gba3end9	2.563	2.276	2.886
max_gba3end9_sq	1.042	1.036	1.048
SNPsite_struct	0.967	0.959	0.976
Ampl_End_struct	0.969	0.962	0.977
x_count	1.334	1.288	1.381

Output 4

The SAS System

12-plex Primer Scores --- External and Internal Scores

The LOGISTIC Procedure

Odds Ratio Estimates

Effect	Point Estimate	95% Wald Confidence Limits	
GC_Max	3.130	2.165	4.524
RealTm_Dif	1.014	1.006	1.023
Ampl_RepNum	1.086	1.070	1.103
Ext_TDN_min	0.980	0.972	0.989
GBAGBA_cumSum	1.061	1.028	1.095
gbagba_cumsum_sq	1.000	1.000	1.000
gbagba_cumsum_cb	1.000	1.000	1.000
Ext_PCRTDN_min	0.953	0.938	0.968
Ext_AmpTmdifSum	1.005	1.003	1.007
Ext_AmpComplexitydif	1.000	1.000	1.000
AT_AA	3.483	1.910	6.354
AT_AC	4.001	1.892	8.460
AT_AT	4.789	2.186	10.489
AT_CA	3.009	1.558	5.812
AT_CC	3.606	1.736	7.488
AT_CG	16.918	1.992	143.699
AT_CT	5.397	1.775	16.412
AT_GA	2.608	1.352	5.031
AT_GC	3.901	1.717	8.861
AT_GG	11.313	4.285	29.866
AT_GT	3.604	1.471	8.830
AT_TA	2.511	1.123	5.614
AT_TC	7.230	3.090	16.915
AT_TG	2.299	1.093	4.836
AT_TT	4.089	1.970	8.484
CA_AA	0.428	0.289	0.634
CA_AT	0.511	0.298	0.877
CA_GA	0.440	0.268	0.721
CA_GC	1.795	1.192	2.703
CA_GG	1.775	1.204	2.617
CA_TA	0.374	0.230	0.609
CG_AC	1.653	1.105	2.473
CG_AG	1.997	1.447	2.757
CG_CC	3.074	2.124	4.449
CG_GC	2.064	1.241	3.432
CG_GG	2.268	1.549	3.321
CG_GT	1.591	1.009	2.508
CG_TC	3.332	1.948	5.701
CG_TT	1.679	1.217	2.316
CT_AC	1.738	1.415	2.134
CT_AG	2.269	1.851	2.783
CT_CC	1.677	1.384	2.032
CT_CG	2.089	1.269	3.440
CT_GC	1.542	1.218	1.951
CT_GG	1.870	1.500	2.330

the odds ratio of primer failure by 0.5%, with the 95% CI between 1.003 and 1.007.

Output 5 shows the model discrimination statistics. The *c*-statistic of 0.729 demonstrates that the model has a reasonably good discrimination power.

The results of Hosmer–Lemeshow goodness-of-fit test are also shown in **Output 5**. There is some evidence that this particular model does not fit the data quite well, as evidenced by $p = 0.0008$. A well-fitted model should generate predicted fail rates that are very similar to observed fail rates across all ranges of fail rates, thus a p of at least 0.05 or higher. Further examination of the model may improve the goodness of fit.

3.5. What Is the Predicted Probability of Success for a Given Primer Design?

3.5.1. Utilize Output Data from Logistic Regression

After the logistic regression, statistics based on the regression model, such as predicted probability of fail, can be output into an output data set. The OUTPUT statement generates the data set that can be named in the "OUT=" option. The "PREDICTED=" option specifies the variable name for the predicted probability of fail to be included in the output data set.

3.5.2. Use of Predicted Probability of Primer Success

After weights are calculated from the model, probability of fail, p, for any new primers can then be calculated as Eq. 5

For existing markers, predicted probability of fail, p, is included in the output data set from the PROC LOGISTIC procedure and named with the "P= *newname*" option, where *newname* is the name for the predicted probability given by the user.

3.6. Other Use of Statistical Model to Monitor Primer Success

For quality control and quality improvement purposes, overall fail rate for various batches, operators, samples, oligo, genotyping equipment, and order of plates in the existing data can also be measured and monitored after adjusting for primer properties.

In the case of operator-specific fail rate, the observed fail rate can be calculated simply as number of failed reactions divided by total number of reactions. Unfortunately, this observed fail rate is not a complete measure of the quality of an operator, because it does not account for the property of primers in the assays. If one operator has considerably more difficult primers than another operator,

Output 5

The SAS System
12-plex Primer Scores --- External and Internal Scores

The LOGISTIC Procedure

Odds Ratio Estimates

Effect	Point Estimate	95% Wald Confidence Limits	
CT_GT	1.330	0.999	1.771
CT_TA	0.734	0.605	0.891
CT_TC	2.345	1.946	2.826
CT_TG	1.990	1.641	2.413
CT_TT	1.808	1.506	2.170
GA_AA	0.771	0.673	0.882
GA_AT	0.529	0.452	0.619
GA_CA	0.860	0.757	0.977
GA_CT	0.619	0.534	0.719
GA_GA	0.749	0.643	0.872
GA_GT	0.622	0.525	0.739
GA_TA	0.552	0.462	0.661
GA_TT	0.530	0.442	0.636
GT_AG	3.306	1.737	6.294
GT_GA	2.468	1.079	5.646
GT_GG	1.996	1.144	3.482
GT_TG	2.926	1.411	6.067
GT_TT	3.791	2.050	7.011

Association of Predicted Probabilities and Observed Responses

Percent Concordant	72.8	Somers' D	0.458
Percent Discordant	27.0	Gamma	0.459
Percent Tied	0.3	Tau-a	0.214
Pairs	188678688	c	0.729

Partition for the Hosmer and Lemeshow Test

Group	Total	fail = 1 Observed	fail = 1 Expected	fail = 0 Observed	fail = 0 Expected
1	2848	388	352.40	2460	2495.60
2	2846	536	512.66	2310	2333.34
3	2845	594	632.61	2251	2212.39
4	2846	743	746.21	2103	2099.79
5	2846	773	870.09	2073	1975.91
6	2844	1013	1010.59	1831	1833.41
7	2847	1207	1180.48	1640	1666.52
8	2848	1405	1391.71	1443	1456.29
9	2845	1710	1679.72	1135	1165.28
10	2829	2167	2158.45	662	670.55

Hosmer and Lemeshow Goodness-of-Fit Test

Chi-Square	DF	Pr > ChiSq
26.8075	8	0.0008

his or her observed fail rate would be somewhat higher. So it would not be fair to measure operators' performance solely on the basis of the percentage of their failed reactions. For example, if one operator's primer assays contain high TDN, his or her PCRs are more likely to fail compared with others', even when the quality of two operators is the same. To make a more fair comparison, the fail rate for each operator needs to be adjusted for how "difficult" the assays are.

First, predicted fail rate for each operator can be calculated by summing up predicted probability (*see* Eq. 5) from all his or her PCRs and divided by total number of his or her PCRs. Second, divide observed fail rate of each operator by his or her predicted fail rate to obtain O/E ratio. If the O/E ratio is more than 1, the operator has higher fail rate than expected. If the O/E ratio is less than 1, the operator has lower fail rate than expected.

The CI for the O/E can be calculated as given below ***(22)***:

$$\text{Lower CI} = \frac{f\{1-[1/(9f)]-[1.96/(3f^{1/2})]\}^3}{p} \tag{7}$$

$$\text{Upper CI} = \frac{(f+1)\left(1-\{1/[9(f+1)]\}+\left\{1.96/\left[3(f+1)^{1/2}\right]\right\}\right)^3}{p} \tag{8}$$

where f is the observed number of failed reactions and p is the predicted number of failed reactions.

If the lower CI is >1, then the operator has a significantly higher than average fail rate. If the upper CI is <1, then the operator has a significantly lower than average fail rate.

The same analysis can be applied to batches, samples, oligo, genotyping equipment, order of plates, and any other factors that may affect PCR primer success.

4. Notes

1. For example, in order for our PC to be connected to the *snpmarkers* table in *gbalab* database from *gbaserver* server, we write:

```
SIGNON gbaserver;
LIBNAME gba REMOTE'gbalab' SERVER=gbaserver
USERID="myuserid" PASSWORD="mypassword";
```

This is just a simple example of connecting an SAS session on a PC to an Oracle server. For more complex connections, please refer to SAS Connect user's guide ***(9)***.

2. Make sure that there are no duplicated records in at least one of the data sets with the same merge key variable, markerid (for instance, gba.pcrproducts in our

example here). If there are duplicated records in the data set, we have to get rid of the duplicated records before we can correctly merge the data sets by using the nodupkey option in the PROC SORT statement (see example below). In addition, always store the output data after the sort in a temporary data set by using the "OUT=" option in the PROC SORT statement so we will not accidentally damage the original data file.

```
PROC SORT DATA=gba.snpmarkers OUT=snpmarker;
By markerid;
PROC SORT DATA=gba.pcrproducts OUT=pcrproducts NODUPKEY;
BY markerid;
DATA gba.analysis;
MERGE snpmarkers pcrproducts (KEEP=markerid melttm annealtm deltatm);
By markerid;
```

3. For example, the following program codes extension mix into six binary variables: *AC, GC, AT, GT, CT*, and *AG*.

```
ARRARY ext_mix(6) AC, GC, AT, GT, CT, AG;
DO i=1 to 6;
Ext_mix(i)=0;
END;
IF extension_mix='AC' THEN AC=1;
ELSE IF extension_mix='GC' THEN GC=1;
ELSE IF extension_mix='AT' THEN AT=1;
ELSE IF extension_mix='GT' THEN GT=1;
ELSE IF extension_mix='CT' THEN CT=1;
ELSE IF extension_mix='AG' THEN AG=1;
```

4. Box plots along with histograms can be obtained in SAS with the following PROC UNIVARIATE procedure with plots option:

```
PROC UNIVARIATE DATA=mydata PLOTS;
VAR original_score1-original_scoren;
RUN;
```

5. `IF 0<=transformed_score <10 THEN score10=1;`

```
ELSE IF 10<=transformed_score<20 THEN score10=2;
ELSE IF 20<=transformed_score<30 THEN score10=3;
ELSE IF 30<=transformed_score<40 THEN score10=4;
ELSE IF 40<=transformed_score<50 THEN score10=5;
ELSE IF 50<=transformed_score<60 THEN score10=6;
ELSE IF 60<=transformed_score<70 THEN score10=7;
ELSE IF 70<=transformed_score<80 THEN score10=8;
ELSE IF 80<=transformed_score<90 THEN score10=9;
ELSE IF 90<=transformed_score<100 THEN score10=10;
```

6. Output data from PROC FREQ needs first rearranged to be directly used by a spreadsheet to make graph. The following program does the transformation:

```
PROC TRANSPOSE DATA=test (KEEP=Score10 fail
count) OUT=test2 PREFIX=fail;
BY score10;
ID fail;
VAR count;
DATA test2;
SET test2 (DROP=_name_);
rate=100*fail0/(fail0+fail1);
PROC EXPORT DATA=test2 OUTFILE='myoutputfile.xls';
RUN;
```

1 where *test* is the output data from PROC FREQ procedure and the input data for PROC TRANSPOSE, *score10* is the transformed primer score, *fail* = 1 for fails and *fail* = 0 for success, *count*= the number of either successes if *fail* = 0 or the number for failures if *fail* = 1, and *test2* is the output data set from PROC TRANSPOSE.

2 The PROC TRANSPOSE procedure rearranged the output data set into a new data set that looks like **Table 2**, where for every transformed primer score, there are the number of successes (named *fail0* by the procedure with the PREFIX=*fail* option and the ID *fail* statement) and number of failures (named *fail1*) in the same data record. The rate of primer fail is then calculated and named *rate*. PROC EXPORT exports the SAS data into an Excel file named *myoutputfile.xls*. Destination of the file can be included in front of the *myoutputfile.xls*.

Table 2
An Example of Genotyping Success and Failure by Transformed Primer Score

Transformed primer score	Genotyping success	Genotyping failure
1–10	240	960
10–20	390	1110
20–30	341	759
30–40	324	576
40–50	315	435
50–60	427	463
60–70	583	517
70–80	530	400
80–90	510	340
90–100	447	263

7. For example, **Table 2** lists the number of successful and failed genotyping for each group of the primer score. In this example, $\chi^2 = 860.63$, $df = (10-1)(2-1) = 9$, $p<0.0001$. Therefore, we reject the hypothesis that genotyping success/fail rates are the same for all primer scores.
8. The following SAS program randomly splits the data into a development data set and a validation data set.

```
DATA sucess fail;
SET gba.mydata;
IF fail=0 THEN OUTPUT success;
ELSE OUTPUT fail;
DATA success1 success2;
SET success;
IF 0<=RANUNI(123)<0.5 THEN OUTPUT success1;
ELSE OUTPUT success2;
DATA fail1 fail2;
SET fail;
IF 0<=RANUNI(123)<0.5 THEN OUTPUT fail1;
ELSE OUTPUT fail2;
DATA gba.development;
SET success1 fail1;
DATA gba.validation;
SET success2 fail2;
RUN;
```

1 This program first separates the data set *gba.mydata* into the *success* data set and the *fail* data set. Then the random uniform function RANUNI draws half of the *success* data into *success1* and the other half *success2*. Half of the *fail* data are drawn to *fail1* and the other half *fail2*. Later, *success1* and *fail1* are combined into *development*, whereas *success2* and *fail2* are combined into *validation*.

9. SAS codes for model development:

```
PROC LOGISTIC DATA=development data-set-name DESCENDING;
MODEL fail=score1 score2 ...scorek /
SELECTION=BACKWARD SLSTAY=0.10 LACKFIT RISKLIMITS;
RUN;
PROC LOGISTIC DATA=validation data-set-name DESCENDING;
MODEL fail=score1 score2 ...scoreka /
SELECTION=BACKWARD SLSTAY=0.10 LACKFIT RISKLIMITS;
RUN;
PROC LOGISTIC DATA=whole data-set-name DESCENDING;
MODEL fail=score1 score2 ...scorekb /
SELECTION=BACKWARD SLSTAY=0.05 LACKFIT RISKLIMITS;
OUTPUT OUT=output-data-set-name
```

PREDITED=*predicted-probability-name*;
RUN;

10. SAS codes to detect multicolinearity.

PROC CORR DATA=*data-set-name*;
VAR $score_1$ $score_2$ ...$score_k$;
RUN;
PROC REG DATA=*data-set-name*;
MODEL $score_1$=$score_2$ ...$score_k$;
MODEL $score_2$=$score_1$ $score_3$ ...$score_k$;
...
MODEL $score_{k-1}$=$score_1$ ...$score_{k-2}score_k$;
MODEL $score_k$=$score_1$ ...$score_{k-2}score_{k-1}$;
RUN;

References

1. Yuryev, A., Huang, J., Pohl, M. et al. (2002) Predicting the success of primer extension genotyping assays using statistical modeling. *Nucleic Acids Research* 30(23), 1–11.
2. SAS Institute Inc. (2002) *SAS 9.1*. SAS Institute Inc., Cary, NC.
3. SPSS Inc. (2005) *SPSS 14.0*. SPSS Inc., Chicago, IL.
4. Stata Corp LP. (2005) *Stata 9*. Stata Corp LP, College Station, TX.
5. SAS Institute Inc. (2005) *JMP 6*. SAS Institute Inc., Cary, NC.
6. Insightful Inc. (2005) *S-PLUS 7*. Insightful Inc., Seattle, WA.
7. Systat Software Inc. (2002) *SYSTAT 11*. Systat Software Inc., Poit Richmond, CA.
8. Spotfire, Inc. (2005) *Spotfire DecisionSite Statistics*. Spotfire, Inc., Somerville, MA.
9. SAS Institute Inc. (2004) *SAS/Connect 9.1 User's Guide*. SAS Institute Inc., Cary, NC.
10. SAS Institute Inc. (2004) *SAS 9.1 SQL Procedure User's Guide*. SAS Institute Inc., Cary, NC.
11. SAS Institute Inc. (2004) *SAS 9.1 Language Reference, Concepts*, Volumes 1 and 2. SAS Institute Inc., Cary, NC.
12. SAS Institute Inc. (2006) *Base SAS 9.1.3 Procedures Guide*, 2nd Edition, Volumes 1–4. SAS Institute Inc., Cary, NC.
13. SAS Institute Inc. (2004) *SAS/STAT 9.1 User's Guide*. SAS Institute Inc., Cary, NC.
14. SAS Institute Inc. (2004) *SAS/GRAPH 9.1 Reference*, Volumes 1–3. SAS Institute Inc., Cary, NC.
15. Keen, O.N. (1995) The log transformation is special. *Statistics in Medicine* 14(8), 811–819.
16. Collett, D. (1991) *Modeling Binary Data*. Chapman and Hall, London.
17. Agresti, A. (1990) *Categorical Data Analysis*. John Wiley & Sons, New York.

18. Cox, D.R. and Snell, E.J. (1989) *The Analysis of Binary Data*, 2nd Edition. Chapman and Hall, London.
19. Hosmer, D.W, Jr. and Lemeshow, S. (1989) *Applied Logistic Regression.* John Wiley & Sons, New York.
20. SAS Institute Inc. (1999) The LOGISTIC procedure, in *SAS/STAT User's Guide*, Version 8. SAS Institute Inc., Cary, NC, pp. 1901–2042.
21. Lewis-Beck, M.S. (1989) *Applied Regression, an Introduction.* Quantities Applications in the Social Sciences Series, Sage University Papers Series. Sage, Newbury Park, CA.
22. Breslow, N.E. and Days, W. (1980) Statistical methods in cancer research, in *The Analysis of Case-Control Studies*, Volume 1, No. 32. IARC Scientific Publication, Lyon, France.

II

Genome-Scale PCR Primer Design

7

GST-PRIME

An Algorithm for Genome-Wide Primer Design

Dario Leister and Claudio Varotto

Summary

The profiling of mRNA expression based on DNA arrays has become a powerful tool to study genome-wide transcription of genes in a number of organisms. GST-PRIME is a software package created to facilitate large-scale primer design for the amplification of probes to be immobilized on arrays for transcriptome analyses, even though it can be also applied in low-throughput approaches. GST-PRIME allows highly efficient, direct amplification of gene-sequence tags (GSTs) from genomic DNA (gDNA), starting from annotated genome or transcript sequences. GST-PRIME provides a customer-friendly platform for automatic primer design, and despite the relative simplicity of the algorithm, experimental tests in the model plant species *Arabidopsis thaliana* confirmed the reliability of the software. This chapter describes the algorithm used for primer design, its input and output files, and the installation of the standalone package and its use.

Key Words: DNA array; GST; PCR; primer design; probe; transcriptome.

1. Introduction

During the last ten years, the increasing accumulation of sequence information has been accompanied by the development of high-throughput methods for the study of mRNA expression levels of hundreds to thousands of genes in parallel. Techniques based on the hybridization of reverse-transcribed representations of the whole transcriptome of cells, tissues, organs, or even whole organisms to immobilized gene probes are currently widely used. These techniques, collectively designated as DNA-array technology, are based on the use of either short oligomers *(1)* or DNA fragments (>100 bp) as

From: *Methods in Molecular Biology, vol. 402: PCR Primer Design*
Edited by: A. Yuryev © Humana Press, Totowa, NJ

immobilized probes representing parts of transcribed regions of genomes (e.g., PCR-amplified cDNAs, expressed-sequence tag (EST) clones, and full-length cDNAs) *(2)*. The accumulation of extensive sequence information from both genomic DNA (gDNA) and EST sequencing projects made it possible to use amplicons from genic DNA regions [gene sequence tags (GSTs)] as immobilized probes on arrays. Because many genes are not represented in EST databases due to their low expression and/or their tissue specificity, GSTs amplified from gDNA based on gene predictions are the only possibility to study their expression. The direct amplification of GSTs from gDNA, however, requires the design of at least one pair of specific primers per GST, requiring the design of hundreds to thousands of primers for an average-sized DNA array. The designed GST primers should ideally be also usable for amplification from cDNA to allow their use in the validation of array data by quantitative reverse-transcriptase PCR (QRT-PCR) or real-time PCR. Because manual design of primers, or its design by one-by-one programs, is not feasible for hundreds or thousands of primer pairs, we have developed GST-PRIME for the large-scale design of GST primers *(3)*. GST-PRIME is a ready-to-use and standalone program package that requires no additional code writing from the user. This chapter describes the algorithm used for primer design, the input and output files, and the standalone package installation and its use. Finally, a comparison between GST-PRIME and other high-throughput primer-design programs is provided, offering the reader a broad choice of programs for specific needs.

1.1. GST-PRIME Overview

The GST-PRIME package has been developed to design, on a large scale, primers suited for highly efficient, direct amplification of GSTs from both DNA and cDNA. As such, it can be used for the generation of collections of GSTs from species for which genomic sequence information is available. It is also useful for species with ongoing sequencing projects, as the automatic retrieval of bacterial artificial chromosome (BAC) clones does not require the availability of fully assembled pseudo-chromosomes.

Because the functional classification of genes is mostly based on homology of their product to polypeptides of known function, the logical starting point for the selection of genes for DNA-array design is usually a collection of protein sequences or their corresponding GIs (*see* **Note 1**). For the design of suitable PCR primers for GST amplification, however, collections of DNA sequences are required (*see* **Fig. 1**). Therefore, in connection with Batch Entrez at National Center for Biotechnology Information (NCBI), a "List of protein Gis" is used to obtain a second list, called "Annotated List," which contains the annotation

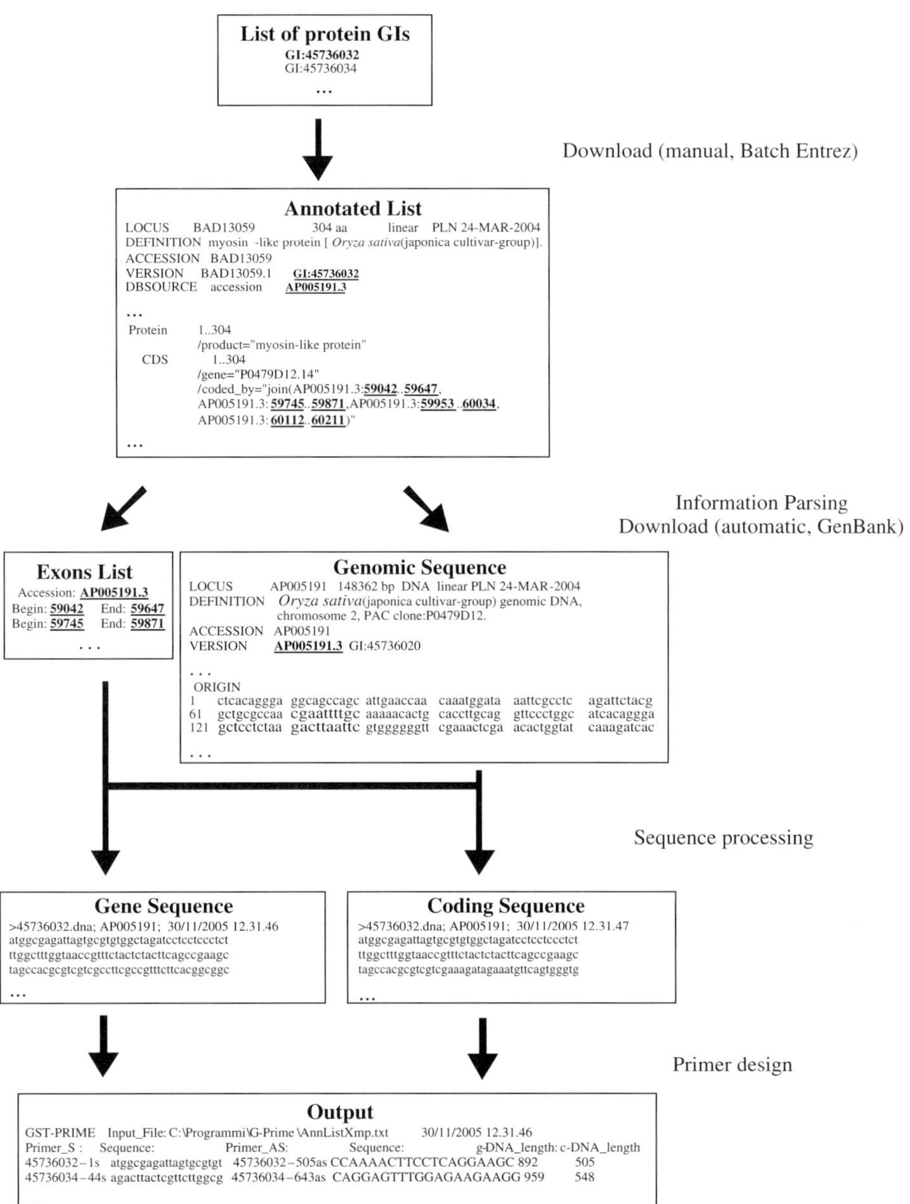

Fig. 1. Overview of the GST-PRIME workflow.

of the corresponding genomic sequence (accession number and position of the coding regions). This file can be then used as input for GST-PRIME. This step has to be performed manually by the user (*see* **Subheading 2.2.**).

GST-PRIME automatically parses the information contained in the Annotated List to extract the pieces of relevant information for the next steps of the primer design: the accession number of the genomic sequence corresponding to each protein and the start and stop positions (base pair number) of each exon coding for the protein in the same genomic sequence. This information is separately assembled for each protein in a text file (called "Exons List" in **Fig. 1**), named according to the polypeptide ID.

The information contained in the single Exons Lists is then used by GST-PRIME to perform automatically two sequential tasks: (1) the downloading from the NCBI database of the genomic nucleotide sequences according to the ID contained in each Exons List and (2) the extraction and reformatting in FastA format of both the whole gene sequence (from start to stop including introns, if present) and the corresponding coding regions.

The program finally starts the iterative process of primer design from the pair of files ("Gene Sequence" and "Coding Sequence" FastA files) that contain the relevant nucleic acid information for each protein of the original "List of protein GIs." At the end of the calculation, the output file "PrimerList.res" contains the list of primers, whereas the "Warning.res" and other files collected in the folder "Results" give additional information concerning the processing of the sequences analyzed (*see* **Subheading 2.7**. for details). In the following sections, the single steps necessary for the operation of GST-PRIME are described in detail.

2. Step-by-Step Protocol for the Use of GST-PRIME

2.1. Definition of the Initial Set of Genes

The selection of genes to be used for DNA-array analyses depends on the biological questions one is interested to answer. In our group, GST-PRIME was used for the rapid and cost-effective amplification of GSTs representing all *Arabidopsis* genes that were predicted to encode proteins imported in the chloroplast *(4,5)*. When a set of proteins has been selected, the first thing to do is to collect all corresponding GenBank GIs (*see* **Note 1**). Depending on which database is used for the retrieval of protein sequences, cross-references to GenBank GIs can be already embedded in the original file format or can be obtained by blasting the peptides of interest to the NCBI database. All retrieved GIs should then be pasted in a text editor and saved as "simple text" file, in which each of the GIs has to be listed in a single line (*see* **Note 2**).

2.2. Obtaining Annotated Lists from List of Protein GIs

The Batch Entrez service at NCBI is used to retrieve a list of annotations for the nucleotide sequences that correspond to the protein accessions chosen (http://www.ncbi.nlm.nih.gov/entrez/batchentrez.cgi?db=Nucleotide).

Necessary steps are

1. Select "protein" from the "Database" drop-down list on the top left.
2. Enter the name of the "List of protein GIs" file or browse to choose the file from your system directory pressing the "Browse" button.
3. Press the "Retrieve" button. A list of document summaries will be shown.
4. Select the format "GenPept" from the "Display" drop-down list on the top left and the number of entries (maximum 500) to be shown at once from the first "Show" drop-down list.
5. Select the option "Text" from the second "Show" drop-down list.
6. Save the text page using the "Save as" option from the "File" menu of the browser in use, giving the file the ".txt" extension. Otherwise, copy the whole page and paste it in an empty document of a text editor.

An "Annotated List" example, correctly formatted, is distributed along with the executable file (file "AnnListXmp.txt," *see* **Note 3**). Annotated lists can be also retrieved by pasting directly a space-separated or tab-separated list of GIs (all of them in one line) into the search text box of the GenBank database, selecting the "nucleotide" subset of sequences. We find, however, this procedure more prone to mistakes.

2.3. Obtaining and Installing GST-PRIME

The package is available free of charge upon request from the corresponding author (Dario Leister). A material transfer agreement form needs to be filled and signed before a compressed version of the executable file (about 2.1 MB), with accompanying documentation and examples, is sent as an e-mail attachment.

Installation of GST-PRIME has been kept as simple as possible to allow also unexperienced users to perform it without any difficulty. It simply requires unzipping of the compressed archive "GST-PRIMEDistr.exe" in a folder created for the installation of the program (*see* **Note 4**). By double-clicking on the "Setup.exe" file, a very simple installation procedure will guide the user through the installation.

2.4. The GST-PRIME User Interface

The main window of GST-PRIME is divided into two parts (*see* **Fig. 2**). On the left side, there are three boxes allowing the user to browse the local file system and an active button to start or stop the primer design. On the right side, there are two text boxes for the visualization of small text files (e.g., to singularly check the "*.dna" or "*.rna" files generated by the program, *see* **Note 5**). At the top of the window, there are three menus: "File," "Options," and "Tools." The window cannot be minimized to remind the user of the presence of the running program (that can be left running in the background, but, in this case, one has to pay attention to avoid running out of memory by using too many other programs simultaneously). Resizing of the window is performed by positioning the mouse at the window edge and dragging it while pressing the left button. The window has a minimal size below which resizing is not allowed. To get a full screen window, the title bar or any gray free space on it has to be double-clicked. A second double-click will restore the window to its original size.

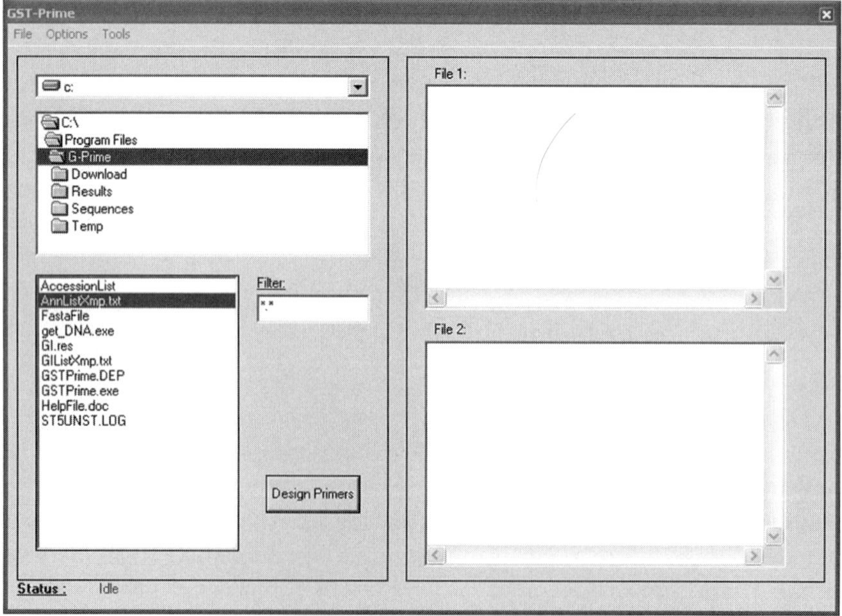

Fig. 2. The main window of GST-PRIME.

2.5. Setting GST-PRIME Parameters Before Primer Design

After the annotated list has been retrieved and saved locally in the folder containing also the GST-PRIME executable, the primer design routine can be started. If the routine for primer design is directly started, the default parameters are used. Otherwise, the user can decide to modify some of the parameters to obtain GSTs more suited for special purposes. In the "Options" menu, the options "Amplification Length," "Sequence Orientation," "Delayed Start," and "Remove BAC Files" can be selected.

The option "Amplification Length" allows the user to set the preferential amplification length at the cDNA level. As discussed in more detail in **Subheading 3**, this parameter is very important to determine the effective probe size (EPS) of the GST, i.e., the number of base pairs of the amplified GST corresponding to coding sequences. Low EPS values require low-hybridization stringencies to obtain sufficient signal intensities on the filter. On the contrary, a high EPS corresponds to larger amplicons, which in turn might require the use of sophisticated polymerase mixes instead of normal Taq polymerase for probe amplification. We found that an EPS of 500 bp, the default value used by GST-PRIME, is a good compromise for model organisms like *Arabidopsis* and *Drosophila*.

The option "Sequence Orientation" allows to choose the orientation of reformatted sequences by keeping the original orientation (default) or reformatting all sequences in "sense" (from start to stop) or "antisense" (from stop to start) orientation. Because the primers are always designed starting from the leftmost (5′) part of the sequences, this option thus allows to select the region of genes from which primers are preferentially designed. Alternatively, it is possible to design different primer pairs by running twice the program on sets of sequences first in sense and then in antisense orientation (*see* **Note 6**).

The option "Only Primer Design" forces the execution of the primer design subroutine without the sequence download and reformatting steps (*see* **Note 7**). This option can be used if one wishes, for example, to design primers to amplify GSTs with different predicted EPS values for the same gene. This can be useful for organisms where a high or low GC content makes it difficult to design primers with the default EPS of 500 bp. If it has not been possible to design suitable primer for some genes within the first run of the program, we suggest repeating the run with a reduced EPS and selecting the option "Only Primer Design" to avoid the useless download and reformatting of the same data set (*see* **Note 8**).

The "Delayed Start" option allows to start the program with a certain delay. It is particularly useful for large sets of sequences or to let the program run during the period of the day or night with less traffic to access the NCBI database. It sets the delay time (in minutes), after which the program will automatically start the primer design (*see* **Note 9**).

If the "Remove BAC Files" option is checked, at the end of the primer design, the whole directory "Download" containing all the original nucleotide sequences downloaded from NCBI will be deleted to save disk space (*see* **Note 10**).

2.6. Primer Design with GST-PRIME

Necessary steps are

1. Select the annotated list file by browsing the drives and the directories of our computer.
2. Highlight the name of the file containing the annotated list by clicking on it once.
3. Click on the button "Design Primers" (*see* **Note 11**).
4. Follow the progress of the subroutines of the program as shown in **Fig. 2** in the "status" line at the bottom of the window.

2.7. Output of GST-PRIME

After each step, GST-PRIME generates output files. Some of them are temporary files, deleted at the end of each run of the program, whereas others are permanent files (*see* **Note 12**). The permanent files are collected in three subfolders generated automatically by the program in the folder containing the Annotated List, "GST-PRIME.exe," and "Get_DNA.exe" files (the three of them have to stay together, otherwise the program will not work). Chronologically, the first files to be generated are the Exons List files (*see* **Fig. 1**, step 2). These files contain the annotation of the coding sequences and are named according to the protein GI, followed by the extension ".list." These files are collected in the directory "Sequences."

In the next step (*see* **Fig. 1**, step 3), the downloaded files containing the original nucleotide sequences are saved in the directory "Download." These files are then reassembled by GST-PRIME in two sets of related files: the "*.dna" files containing the genomic sequence from start to stop (including introns, if present) and the "*.rna" files containing only the coding regions corresponding to each gene. Both file sets are designated according to the protein GIs (*see* **Note 13**).

"Troubles" encountered during step 2 (e.g., wrong annotation or an empty "*.list" file) and step 3 (e.g., sequence with missing start and/or stop) are listed in the "WARNING.res" file (folder "Results").

In the same folder, the file "PrimerList.res" is generated during step 4 (*see* **Fig. 1**) and contains for every gene the sense and antisense primer, the predicted amplification length from both gDNA and cDNA, the protein GI, the corresponding nucleotide accession number, and the primer that can be used for first-strand cDNA synthesis ("sense" or "antisense").

Additional files contained in the folder "Results" are the "DNALength.res" (containing the length in base pairs of the sequence in every ".dna" file), "GI-Accession.res" (a list of all the GIs and the corresponding Accessions Numbers used for the primer design), "LastDownload.res" (a list of the nucleotide sequences downloaded during the last run of the program), and "Statist.res" (predicted amplification length from cDNA, gDNA, and orientation of the primer to be used for first-strand cDNA synthesis).

3. Algorithm Used by GST-PRIME for Primer Design

The task of amplifying GSTs for DNA arrays imposes some constraints to the features that each amplicon should have. The first requirement is that they should contain a minimal amount of coding region to allow hybridization to the reverse-transcribed RNA sample to be assayed. We call the amount (in base pairs) of coding regions associated with each GST EPS (*see* **Subheading 2.5.**). For instance, amplifying a GST containing only intronic sequences would correspond to an EPS of 0, making it therefore useless as a probe. Therefore, the primers are designed to anneal exclusively in exons, and they amplify GSTs with an EPS preferentially of 500. This default EPS value, chosen to allow a good hybridization even in the presence of introns in the probe, can be changed by the user. We would recommend, in general, using high EPS values for species and/or gene sets known to have short exons and introns; on the contrary, a low EPS could be set to amplify GSTs from genes with long introns that could hinder the amplification efficiency. The second requirement is that the primers designed by GST-PRIME should work on both gDNA and cDNA templates to allow their use not only for probe amplification but also for downstream validation of the hybridization results by reverse-transcriptase (RT)-PCR assays. This implies that primers should be complementary to coding regions, and at the same time, they should not span splicing regions. Finally, to contain the costs associated to primer synthesis and ensure robust amplification, primers are designed by default as 20 nt long with a 50% GC content. In the case that no primer pair can be designed to fulfill the above criteria, the procedure is iteratively repeated by increasing the primer length by 1 nt up to a maximal length of 22 nt.

First of all, GST-PRIME calculates for each gene the length of the coding sequence, and according to it, it determines the two sequence regions where the search of a suitable sense and antisense primer, respectively, is carried out. The corresponding sequences are then extracted to start the primer design:

```
Pseudocode:
Open the file containing the CDS in FastA format
# length in bp of the CDS sequence for the gene under examination
Calculate Sequence_Length
# according to sequence length, fix the sequence regions where to search
for a suitable sense and antisense primer, indicated by SenseRegion and
AntisenseRegion
# No primer design performed for this gene (too short)
If Sequence_Length<120 then Discard gene
# try to design Sense primer in the first half of the sequence and Antisense
primer in the second half
Else if 120<Sequence_Length<720 then
    SenseRegion is between 1 and Sequence_Length/2
    AntisenseRegion is between Sequence_Length/2+1 and Sequence_Length
End if
# try to design SensePrimer in the first 180 bp and AntisensePrimer in the
240 bp at a distance of about EPS bp from the SenseRegion
Else if Sequence_Length>720 then
SenseRegion is between 1 and 180 bp
AntisenseRegion is between (EPS-20) and (EPS-20)+240
End if
Extract sequences corresponding to SenseRegion and AntisenseRegion
```

The first primer to be designed is the sense primer. GST-PRIME iteratively scans the Sense Region for putative primers having 10 G + C by means of a sliding window of size 20 bp. If no such primer is found, the scanning is repeated for window sizes of 21 and, if necessary, 22 bp. Each putative primer having the desired GC content is checked for overlaps with exon–intron junctions with the help of the file "Exons List" for the gene under examination. This procedure ensures that the designed sense primer will be the shortest, "leftmost" primer with the desired GC content within the selected region:

```
#sliding window scan for GC content calculation
Calculate GC_Content of each 20 bp oligo within SenseRegion starting
from5′
If GC_Content<>10 for every oligo then
```

```
  repeat for oligos 1 bp longer up to a max of 22 bp
# oligo has the requested GC content
Else
Open the file containing the annotated list
Compare the position of oligo to that of exon–intron junctions
    If any exon–intron junction falls within oligo then
# oligo is on junction
        Discard oligo and resume search
# sense primer found
    Else SensePrimer=oligo
End if
End if
```

GST-PRIME then searches for the "AntisensePrimer" in the "AntisenseRegion" in a way similar to that used for the "SensePrimer": the "AntisenseRegion" is scanned from 3′ to 5′ and a "TemporaryAntisensePrimer" is found as soon as an antisense primer fulfills the GC content constraint. In contrast to the procedure for the design of the sense primer, the search for additional antisense primers proceeds, and if another primer is found that amplifies a cDNA fragment closer in size to the selected EPS than the first one, the former "TemporaryAntisensePrimer" is substituted by the latter primer. This method ensures that the "AntisensePrimer" selected by GST-PRIME is the shortest antisense primer able to guarantee the best approximation to the selected EPS.

```
#design AntisensePrimer
Calculate the GC_Content of each 20 bp oligo within AntisenseRegion
starting from 3′
If GC_Content<>10 for every oligo then
  repeat for oligos 1 bp longer up to a max of 22 bp
# oligo has requested GC content
Else
Open the file containing the annotated list
Compare the position of oligo to that of exon–intron junctions
If an exon–intron junction falls within oligo then
  #oligo is on junction
Discard oligo and resume search
Else TempAntisensePrimer=oligo
  # look for oligo with better EPS
Resume search
```

```
End if
End if
If (EPS-AmplificationLength(TempAntisensePrimer – SensePrimer))<0 then
   #found antisense primer
   AntisensePrimer=TempAntisensePrimer
End if
```

4. Comparison of GST-PRIME with Other Programs for Primer Design

Currently, several programs have been released with the potential to assist in probe design for DNA-array construction. In **Table 1**, a comparison of the packages available is provided to help the reader to choose the package that suits his/her needs the best. Particular emphasis has been given to those features that allow high-throughput amplification of probes from gDNA for DNA-array analyses. The features that, in our opinion, are most important to this purpose are in the following order: (1) the possibility to consider the gene structure during design ("gene structure"), (2) the possibility to check for specificity and uniqueness of the amplified probes to avoid cross-hybridization with other genomic or transcribed regions because of sequence homology ("homology checks"), (3) the possibility to offer retrieval facilities for the input sequences ("retrieval of gene sets"), (4) the number and types of checks performed on the primer quality (e.g., length, GC content, hairpin loops, bulge loops, dimers, GC clamps, and thermodynamic stability) ("quality checks on primers"), (5) the possibility to be used on any set of sequences ("general applicability"), and (6) the possibility to be able to design oligomeric probes in addition to primers ("oligos"). On the basis of these parameters, we provide a score for the general applicability of these programs for GST design for DNA arrays. According to this classification, the program PrimeArray *(6)* scores relatively low because it lacks several features. Primer3 (original package without its incorporation in dedicated array-oriented applications *(7)*) is slightly better suited: even though not created with the purpose of designing probes for DNA-array construction, it performs robust quality checks on the primers and can be easily incorporated (by experienced computer programmers) in other programs to design primers at a high throughput. PRIMEGENS *(8)* and Probewiz *(9)* use the same general approach and perform similarly on the basis of the criteria considered. Both of them do not include a "Gene structure" feature and cannot ensure that primers designed on gDNA will work for the amplification of cDNA and vice versa. Mprime *(10)*, despite being an extremely complete program specifically created for large-scale primer design for customized microarrays and providing elaborate sequence retrieval facilities, also lacks the "Gene structure"

Table 1
Comparison and Scoring of Different Programs for Large-Scale Primer Design with Respect to Their Applicability GST Designs

Program	Gene structure	Homology checks	Retrieval of gene sets	Quality checks on primers	General applicability	Oligos	Score
Genome PRIDE	+	+	−	Medium	+	+	11
Spads	+	+	−	High	+	−	11
GST-PRIME	+	−	+	Low	+	−	8
Medusa	+	−	−	High	+	−	8
PCR-Suite (Primer3)	+	−	−	High	+	−	8
Mprime	−	+	+	Medium	−	+	8
PRIME GENS/ Primer3	−	+	−	High	+	−	7
ProbeWiz	−	+	−	High	+	−	7
Primer3	−	−	−	High	+	+	5
PrimeArray	−	−	−	Medium	+	−	3

GST, gene-sequence tag.
For scoring of the programs, the different features were weighted: 4 for column "Gene structure," 3 for "Homology checks," 2 for "Retrieval of gene sets," and 1 for "Quality checks on primers," "General applicability," and "Oligos." Presence of a feature corresponds to 1 and absence to 0, except "low," "medium," and "high" in column "Quality checks on primers," which correspond to scores of 1–3, respectively. The high relevance of the first two criteria reflects their essential character in designing non-cross-hybridizing GSTs from genomic DNA. "Retrieval of gene sets" is weighted higher than "Quality checks on primers," because even simple quality checks on simple features like primer length and GC content can result in a relatively high robustness of primer design.

feature. All remaining programs described in **Table 1** consider, to different extents, gene structures in primer design. The PCR-Suite *(11)* offers, through the interface "Genomic Primers," the possibility to design primers around exons (a task that finds application in mutation analysis). This program, however, is not suitable for large-scale primer design. Medusa *(12)* is an interface for different programs for primer design. By integrating feature information (exon start and stop position) and PCR primer criteria, Medusa can be used for the design of primers for gDNA, but the information about gene structures is not automatically parsed from, for example, GenBank files and has to be provided separately by the user. The program, moreover, does not perform any homology check. Also GST-PRIME, even though dedicated to DNA-array construction, has its strongest limitation in not performing homology checks at both primer and GST level. This choice was motivated, at the time of its design, by the fact that the incompleteness of the target genome (of *Arabidopsis thaliana*) made this feature obsolete. Moreover, the fact that primers can be designed preferentially for the individual 3′ region of genes was considered to guarantee a reasonable degree of specificity in the amplification of GSTs. Despite a very simple algorithm for primer design, the amplification robustness has been experimentally found to be satisfactory ***(3)***. The last two programs, GenomePRIDE ***(13,14)***, an update of PRIDE, ***(15)*** and Spads ***(16)***, lack facilities for retrieval of large sets of sequences but provide stringent checks on primer quality and homology checks. In particular, both the specificity of primer pairs and the lack of cross-hybridization of the resulting GST are checked. Both programs also incorporate highly refined procedures to keep into account gene structure and use it to explicitly define the maximal fraction of intronic sequences contained in the GST of a defined length. This feature approximates the EPS parameter calculated by GST-PRIME. Spads perform more accurate quality checks on the primer designed, whereas GenomePride (like also Mprime and Primer3; *see* **Table 1**) offers the possibility to design primers for both PCR-based probe amplification and oligonucleotide probes (40–70 nt) (*see* column "Oligos" in **Table 1**). For oligo-probe design, also other programs have been specifically developed (e.g., OligoWiz, OligoArray 2.0, OligoPicker, ROSO, and PROBE ***(17–23)***).

Taken together, all the programs presented in **Table 1** are suitable—to different extents—for large-scale primer design. Among them, however, only five programs incorporate information on gene structure in primer design. Among the three programs specifically developed for primer design for GST probe amplification, the best performance is provided by GenomePride and Spads. The suitability of these programs in the realization of DNA arrays is reflected by the increasing number of groups that use either of them to this purpose.

5. Notes

1. GI numbers (which are unique identifiers to DNA and protein sequences) can be obtained from the GenBank flat file corresponding to each sequence as nucleotide ID (NID for DNA sequences) or protein ID (PID for protein sequences). The GI can be obtained from the PID by removing the letter prefix entries. Unlikely

the accession numbers, which are the original names given to the sequences at the moment of submission, GI numbers can change without notice. We therefore recommend compiling the list of GIs shortly before its use by GST-PRIME.

2. The use of simple text editors like NotePad automatically guarantees the lack of formatting that could hinder the retrieval of the corresponding sequencing from NCBI. If using more complex text editors like WordPad or MS Word, chose the option "text only with line breaks" while saving the file (if the version of the text editor used should not have this format available, any format including at the end of each line both a "linefeed" and a "carriage return" character will work).
3. If the browser is not correctly configured, the text file corresponding to the annotated list could have the wrong format. GST-PRIME requires the presence of both the "carriage return" and the "linefeed" characters at the end of each line. MS Word can be used to convert the file to one of the formats read by GST-PRIME:

   - select the "Save as" option from the "File" menu and
   - choose the format "Text with line breaks" or an analogous format from the "Save as type" drop-down list.

4. The folder for the installation cannot be in the directory containing the Setup source files. We suggest installing the program in the folder "Program files" normally present in the drive "C" in Windows.
5. If files are too large to be displayed, a message prompts the user to use an external editor such as MS Word, NotePad, or WordPad.
6. The reverse transcription and concomitant labeling of mRNA for hybridization is normally performed with oligo-dT oligomer primers. In this case, by selecting the option "antisense," the GST will be preferentially amplified from gene regions close to their predicted 3′-end. This solution can be useful to assess the expression level of genes that are particularly long or rich in secondary structures.
7. The "Only Primer Design" option becomes available only after the first run of the program and requires the use of the annotated list obtained from the previous run. Its selection automatically disables the option "Sequence Orientation."
8. We suggest the use of this option as often as possible to avoid useless load on the NCBI database. Ideally, sequences should be downloaded only once and then re-used in following runs of the program if the user wants to experiment the change of some parameters on the output.
9. Delay times up to 48 h (2880 min) are possible. Select the annotated list to be processed before checking this option.
10. We recommend the use of this option only when running low on hard-disk space. This will avoid repeated downloads of the same sequences from the NCBI database.
11. Please make sure that the annotated list is in the same folder as "GST-Prime.exe" and "Get_DNA.exe."
12. If one of the files employed as input for primer design by GST-PRIME is in use by another software (e.g., Word or Excel), an error is generated and the execution

of GST-PRIME is terminated. WordPad represents an exception, allowing GST-PRIME to finish its execution without generating any error.

13. Depending on the version of Windows or MS Office used, the format of the output files (in particular those contained in the "Results" folder) might not be read correctly. To read them, we suggest using WordPad or Excel. In WordPad, select the whole text and convert it into a non-proportional font ("Courier" or "Courier New"). If using Excel, select the "delimited" option and "space" as delimiter during the file import.

References

1. Lipshutz RJ, Fodor SP, Gingeras TR, Lockhart DJ. (1999) High density synthetic oligonucleotide arrays. *Nat. Genet.* **21**, 20–24.
2. Stoughton RB. (2005) Applications of DNA microarrays in biology. *Annu. Rev. Biochem.* **74**, 53–82.
3. Varotto C, Richly E, Salamini F, Leister D. (2001) GST-PRIME: a genome-wide primer design software for the generation of gene sequence tags. *Nucleic Acids Res.* **29**, 4373–4377.
4. Abdallah F, Salamini F, Leister D. (2000) A prediction of the size and evolutionary origin of the proteome of chloroplasts of Arabidopsis. *Trends Plant Sci.* **5**, 141–142.
5. Kurth J, Varotto C, Pesaresi P, Biehl A, Richly E, Salamini F, Leister D. (2002) Gene-sequence-tag expression analyses of 1,800 genes related to chloroplast functions. *Planta* **215**, 101–109.
6. Raddatz G, Dehio M, Meyer TF, Dehio C. (2001) PrimeArray: genome-scale primer design for DNA-microarray construction. *Bioinformatics* **17**, 98–99.
7. Rozen S, Skaletsky H. (2000) Primer3 on the WWW for general users and for biologist programmers. *Methods Mol. Biol.* **132**, 365–386.
8. Xu D, Li G, Wu L, Zhou J, Xu Y. (2002) PRIMEGENS: robust and efficient design of gene-specific probes for microarray analysis. *Bioinformatics* **18**, 1432–1437.
9. Nielsen HB, Knudsen S. (2002) Avoiding cross hybridization by choosing nonredundant targets on cDNA arrays. *Bioinformatics* **18**, 321–322.
10. Rouchka EC, Khalyfa A, Cooper NG. (2005) MPrime: efficient large scale multiple primer and oligonucleotide design for customized gene microarrays. *BMC Bioinformatics* **6**, 175.
11. van Baren MJ, Heutink P. (2004) The PCR suite. *Bioinformatics* **20**, 591–593.
12. Podowski RM, Sonnhammer EL. (2001) MEDUSA: large scale automatic selection and visual assessment of PCR primer pairs. *Bioinformatics* **17**, 656–657.
13. Haas SA, Hild M, Wright AP, Hain T, Talibi D, Vingron M. (2003) Genome-scale design of PCR primers and long oligomers for DNA microarrays. *Nucleic Acids Res.* **31**, 5576–5581.
14. Xue Y, Haas SA, Brino L, Gusnanto A, Reimers M, Talibi D, Vingron M, Ekwall K, Wright AP. (2004) A DNA microarray for fission yeast: minimal changes in global gene expression after temperature shift. *Yeast* **21**, 25–39.

15. Haas S, Vingron M, Poustka A, Wiemann S. (1998) Primer design for large scale sequencing. *Nucleic Acids Res.* **26**, 3006–3012.
16. Thareau V, Dehais P, Serizet C, Hilson P, Rouze P, Aubourg S. (2003) Automatic design of gene-specific sequence tags for genome-wide functional studies. *Bioinformatics* **19**, 2191–2198.
17. Wernersson R, Nielsen HB. (2005) OligoWiz 2.0—integrating sequence feature annotation into the design of microarray probes. *Nucleic Acids Res.* **33**, W611–W615.
18. Nielsen HB, Wernersson R, Knudsen S. (2003) Design of oligonucleotides for microarrays and perspectives for design of multi-transcriptome arrays. *Nucleic Acids Res.* **31**, 3491–3496.
19. Rouillard JM, Zuker M, Gulari E. (2003) OligoArray 2.0: design of oligonucleotide probes for DNA microarrays using a thermodynamic approach. *Nucleic Acids Res.* **31**, 3057–3062.
20. Rouillard JM, Herbert CJ, Zuker M. (2002) OligoArray: genome-scale oligonucleotide design for microarrays. *Bioinformatics* **18**, 486–487.
21. Wang X, Seed B. (2003) Selection of oligonucleotide probes for protein coding sequences. *Bioinformatics* **19**, 796–802.
22. Reymond N, Charles H, Duret L, Calevro F, Beslon G, Fayard JM. (2004) ROSO: optimizing oligonucleotide probes for microarrays. *Bioinformatics* **20**, 271–273.
23. Li F, Stormo GD. (2001) Selection of optimal DNA oligos for gene expression arrays. *Bioinformatics* **17**, 1067–1076.

8

Genome-Scale Probe and Primer Design with PRIMEGENS

Gyan Prakash Srivastava and Dong Xu

Summary

This chapter introduces the software package PRIMEGENS for designing gene-specific probes and associated PCR primers on a large scale. Such design is especially useful for constructing cDNA or oligo microarray to minimize cross-hybridization. PRIMEGENS can also be used for designing primers to amplify a segment of a unique target gene using reverse-transcriptase (RT)-PCR. The input to PRIMEGENS is a set of sequences, whose primers need to be designed, and a sequence pool containing all the genes in a genome. It provides options to choose various parameters. PRIMEGENS uses a systematic algorithm for designing gene-specific probes and its primer pair. For a given sequence, PRIMEGENS first searches for the longest gene-specific fragment and then designs best PCR product for this fragment. The 2.0 version of PRIMEGENS provides a graphical user interface (GUI) with additional features. The software is freely available for any users and can be downloaded from http://digbio.missouri.edu/primegens/.

Key Words: PCR primer design; cross-hybridization; cDNA microarray; oligo arrays; qRT-PCR; sequence alignment; dynamic programming.

1. Introduction

Various genome-scale sequencing project have generated vast amounts of sequence data. High-throughput data analysis and its study are one of the primary focuses for molecular biologist. Microarray is one of the most common tools for studying gene expressions on a large scale ***(1,2)***. In cDNA microarray, typically each spot on the array contains sequence segment of a specific gene, which is amplified by PCR. The segment is expected to be gene specific to

From: *Methods in Molecular Biology, vol. 402: PCR Primer Design*
Edited by: A. Yuryev © Humana Press, Totowa, NJ

avoid cross-hybridization among genes sharing significant sequence identity. In another case, researchers may simply want to amplify gene-specific segments for a selected group of genes using reverse-transcriptase (RT)-PCR. In both cases, the problem can be formulated to choose a gene-specific segment for a gene in a genome and then design PCR primers according to some specifications. Such an objective is often achieved manually, e.g., using Primer3 *(3)* for primer design for a given sequence. Primer3 designs many possible primer pairs for a given sequence, but it does not guarantee their uniqueness in the whole genome. Therefore, a user has to manually run BLAST *(4)* for each PCR product against the genome to search to avoid cross-hybridization. Such manual approach cannot be applied to a large scale. PRIMEGENS *(5)* does not only fulfill this task but also automate the primer generation on the large scale. Furthermore, PRIMEGENS has a rigorous formulation, which has a much better chance to find gene-specific segment than a manual process.

Figure 1 gives some general idea about PRIMEGENS. The essence of PRIMEGENS is based on searching the sequence-specific fragment for any particular sequence. PRIMEGENS implements this task by finding the fragment of a given DNA sequence, which does not have high-sequence similarity with any other sequence in the given sequence pool (whole genome in general). If the given sequence is unique, then the whole sequence is considered as the

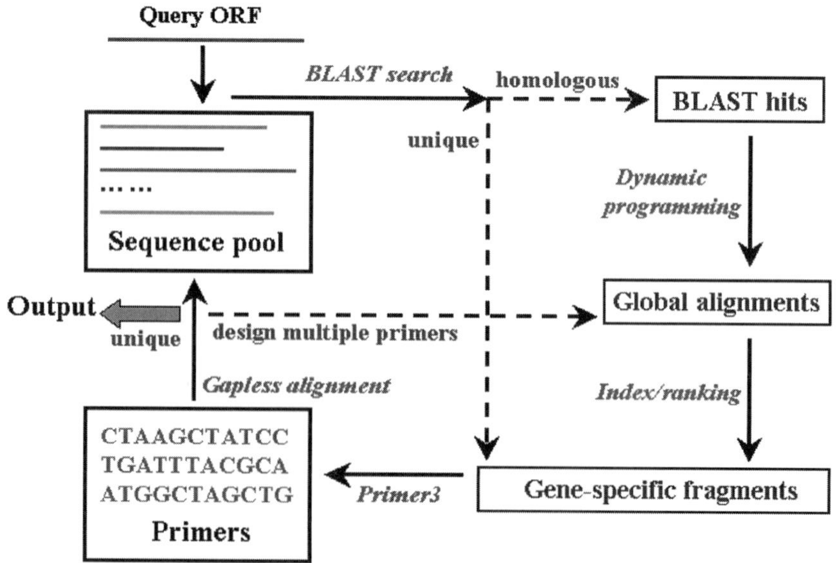

Fig. 1. Basic PRIMEGENS model.

sequence-specific fragment. Otherwise, PRIMEGENS searches for the unique fragment based on the BLAST result for the query sequence. The optimal global alignment between the query sequence and each of its significant BLAST hits is performed ***(6)***. Based on the alignment, PRIMEGENS searches for the longest unique segment for the query sequence. Finally, it designs primers on the selected gene-specific fragment using Primer3. **Figure 2** describes the detailed algorithm of the PRIMEGENS implementation.

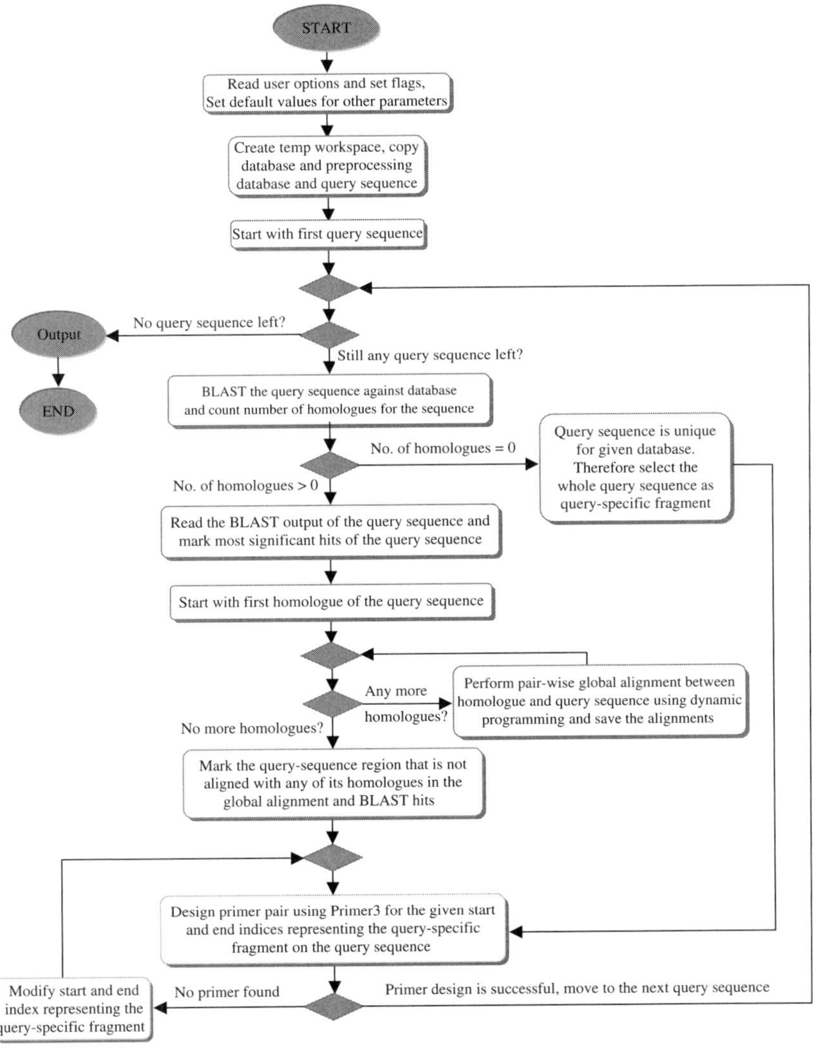

Fig. 2. Flowchart for PRIMEGENS implementation.

We recently developed PRIMEGENS version 2.0. In this new version, we improved the main algorithm and added a number of new features. In particular, we developed a Java-based graphical user interface (GUI), which can be used under both Windows and Linux platforms.

2. Description of PRIMEGENS

This section will explain what a user needs to specify for running PRIMEGENS. **Subheading 2.1** describes compositions of PRIMEGENS. **Subheading 2.2** shows various types of inputs. **Subheading 2.3** covers different types of execution features supported by PRIMEGENS. **Subheading 2.4** explains about the *append file (3)*, which is the input for primer design that controls the specifications of primers. **Subheading 2.5** describes the format of primer design results. More detailed description about PRIMEGENS is provided in the software documentation, which comes along with the PRIMEGENS package.

2.1. Composition of the Software Package

PRIMEGENS 2.0 is available in the form of compressed format as *PRIMEGENSv2.zip* for Windows and *PRIMEGENSv2.tgz* for Linux. These packages are freely available for any users and can be downloaded from http://digbio.missouri.edu/primegens/. For installation, user should specify the location of PRIMEGENS folder by setting the environment variable *PRIMEGENS_PATH* with the absolute path of the software location. More detailed description about software installation can be found in *README.txt*.

PRIMEGENS 2.0 (PRIMEGENS as the main folder) consists of following major directories and files:

1. bin/: console application executables
2. blast/: BLAST executables
3. doc/: documentation
4. include/: supporting resources
5. output/: output results
6. primer3/: primer3 executables
7. primerdesign/: graphical interface files
8. test/: testing resources
9. README.txt: instruction manual
10. primerdesign.jar: main Java executable for graphical interface

2.2. Inputs to PRIMEGENS

PRIMEGENS supports various input features according to the user requirements. This section describes each type of input.

2.2.1. Sequence Pool

To start primer design, a user needs to create a database file consisting of all the sequences in the FASTA format. The content format of the database file should look like as shown in **Fig. 3**.

2.2.2. Sequence of Interest

By default, PRIMEGENS searches for the unique sequence-specific fragment and primer pair relative to all the sequences present in database file. Alternatively, if the user is interested in a set of those sequences, a list of these sequences should be provided in to a separate file. This subset file can be either in the FASTA format or a list in which each line gives the name of the gene.

2.2.3. Saving Result Files

PRIMEGENS generates various types of results in different files. This information may be useful subsequently; therefore, a user can specify any location on the local computer to save all the result files.

2.3. Execution Features (Command-Line Options)

Once the input files are selected, PRIMEGENS provides various execution features. Following are some of the useful options provided by PRIMEGENS.

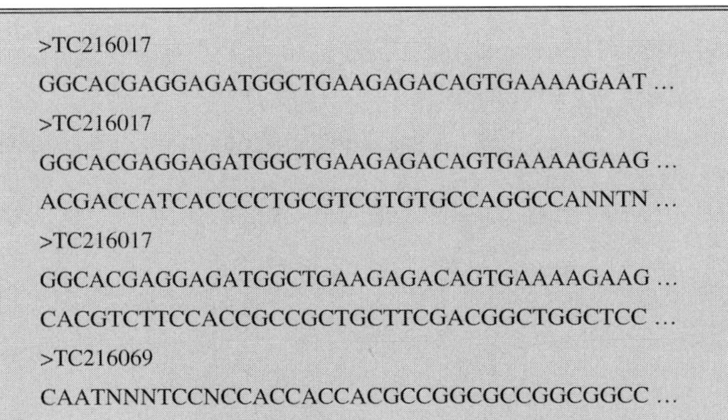

```
>TC216017
GGCACGAGGAGATGGCTGAAGAGACAGTGAAAAGAAT ...
>TC216017
GGCACGAGGAGATGGCTGAAGAGACAGTGAAAAGAAG ...
ACGACCATCACCCCTGCGTCGTGTGCCAGGCCANNTN ...
>TC216017
GGCACGAGGAGATGGCTGAAGAGACAGTGAAAAGAAG ...
CACGTCTTCCACCGCCGCTGCTTCGACGGCTGGCTCC ...
>TC216069
CAATNNNTCCNCCACCACCACGCCGGCGCCGGCGGCC ...
```

Fig. 3. Input database format.

2.3.1. Keeping All the BLAST Results (from GUI)

This option can be controlled by the user to keep all the BLAST results (which otherwise will be deleted by default) for each query sequence against the database. The BLAST result for each query sequence against the database is stored in the TMP directory in the main PRIMEGENS workspace. The file nomenclature is organized in such a way that query sequence name appears as follows *<Query-name>_<PID>_primegens_log*, where "PID" is the process ID of this computational job. The user can open these files in any text editor like Microsoft Word Pad.

2.3.2. User-Defined Expectation Value for BLAST

This optional feature allows user to set any user-defined expectation value ***(4)*** for running the BLAST program. In case the expectation value is not specified by the user, PRIMEGENS sets its default value to *1e-5*.

2.3.3. Using Subset File

One can choose a subset file for designing primers, where the entries in the subset files are selected from the search database file. In this case, a user needs to specify if the file is in the FASTA format or in the non-FASTA format using the command-line execution as follows:

$> *primegens.exe –lf <fasta-subset-file-name> <database-file-name>*

or

$> *primegens.exe –l <non-fasta-subset-file-name> <database-file-name>*

2.3.4. User-Defined PCR Product Size

A user can define its own PCR product-size range for the primer pair design. These values can be specified in the form of maximum and minimum value for the product size using the *–fz* flag on the command line. For example, the command line defining the product size range of 80–120 is as follows:

$> *primegens.exe –fz 80 120 –lf <subset-file-name> <database-file-name>*

In case a user wants to specify product size as a function of sequence length, the command to specify product size should be

$> *primegens.exe –f <fraction> -lf <subset-file-name> <database-file-name>*

Here, *fraction* is a value between 0 and 1, which represents the ratio between the minimum length of the fragment and the whole sequence length. By default, PRIMEGENS assumes the maximum product size as the sequence length itself.

2.3.5. Uniqueness of Primers

Primer pair designed for any gene-specific segment still has non-zero probability to amplify a different gene, as the short primer pair itself may not be gene specific. In particular, if the first *K* bases in the left primer and the last *K* bases of the right primer match exactly in some region of another gene in the whole sequence pool, then the designed primer pair may amplify both genes. PRIMEGENS allows user to set a parameter *K*, whose default value is 10. PRIMEGENS uses this option to make sure that primer pair is unique for the first *K* bases (left primer) and the last *K* bases (right primer) for associated query sequence. A user can choose the option like the following command

$> *primegens.exe –pterm K –lf <subset-file-name> >database-file-name>*

2.4. Primer3 Parameter Input (Append File)

All the optional parameters for Primer3 are stored in a file with a reserved name called *append.txt* **(3)**. If a user wants to specify his or her own parameter values, he or she needs to modify this file. A sample format for *append.txt* is shown in **Fig. 4**. The location of this file should be the same as the location of the input database file and the optional subset file (for the console version rather than the GUI version). If PRIMEGENS cannot find this file, it will use the default values provided in the "include" directory of the software package.

2.5. Output of PRIMEGENS

PRIMEGENS supports permanent storage of primer design results. To generate organized results, PRIMEGENS creates various files and directories. Here is a brief description of all types of generated files and directories. For generality, it is assumed that a user has selected both *Database.txt* as the database file and *subset.txt* as the subset file for PRIMEGENS. If *subset.txt* is not specified, it will select all the entries from *Database.txt* without sequences (only sequence IDs).

- Database_nohit.txt: This file contains those query sequences from *subset.txt* that are unique in *Database.txt* (no hit is found in BLAST). The file is in the FASTA format with header line containing the sequence name and length.
- Database_nohit.txt_back: This is the backup file of the *Database_nohit.txt* result for the previous run.

```
PRIMER_EXPLAIN_FLAG=0
PRIMER_OPT_SIZE=20
PRIMER_MIN_SIZE=19
PRIMER_MAX_SIZE=23
PRIMER_MIN_TM=56.0
PRIMER_OPT_TM=60.0
PRIMER_MAX_TM=66.0
PRIMER_MAX_DIFF_TM=10
PRIMER_MAX_GC=60.0
PRIMER_MIN_GC=40.0
PRIMER_SALT_CONC=50.0
PRIMER_DNA_CONC=50.0
PRIMER_NUM_NS_ACCEPTED=0
PRIMER_SELF_ANY=8.00
PRIMER_SELF_END=3.00
PRIMER_FILE_FLAG=0
PRIMER_MAX_POLY_X=5
PRIMER_LIBERAL_BASE=0
PRIMER_FIRST_BASE_INDEX=1
=
```

Fig. 4. Format of *append file* to PRIMEGENS.

- Database_nohit_primer.txt: This file contains all the primers successfully designed by the PRIMEGENS for sequences in *Database_nohit.txt*. The file includes the left and right primer sequences along with their locations on the query sequence.
- Database_nohit_primer.txt_back: This backup file contains all the record of *Database_nohit_primer.txt* for the previous primer design results created by PRIMEGENS.
- Database_primer.xls: This plain text file can be opened in an excel spreadsheet, and it contains all the designed primers along with various other information about the primer pairs.
- Database_primer.xls_back: This is the backup file of *Database_primer.xls* for the previous primer design.
- Database_primer_undo.xls: This plain text file can be opened in an excel spreadsheet, and it contains all the query sequences whose primer pairs could not be designed by PRIMEGENS based on the specified parameters.
- Database_primer_undo.xls_back: This is the backup file of *Database_primer_undo.txt* for the previous primer design.
- Database_seg.txt: This file contains those query sequences that are not completely unique in the database but have some sequence-specific fragments that are unique in whole database. It includes the name of the query sequence along with the longest sequence-specific fragments and its location on the query sequence.
- Database_seg.txt_back: This is the backup file of *Database_seg.txt* from the previous execution of PRIMEGENS.

- Database_seg_primer.txt: This file contains all the successfully designed primers for the sequences in *Database_seg.txt*. The file includes the query-specific fragments along with the designed primer pairs and their locations on the original sequences.
- Database_seg_primer.txt_back: This is the backup file of *Database_seg_primer.txt* for the previous run.
- Database_sim.txt: This file contains information about the query sequences that are not unique in database. This file also shows which sequence in the database is the closest to the query sequence along with the sequence identity.
- Database_sim.txt_back: This is the backup file of *Database_sim.txt* for the previous run.
- Primer_plate_left [index]: This file contains query sequence name and its left primer, which should be kept in a 96-well primer plate. The plate number is represented by the index digit.
- Primer_plate_right [index]: This file contains query sequence name and its right primer, which should be kept in a 96-well primer plate. The plate number is represented by the index digit.

Besides the various files, there are some temporary directories that are created by PRIMEGENS. Unlike the above files, these directories are cleaned before and after the software execution. A user may save them in different folders to keep them from deleting. These directories are described below.

- TMP/directory: This directory contains the BLAST output for each query sequence in *subset.txt* against *Database.txt*. The file name is selected on the basis of query sequence name. This directory can also be saved through an option from GUI (*see* **Subheading 2.3.1**.).
- LOGS/directory: This directory contains various logs, which are created during the primer design process when debugging mode is used for execution. The purpose of this option is to provide run-time information of the whole process from the user perspective.
- SEQ/directory: This directory contains individual sequence file for each query sequence in the FASTA format. The purpose of this directory is to retrieve sequence of interest without searching into the database.
- ALIGNMENT/directory: This directory contains global alignments performed for those sequences, which are not unique in the database. The alignment is performed to find the longest sequence-specific fragment for the query sequence.

3. PRIMEGENS Execution

This section will explain how to run PRIMEGENS. We will describe it for both command line and GUI version.

3.1. Using PRIMEGENS as Console Application

To use the software for designing primers, the following basic procedure should be followed.

1. Create a new directory to save all the input and output of the software.
2. Prepare a database file (i.e., *Database.txt; see* **Subheading 2.2.1**.) consisting of all the potential sequences with cross-hybridization in the new directory.
3. If necessary, select a subset of sequences (i.e., *subset.txt; see* **Subheading 2.2.2**.) from above database, whose primers should be designed, and keep it in the new directory.
4. Copy *append.txt* file from include/of the PRIMEGENS folder in to the new location. Modify the primer parameters according to the experiment and save it. If not specified by the user, PRIMEGENS will use the default parameters.
5. Change the directory to the newly created directory and give following command,

   <PRIMEGENS_PATH>/bin/primegens.exe –option subset.txt database.txt

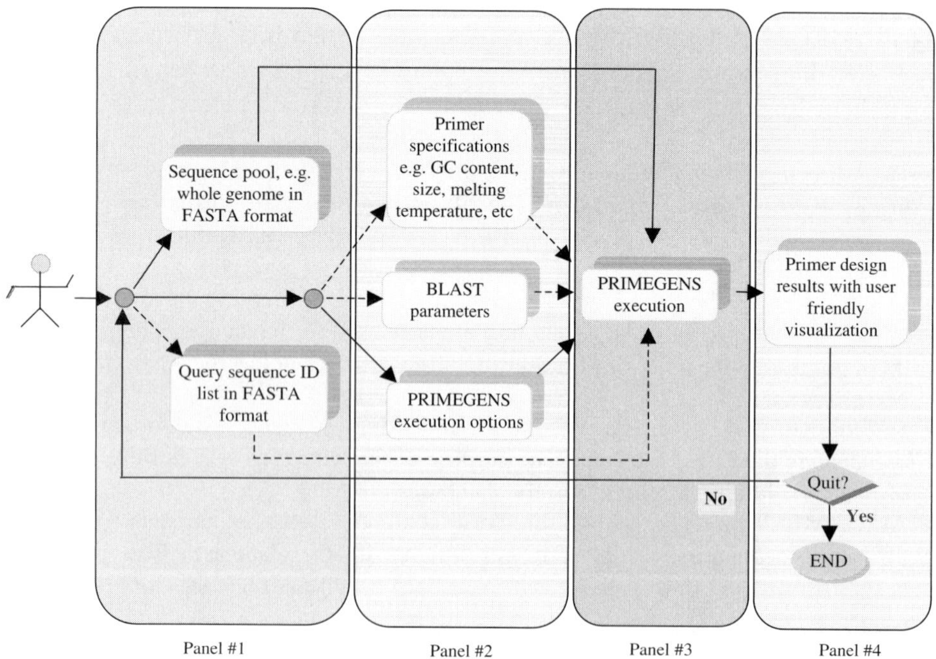

Fig. 5. Workflow for primer design. Database input (panel #1), design specification (panel #2), primer design process (panel #3), and results view (panel #4). Solid arrows represent required inputs, whereas dashed arrows show optional inputs.

More information about various options and command format is described in **Subheading 2.3**.

3.2. Running PRIMEGENS from GUI

To use the GUI version of PRIMEGENS, the database file also has to be prepared first. The user can run the software by double clicking on the executable. **Figure 5** provides a systematic workflow from the user perspective for primer design. The details for using the GUI version are explained in the following subheadings.

3.2.1. Database Input

Figure 6 shows the first panel, which will be visible when a user clicks to run PRIMEGENS. User should input the database, subset file (optional), and result storage location (optional). Once completed with input, the "next" button should be pressed for execution options.

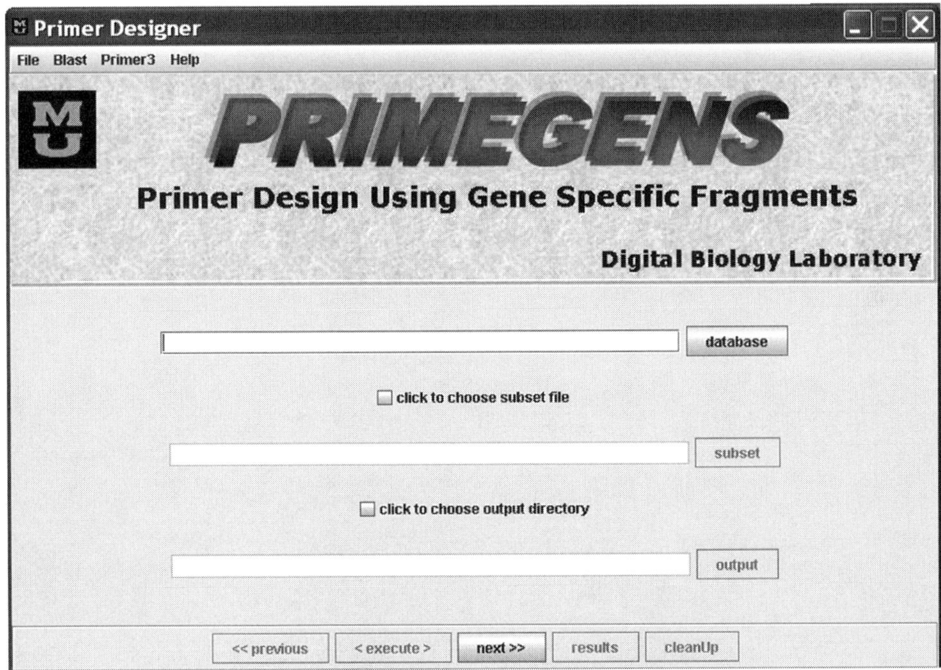

Fig. 6. PRIMEGENS graphical user interface (GUI).

3.2.2. Execution Option Window

This panel contains various execution options as explained in **Subheading 2.3**. The execution features supported by the GUI map to all command-line features correspondingly. **Figure 7** shows the option panel. Once user selects any attribute, the optional attribute value field shows the default attribute value, which can be modified then.

3.2.3. Primer Pair Specification

In case a user has specific parameter requirements for primer pairs, he or she can specify those values in the primer-specification window. The user can click on the *Primer3* menu to modify primer specifications. These modifications correspond to changes in the *append.txt* file in the command-line version.

3.2.4. Execution-Display Window

After specifying inputs and options, the software allows a user to open the execution window. Once the primer design is completed, the *result* and the

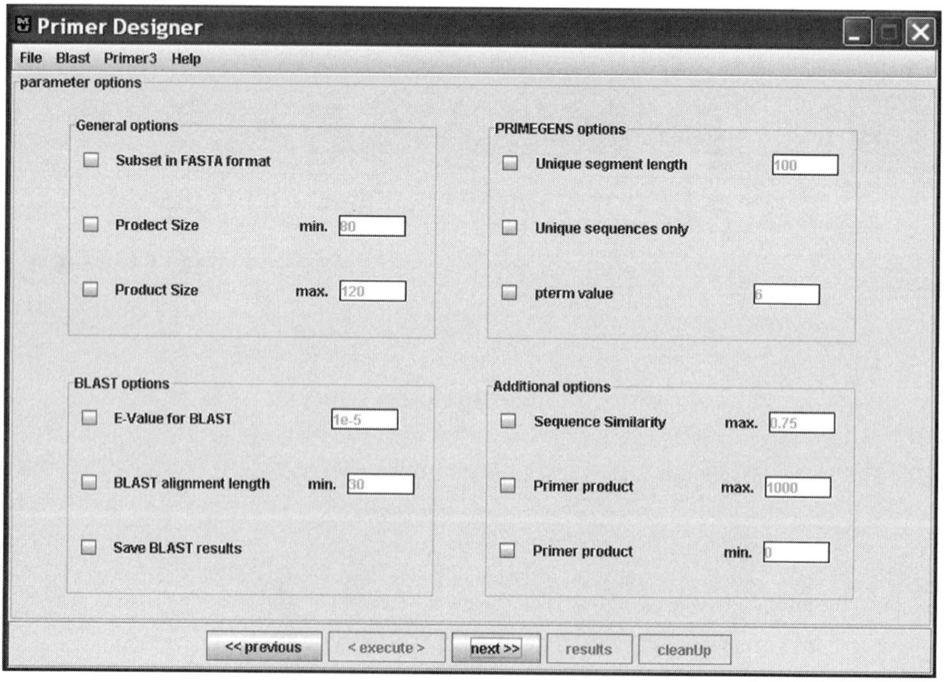

Fig. 7. Execution option window.

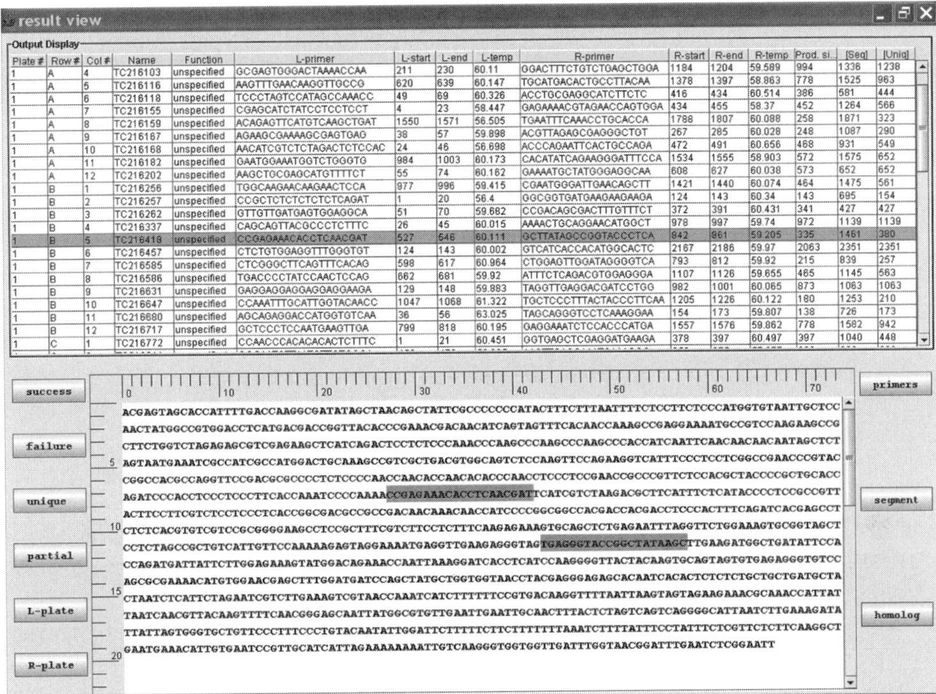

Fig. 8. Primer design result display window.

clean buttons are activated. User can click on the *result* button to see the results and *clean* to remove all the temporary results from buffer and reset the software to the first window.

3.2.5. Result Analysis and Visualization Window

This window displays the primer design results generated by PRIMEGENS for all the sequences whose primers are found, including the gene-specific fragment and the global alignment between the sequence and its BLAST hits. The various buttons are self-explanatory. **Figure 8** shows a sample result visualization window.

4. Application of PRIMEGENS in the Quantitative RT-PCR Primer Design for the Transcription Factors in the Soybean Genome

As an example of successful PRIMEGENS application, the following shows some details for an actual research project. The aim of this project is to use gene-specific primers to isolate transcriptional factors (TFs) in the soybean *Glycine max* through quantitative (q)RT-PCR. We have used several approaches toward identification of putative TFs in soybean. Data were acquired from multiple resources (as below) and combined to generate the non-redundant final list of 734 putative TFs.

1. Soybean sequences homologous to the ~ 2300 *Arabidopsis thaliana* TFs were identified by searching the National Center for Biotechnology Information (NCBI) *Glycine max* Unigene database at ftp://ftp.ncbi.nih.gov/repository/UniGene/Glycine_max. To assure a high homology between *Arabidopsis* and *Glycine max* sequences, an E-value ≤ 1e-12 was considered as a cutoff. These analyses yielded 238 putative TFs in *Glycine max*, which fulfill these requirements.
2. We searched the listing of several TF domains and families already registered at the TIGR Gene Indices database (http://www.tigr.org/tigr-scripts/tgi/T_index.cgi?species=soybean). So far, 1321 tentative consensus (TC) sequences are available on this Web site.
3. To expand the above list, the expressed-sequence tag (EST) soybean database (dbEST, http://www.ncbi.nlm.nih.gov/dbEST) was also screened for TF sequences. An initial search identified 1043 ESTs.
4. All putative TFs identified from TC sequences and ESTs (from **steps 2** and **3** above) were combined to form a data set, and were compared against the *Glycine max* Unigene data set using BLAST with an E-value cutoff of ≤ 1e-12. This yielded 496 putative TFs in *Glycine max*, out of which 449 match the putative TF TC sequences and 47 match the putative TF ESTs.
5. The final list (from **steps 1** and **4**) contains 734 putative TFs in *Glycine max*. PRIMEGENS was used to design primers for qRT-PCR experiment for validation of the above 734 TFs. All sequences were compared against the *Glycine max* Unigene database to ensure that the designed primers are unique to the sequence and will give specific PCR amplification.

4.1. Input File

The NCBI unique unigenes database for soybean (15,047 unigenes at the time of the design), acquired from ftp://ftp.ncbi.nih.gov/repository/UniGene/Glycine_max, was used as the database, and 734 TFs identified in soybean were used as the subset file for primer design.

4.2. Execution Option

Product size range = 80–150.

The rest of the parameters were all set to the default values.

4.3. Primer Parameters (Append File)

```
PRIMER_EXPLAIN_FLAG=0
PRIMER_OPT_SIZE=20
PRIMER_MIN_SIZE=17
PRIMER_MAX_SIZE=25
PRIMER_MIN_TM=50.0
PRIMER_OPT_TM=60.0
PRIMER_MAX_TM=70.0
PRIMER_MAX_DIFF_TM=10
PRIMER_MAX_GC=65.0
PRIMER_MIN_GC=35.0
PRIMER_SALT_CONC=50.0
PRIMER_DNA_CONC=50.0
PRIMER_NUM_NS_ACCEPTED=0
PRIMER_SELF_ANY=8.00
PRIMER_SELF_END=3.00
PRIMER_FILE_FLAG=0
PRIMER_MAX_POLY_X=5
PRIMER_LIBERAL_BASE=0
PRIMER_FIRST_BASE_INDEX=1
```

4.4. Machine Configuration

Hardware: X86 64-bit processor.
Memory: 8 GB random access memory (RAM).
Platform: Linux release 2.6.12.

4.5. PRIMEGENS Result Statistics

Primer pair design targets = 734 genes.

Possible gene-specific fragments found = 680 genes [(680/734) = 92.64%].

Successfully designed primer pairs = 670 genes [(670/734) = 91.28%].

Primer design failure = 10 genes [(10/734) = 1.36%].

Unique query genes for the whole sequence = 182 genes [(182/680) = 26.8%].

Genes containing gene-specific fragments = 498 genes [(498/680) = 73.24%].

Total execution time = 4 h 44 min 58 s.

Average time per gene (sequence) = $[(4 \times 3600) + (44 \times 60) + 58]/734 =$ 23.29 s].

Our collaborator Gary Stacey's laboratory at the University of Missouri-Columbia performed qRT-PCR for the first 96 designed primers. Among them, 89 (92.7%) primer pairs yield single bands, which means that unique PCR products are amplified as expected (manuscript in preparation). Given that the soybean whole genome has not been sequenced, that is, the 15,047 unigenes that we used do not cover all the genes; such a successful rate is satisfactory.

5. Conclusion

PRIMEGENS provides an easy-to-use tool for biologist to select gene-specific fragment and to design PCR primers. A biologist with little computer skill can easily use the GUI version, which is available for both Windows and Linux platforms. The primer amplification results of TFs in the soybean genome, results from other projects ***(5,7–9)***, and feedback from users all indicate that the software works successfully and efficiently. We are continuing to refine the software efficiency and develop more features. The new versions of the software together with documentations will be released at the PRIMEGENS Web site (http://digbio.missouri.edu/primegens/).

Acknowledgments

The development of PRIMEGENS was sponsored by the Office of Biological and Environmental Research, U.S. Department of Energy, under Contract DE-AC05-00OR22725, managed by UT-Battelle, LLC, and a startup fund for Dong Xu at the University of Missouri-Columbia. We thank Ying Xu, Jizhong Zhou, Gary Li, Liyou Wu, Trupti Joshi, and Gary Stacey for collaboration and assistance.

References

1. DeRisi, J. L., Iyer, V. R., and Brown, P. O. (1997) Exploring the metabolic and genetic control of gene expression on a genomic scale. *Science*, 278, 680–686.
2. Duggan, D. J., Bittner, M., Chen, Y., Meltzer, P., and Trent, J. M. (1999) Expression profiling using cDNA microarrays. *Nat. Genet.*, 21, 10–14.
3. Rozen, S. and Skaletsky, H. J. (2000) Primer3 on the WWW for general users and for biologist programmers. *Methods Mol. Biol.*, 132, 365–386.
4. Altschul, S. F., Madden, T. L., Schaffer, A. A., Zhang, J., Zhang, Z., Miller, W., and Lipman, D. J. (1997) Gapped BLAST and PSI-BLAST: a new generation of protein database search programs. *Nucleic Acids Res.*, 25, 3389–3402.
5. Xu, D., Li, G., Wu, L., Zhou, J., and Xu, Y. (2002) PRIMEGENS: robust and efficient design of gene-specific probes for microarray analysis. *Bioinformatics*, 11, 1432–1437.

6. Smith, T. F. and Waterman, M. S. (1981) Comparison of biosequences. *Adv. Appl. Math.*, 2, 482–489.
7. Liu, Y., Zhou, J., Omelchenko, M. V., Beliaev, A. S., Venkateswaran, A., Stair, J., Wu, L., Thompson, D. K., Xu, D., Rogozin, I. B., Gaidamakova, E. K., Zhai, M., Makarova, K. S., Koonin, E. V., and Daly, M. J. (2003) Transcriptome dynamics of *Deinococcus radiodurans* recovering from ionizing radiation. *Proc. Natl. Acad. Sci. U. S. A.*, 100, 4191–4196.
8. David, J.-P., Strode, C., Vontas, J., Nikou, D., Vaughan, A., Pignatelli, P. M., Louis, C., Hemingway, J., and Ranson, H. (2005) The *Anopheles gambiae* detoxification chip: a highly specific microarray to study metabolic-based insecticide resistance in malaria vectors. *Proc. Natl. Acad. Sci. U. S. A.*, 102, 4080–4084.
9. Wu, L. Y., Thompson, D. K., Li, G. S., Hurt, R. A., Tiedje, J. M., and Zhou, J. Z. (2001) Development and evaluation of functional gene arrays for detection of selected genes in the environment. *Appl. Environ. Microbiol.*, 67, 5780–5790.

A

Repeat Masking for PCR Primer Design

9

SNPbox

Web-Based High-Throughput Primer Design with an Eye for Repetitive Sequences

Stefan Weckx, Peter De Rijk, Wim Glassee, Christine Van Broeckhoven, and Jurgen Del-Favero

Summary

In this omics era, researchers in biosciences need more than ever user-friendly, fast, and straightforward computational tools for high-throughput research. In this chapter, we present SNPbox and explain the general primer design strategy. We also show how the four modules, Exon, Single-Nucleotide Polymorphism (SNP), Saturation, and Exon + Saturation, exactly apply the primer design strategy, give clear guidance for users of the Web interface at http://www.SNPbox.org, and explain how pre-designed primers for Ensembl genes can be visualized and retrieved.

Key Words: PCR primer design; high throughput; resequencing; Web server; Ensembl; DAS.

1. Introduction

In February 2001, we witnessed a major event in biosciences: the release of the first draft sequence of the human genome ***(1,2)***. It was a notable achievement and the result of years of precise work and major technological improvements. Now, we can say that the human genome was just the beginning of a genomics era, actually of a whole "omics" era. Indeed, many other genome projects for eukaryotes and prokaryotes followed, and new projects are still set up. With all these high-quality genomic sequences available, we now need more than ever fast, high-throughput, and user-friendly computational tools to explore the

From: *Methods in Molecular Biology, vol. 402: PCR Primer Design*
Edited by: A. Yuryev © Humana Press, Totowa, NJ

wealth of genomic information so that this information can be taken back to the laboratory and used and applied in research.

Several techniques are available to discover single-nucleotide polymorphisms (SNPs) and insertion–deletion polymorphisms (INDELs). One of the most common techniques is resequencing the same genomic regions in different individuals. Hereto, primers are needed for PCR and sequencing reactions. During the primer design process, attention has to be paid to repetitive sequences, because almost 50% of the human genome consists of repeats. When these repeats are neglected during the primer design process, PCR can result in ambiguous PCR products. It is obvious that this kind of primers are useless, as the genomic regions cannot be properly resequenced.

Taking repetitive sequences into account while designing primers is not yet a standard procedure in primer design programs. One can repeat mask each sequence separately and subsequently design primers, but this is not manageable for larger projects in which all coding sequences of a gene and/or genomic regions require the design of primers. Furthermore, in this kind of large-scale projects, it is important that all primers are designed to work under the same experimental conditions and that larger genomic regions are covered by overlapping PCR products. Therefore, we developed SNPbox, a high-throughput Web-based program for primer design, based on Primer3 *(3)*, where repeat annotation obtained by RepeatMasker and Sputnik is taken into account during the design process (*see* **Box 1**).

Box 1. Detecting Repetitive Sequences

RepeatMasker

RepeatMasker (http://www.repeatmasker.org) is probably the most commonly used program to detect interspersed repetitive elements and low-complexity stretches in human genomic sequences. RepeatMasker relies on cross_match, an alignment program based on the Smith-Waterman algorithm written by Phil Green at the University of Washington (http://www.phrap.org) and on a collection of known repetitive elements, Repbase Update. This collection comes with the free version of RepeatMasker for non-profit and academic researchers and is actively updated by the Genetic Information Research Institute, GIRI (http://www.girinst.org). Although RepeatMasker is mostly used to replace repetitive sequences by the IUPAC ambiguity base "N," we only use the program to annotate genomic sequences with repetitive elements, leaving the genomic sequence unchanged.

Sputnik

The original version of Sputnik detects microsatellite repeats with repeat elements of 2–5 nucleotides, e.g., trinucleotide repeats, and was developed by Chris Abajian at the University of Washington (http://espressosoftware.com/). We have extended the capabilities of Sputnik so that the program now also finds single-base stretches. The adapted Sputnik program is available from the SNPbox Web site.

2. Primer Design Concept: What Is Behind SNPbox

Many life-science laboratories have entered the high-throughput era, with robot platforms and laboratory equipment conceived for parallel sample processing in 96-well or 384-well plates. To take full benefit from this technological evolution, bioinformatics software must follow this trend, e.g., primer design for PCR and sequencing reactions. In a high-throughput environment, a primer design program should be able to design primers with uniform conditions and with as little interaction from the user as possible. Also, repetitive sequences should automatically be taken into account, because a high-throughput environment does not allow for extensive manual checks *(4)*.

2.1. About Objects and Targets

SNPbox will design primers for exons, SNPs, or genomic regions. Upon selection, they are transformed into objects: parts of the genomic sequence to be included in the PCR product. Because we want to obtain PCR products of similar length, targets for primer design are defined covering the objects. The default target length is 450 bp, but it can be chosen between 100 and 5000 bp. At each side of the target, there is a primer design frame of 70 bp (*see* **Fig. 1**). As a consequence, the minimal PCR product length is 490 bp, the maximum 590 bp. In general, objects that are smaller than the default target length are symmetrically covered by one target. Several small objects like mapped SNPs that are close enough to each other are bundled into the same target, and large objects like large exons or genomic regions will be covered by multiple overlapping targets (*see* **Fig. 1**).

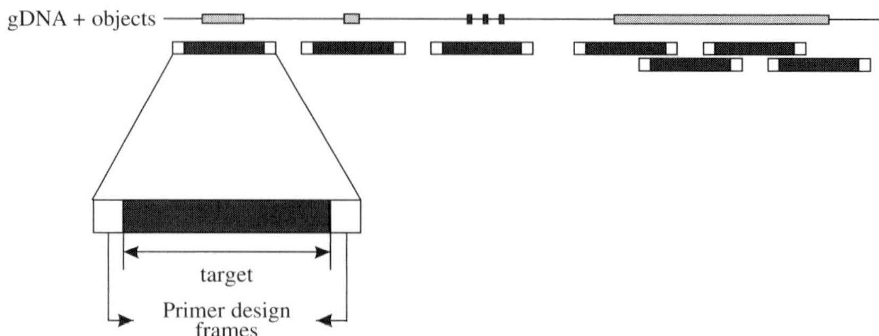

Fig. 1. Concept of objects and targets.

2.2. When Repeats Come in . . .

As almost 50% of the human genome consists of repetitive elements, it is likely that a repeat occurs in the neighborhood of an object.

Repeated sequences can be roughly divided into two classes: tandem repeats that can be polymorphic and therefore variable in length and repeats with a fixed length. The tandem repeats are built up of small units of one or several bases, like a single-base stretch, di-nucleotide, or tri-nucleotide repeat. The di-nucleotide and tri-nucleotide repeats very often are polymorphic, and consequently one individual can be heterozygous for such a repeat resulting in bad sequencing data. Interspersed repeats have a fixed length and occur at several places in the genome *(5)*. As a consequence, these repeats are not problematic as long as PCR and sequencing primers are not designed within such a repeat.

In practice, SNPbox takes repeats into account during target definition, applying these general rules:

- Primers are never selected in a repeat, to avoid multiple hybridization positions on the genome, with ambiguous PCR products and sequences as a result.
- The primer design frame of 70 bp can be decreased to 50 bp in case a non-includable repeat is too close to the object.
- An interspersed repeat can be included in a target if the repeat is closer than 100 bp to the object and if the repeat length is less than 300 bp. Consequently, these repeats are amplified and sequenced. This is however needed to obtain a high-quality sequence for the nearby object, important for variation discovery in exons.

- A repeat will only be included if it belongs to a repeat class with fixed repeat length, like short interspersed repetitive elements (SINEs) and Alu's.
- Variable repeats like variable number of tandem repeats (VNTRs) are non-includable repeats. After all, both alleles can have different repeat numbers, resulting in low sequence quality in case of heterozygosity. Also single-base stretches are avoided because of the risk of slippage of the polymerase enzyme, which also results in bad-quality sequences.
- In primer design for genomic regions, the succession of overlapping targets is interrupted by non-includable repeats. Here, the default target length is dynamically adapted by SNPbox to avoid large differences in PCR product lengths.

2.3. *Illustrations of Strategy*

Figure 2 shows four examples to illustrate the strategy described in **Subheading 2.2**.

In panel A, the object has a 5′-flanking repeat at a distance that is smaller than 100 bp, making it possible to include the repeat in the target. The repeat length is less than 300 bp, and the repeat cannot vary in length. As a consequence, the repeat is included in the target. Therefore, the 5′-primer design frame is positioned at the 5′-end of the repeat, and the target and 3′-primer design frame are positioned accordingly to obtain the optimal target length.

In panel B, the object has a 5′-flanking repeat that cannot be included in a target. The space between repeat and object is large enough to allow the

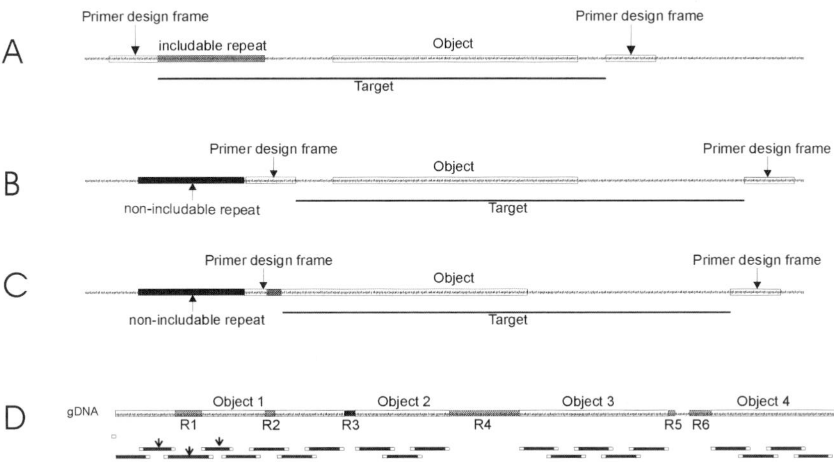

Fig. 2. Examples and special cases of the SNPbox strategy.

positioning of the 5′-primer design frame in that space. The target and 3′-primer design frame are positioned accordingly to obtain the optimal target length.

In panel C, the 5′-flanking repeat cannot be included. The space between repeat and object is too small to position the 5′-primer design frame of 70 bp; it is even too small to position a primer design frame of 50 bp. As a consequence, a primer design frame of 50 bp is positioned toward the 3′-end of the repeat, resulting in a small overlap between the primer design frame and the object, indicated as a shaded area. This might result in a part of the object not covered by the PCR product and not sequenceable if starting from the PCR product. At the 3′-end, there is no limiting repeat, and the 3′-primer design frame together with the target is positioned as expected.

Panel D shows a genomic region that was subject to primer design. R1–6 can be included in a target based on their class; R3 cannot be included. R1 and R2 are effectively included. The targets indicated with an arrow have their length changed to optimally cover the genomic sequence. R4 is not included because of its length; R5 and R6 are considered as one repeat because these are close to one another.

3. How to Use SNPbox to Design Primers for Your Project

SNPbox is accessible at http://www.SNPbox.org *(6)*. The “Web service” link brings the user to a page with an academic user statement. Agreeing on this statement provides access to the SNPbox Web-based user interface. Commercial users should contact the corresponding author (Jurgen Del-Favero) to inquire how they can use SNPbox.

3.1. General Settings and Requirements

The window following the agreement lists the available modules: Exon, SNP, Saturation, and Exon + Saturation. For all modules, a genomic sequence must be provided by pointing to either local data or information that will be obtained from GenBank (*see* **Fig. 3**).

Local data can be provided in FASTA and GenBank file format. This possibility allows the user to provide own curated or private information about a gene and/or exon information to SNPbox. Please be aware that these files must be saved in text format prior to upload, not in Microsoft Word format.

Public data can be directly obtained from GenBank in three ways: by GI identifier, by accession number, or by an Entrez gene name. For the latter, the species name is obligatory and needs selection from the drop-down menu. Also, the user should be aware that one relies on the annotation of Entrez Gene, so it

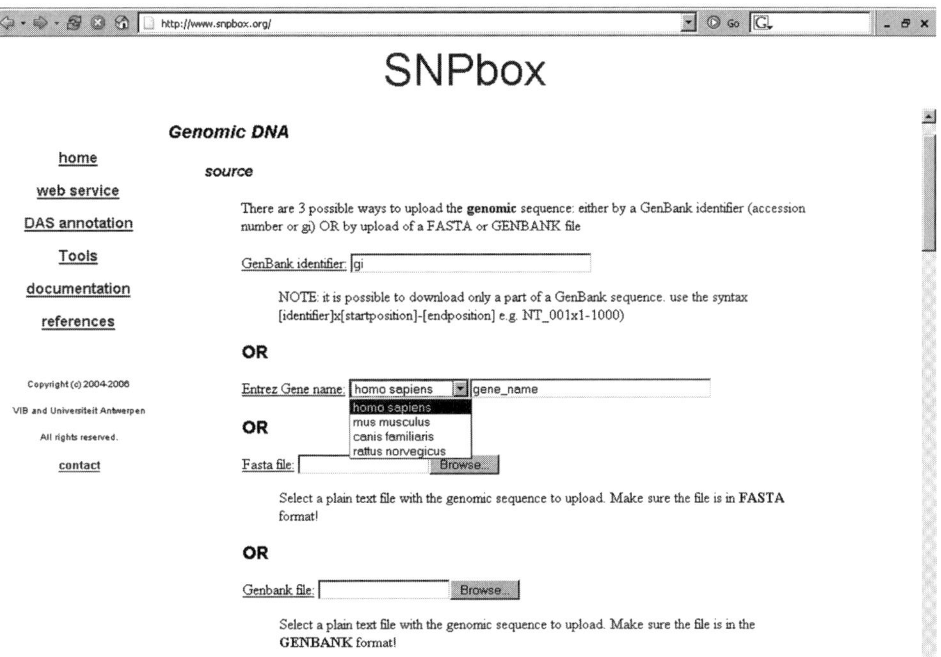

Fig. 3. General settings and requirements: ways to provide information on genomic DNA.

is possible that not all mRNA/CDS information is included in the Entrez Gene overview. In case a GenBank accession number or GI identifier is used, the user is able to define a subsequence by using the syntax [identifier] × [startposition] – [endposition], e.g., NC_000020 × 10001–22000. In this example, SNPbox will only retrieve a subsequence of NC_000020 from GenBank with length 12,000 bp.

One of the strengths of SNPbox is its ability to cope with repetitive sequences near objects, so the option "Mask repeats" is selected by default (*see* **Fig. 4**). Furthermore, the selection of RepeatMasker settings can be changed. The user can select the DNA source, and the level of sensitivity can be decreased resulting in a faster repeat search. Also, the types of repeats that are searched for can be changed: skip or mask only low complexity and simple repeats and mask only interspersed sequences. The integration of Sputnik allows searching for simple repeats and single-base stretches, of which the minimal length can be user defined (default is 10).

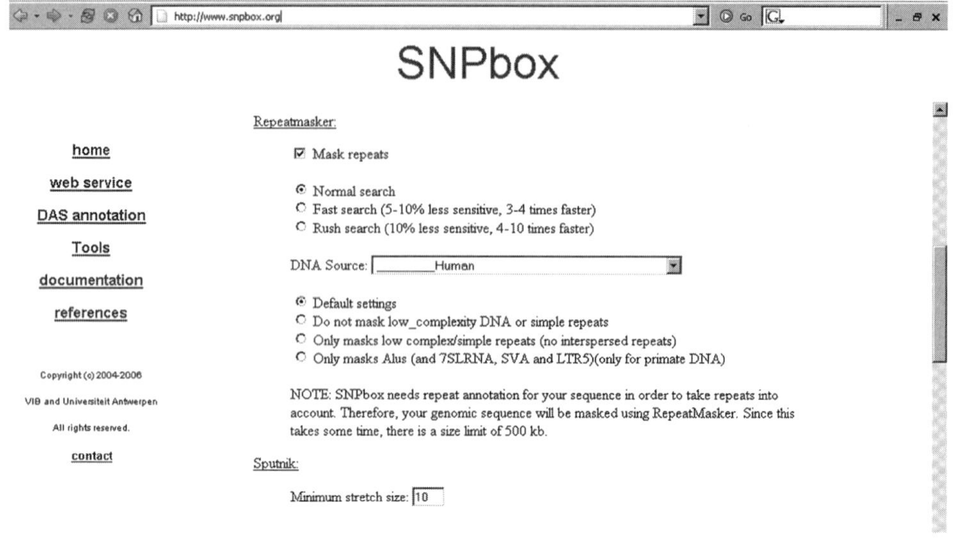

Fig. 4. General settings and requirements: repeat-masking settings.

Note that there is a length limit of 500 kb for the genomic sequence, for reasons of server performance. This limit can be increased to 2 Mb if RepeatMasker is used in "Rush" option.

3.2. Module-Dependent Settings and Requirements

3.2.1. Exon Module

With this module, primers are designed to amplify coding sequences. Here, SNPbox needs an additional stretch of at least 500 bp of genomic sequence before the first and after the last exon to allow an optimal object definition. Especially users who have downloaded GenBank files for a specific gene should pay attention to this recommendation, as the genomic sequence typically starts immediately with the first base of the first exon and ends with the last base of the last exon in GenBank files.

The Exon module performs following steps:

1. Exon annotation of the genomic sequence. Coding sequences are identified within the genomic sequence by aligning cDNA and/or expressed-sequence tag (EST) sequences using the program Spidey *(7)*, or alternatively, the exon annotation is obtained from a GenBank file uploaded as described in **Subheading 3.1**. The latter method is especially useful for genes with numerous exons and/or genes with large introns, as these circumstances may cause incorrect alignments by Spidey.

2. Defining the objects. The start and end positions of the exons are used to define objects. To make sure that the branch points and splicing sites are included in the object, exons are extended 5′ and 3′ with 50 bp. In case a non-includable repeat is near an exon, the 5′ and 3′ extension can be limited to 25 bp or not extended at all. Objects that are closer than 250 bp to each other are joined into one object.
3. Targets for primer design are defined taking the optimal target size into account; primers are designed as described in **Subheading 2**.

Note: For most of the genes, multiple mRNA sequences are available in public databases, sometimes also including splice variants. Multiple mRNA sequences can be provided as input file; the GI identifiers have to be separated by a space. SNPbox will compile all exons into assembled exons so that all exons from the provided mRNA sequences are covered. These assembled exons are used in **step 2**, the object definition (*see* **Fig. 5**).

3.2.2. SNP Module

In this module, primers are designed to amplify SNPs including their flanking sequence. Next steps can be distinguished:

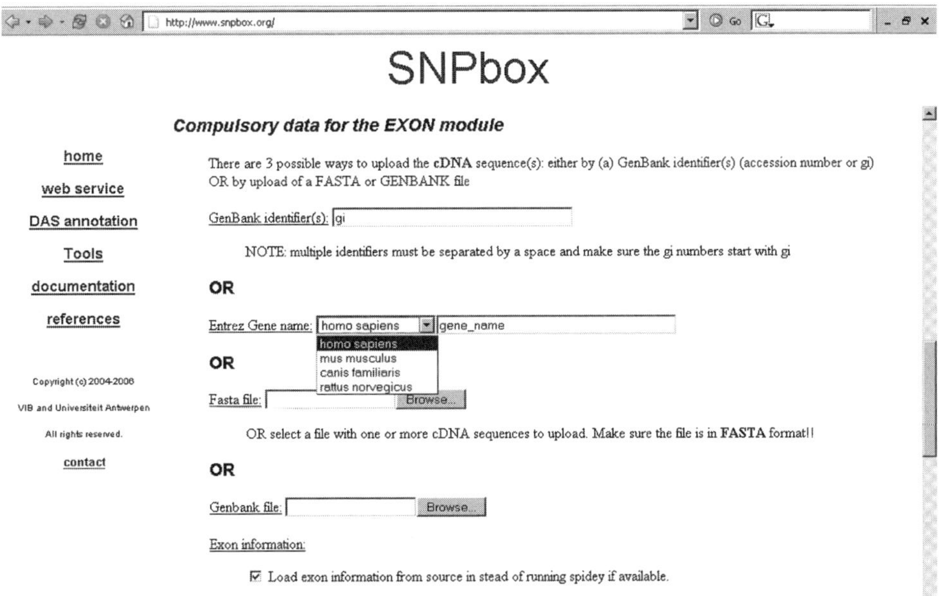

Fig. 5. Overview of the compulsory data for the exon module.

1. SNP mapping. Using the BLAST algorithm, HapMap SNPs *(8)* are aligned to the genomic sequence. Hereto, all HapMap SNPs including 50 bp of flanking sequence at each side were stored on the SNPbox server (*see* **Box 2**).
2. Defining objects. Each mapped region of 101 bp containing a SNP is an object. Multiple objects can be joined into one larger object in case the SNPs are closer than 250 bp to each other.

Targets for primer design are defined taking the optimal target size into account, and primers are designed as described in **Subheading 2**. In most of these cases, the objects of 101 bp are extended to targets of 450 bp. It might be considered to lower the optimal target size in this module.

Box 2. HapMap

HapMap (http://www.hapmap.org) is an international project that aims to collect all sequence variations of the human genome. HapMap uses DNA from different ethnic origins to determine the population frequencies of the haplotypes. The HapMap Public Release #19 (2005-10-24) contains about 5.8 million genotyped single-nucleotide polymorphisms (SNPs) and 3.9 million quality-checked SNPs. It is this collection of public SNPs that can be mapped to the genomic sequences used by SNPbox.

Note: To speed up the mapping process, subsets of HapMap SNPs can be selected according to the chromosomal location of the genomic DNA sequence (*see* **Fig. 6**).

3.2.3. Saturation Module

In the Saturation module, the aim is to design overlapping primers for PCR amplification covering a whole genomic region. Following steps can be distinguished:

1. Genomic region. The only required input is the genomic sequence and the start and end position on that sequence to mark the genomic region that will be considered as object.
2. Defining objects. In first instance, the whole genomic region of interest is considered as one object. Inspection for the presence of repeats is very important: the initial large object is interrupted with non-includable repeats (*see* **Fig. 2**, panel D).
3. Targets for primer design are defined based on the resulting objects, which in most cases result in overlapping targets. Primers are designed as described in **Subheading 2**, where target length can be dynamically adapted to avoid some PCR products with lengths that differ too much from the optimal target length.

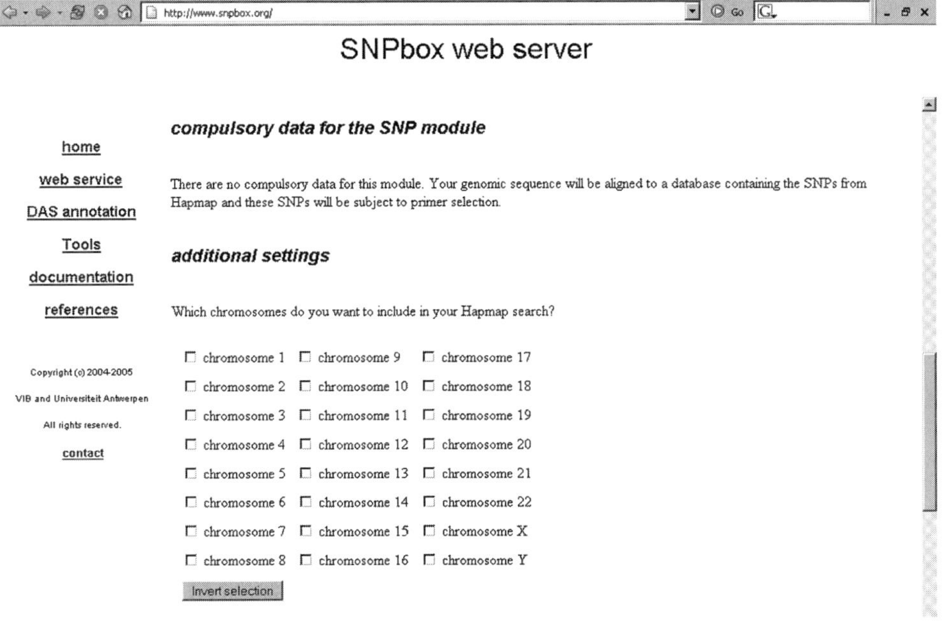

Fig. 6. Overview of additional settings for the SNP module.

3.2.4. Exon + Saturation Module

A commonly used strategy for hunting polymorphisms associated with diseases is to focus first on coding sequences and subsequently on non-coding regions surrounding the coding region, like the upstream region that could contain regulatory elements. In the Exon + Saturation module, both formerly discussed modules are combined in one single module (*see* **Fig. 7**). Following steps can be distinguished:

1. Annotation of the genomic sequence for exons based on sequence alignment of the genomic sequence and coding sequences or information provided in GenBank format (*see* **Subheading 3.2.1., step 1**, for details).
2. Defining objects based on the exons, where the exons are extended 5′ and 3′ with 50 bp (*see* **Subheading 3.2.1., step 2**, for details).
3. Targets for primer design are defined taking the optimal target size into account, and primers for the coding sequences are designed as described in **Subheading 2**. This set of primers is here considered as the first set.
4. Subsequently, the saturation module is applied, where the genomic region that has to be covered by amplicons is considered as one large object. This object is screened for repeats and split where a non-includable repeat is located (*see* **Fig. 2**, panel D).

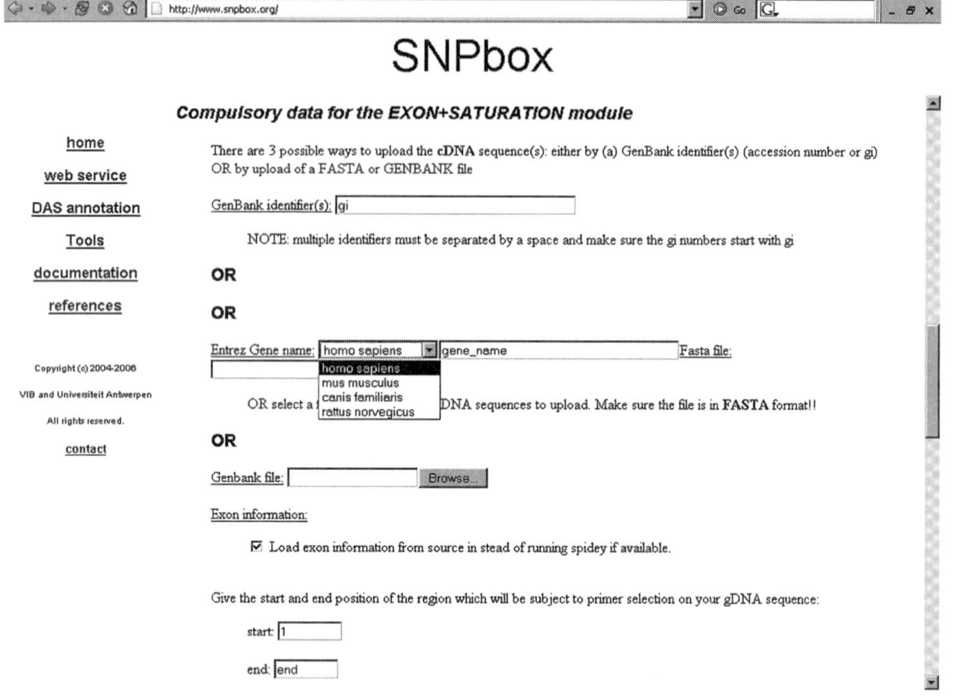

Fig. 7. Overview of compulsory settings for the Exon + Saturation module.

5. In addition, the resulting objects are checked for the presence of sequence regions already covered by the first set of primers. If this is the case, the objects are shortened to avoid extensive overlap.
6. Targets for primer design are defined based on the resulting objects and taking the optimal target length into account. Primers are designed as described in **Subheading 2**.

3.3. Additional Settings

Some additional settings are related to SNP mapping based on HapMap SNPs, Primer3 settings, and feedback to the user. If module Exon, Saturation, or Exon + Saturation is chosen, SNPbox can map SNPs from HapMap SNPs to the genomic sequence just for information. To speed up the mapping process, users can select subsets of HapMap SNPs according to chromosome number (*see* **Fig. 8**).

One of the strengths of SNPbox is that all primers are designed under the same set of standard conditions. However, there might be a need to change

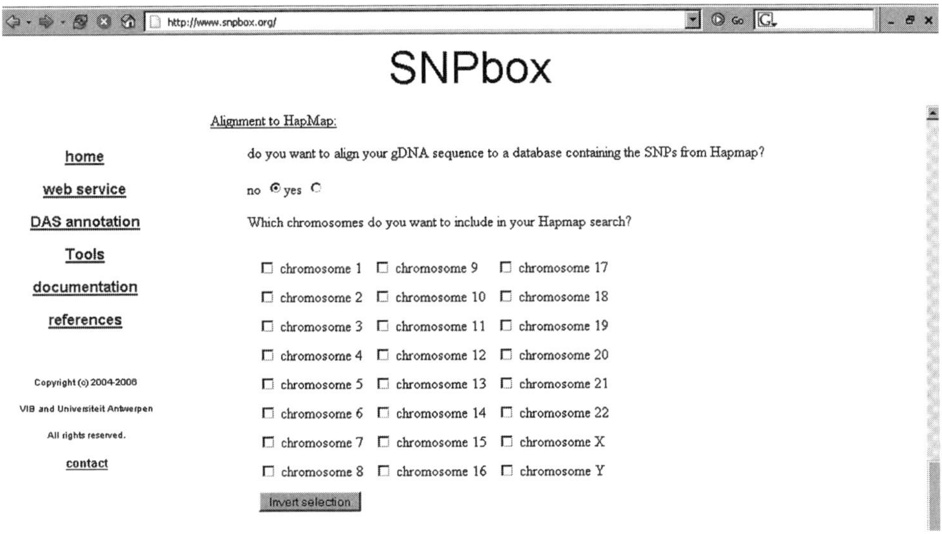

Fig. 8. Additional settings for the Exon and Saturation modules.

these settings. Therefore, the input parameters can be changed. These settings will then be used for all primer sets that will be designed in that particular experiment. Also, users can change the default length of the target. Finally, users can provide an experiment name that will be used in the output files as well as an e-mail address that will be used to send a notification once primer design has been completed (*see* **Fig. 9**).

3.4. SNPbox Output

To illustrate SNPbox output, we used the exon module for human *KCNQ2*. A subsequence of NC_000020 was selected as genomic sequence (NC_000020 × 61500001–61575000). As mRNA sequences, we used NM_004518, NM_172106, NM_172107, NM_172108, and NM_172109. Default settings for repeat masking were used, and as *KCNQ2* is located at 20q13.3, we opted for mapping HapMap SNPs from chromosome 20 only. After the input page, an overview page is shown before the actual job is started (*see* **Fig. 10**). The transition to this overview page might take some time, as the sequences are being retrieved at this stage. Feedback is provided to the user in case one of the sequences cannot be retrieved.

After starting SNPbox, the user is forwarded to a Web page to confirm that the program has been started (*see* **Fig. 11**). The page contains a link to the

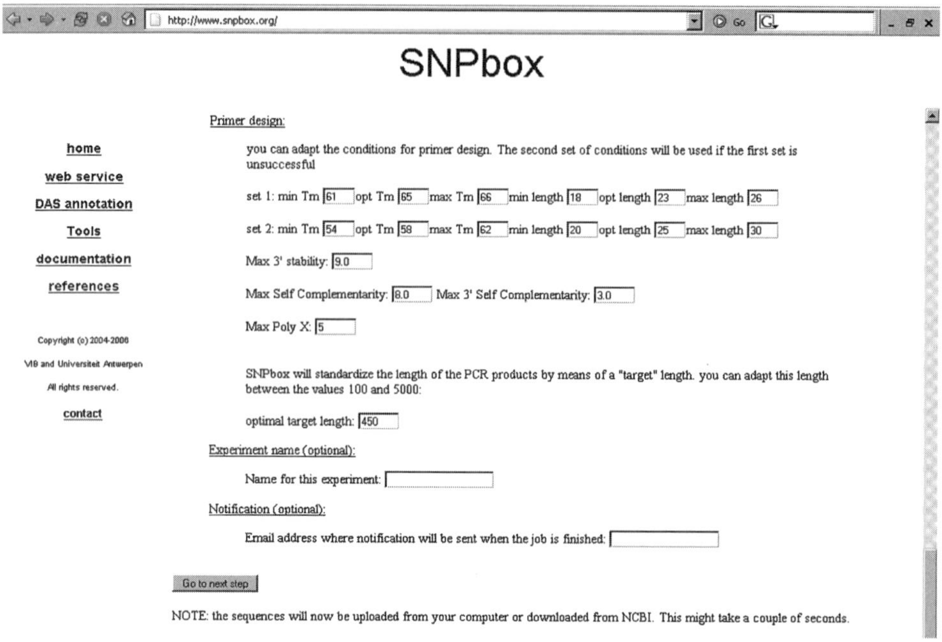

Fig. 9. Additional settings concerning primer design parameters, optimal target length, experiment name, and e-mail address for automated notification.

HTML output page. Following this link, a new browser window opens with a temporary page that refreshes every minute and is replaced once the primer design job is finished. The processing time depends on the sequence length, the mapping of SNPs, and on other jobs in the queue. For privacy reasons, the URL to the output page contains a random generated 8-digit code (TYJLNTFH in **Fig. 11**), to avoid users browsing SNPbox output from other jobs. Furthermore, data on the SNPbox server can be removed from the server by a dedicated tool in the "Tools" section of the SNPbox Web site, using the experiment number and the 8-digit code.

SNPbox' output page consists of two frames. The upper frame shows a graphical representation of the sequence in Scalable Vector Graphics (SVG) format. SVG is supported by Firefox (release 1.5) available at http://www.mozilla.com/firefox/. In other cases, a free SVG viewer has to be installed, like the one of Adobe, downloadable from http://www.adobe.com/svg/. The genomic sequence is represented by black lines, repeats by red lines, amplicons by blue lines, and SNPs by green lines

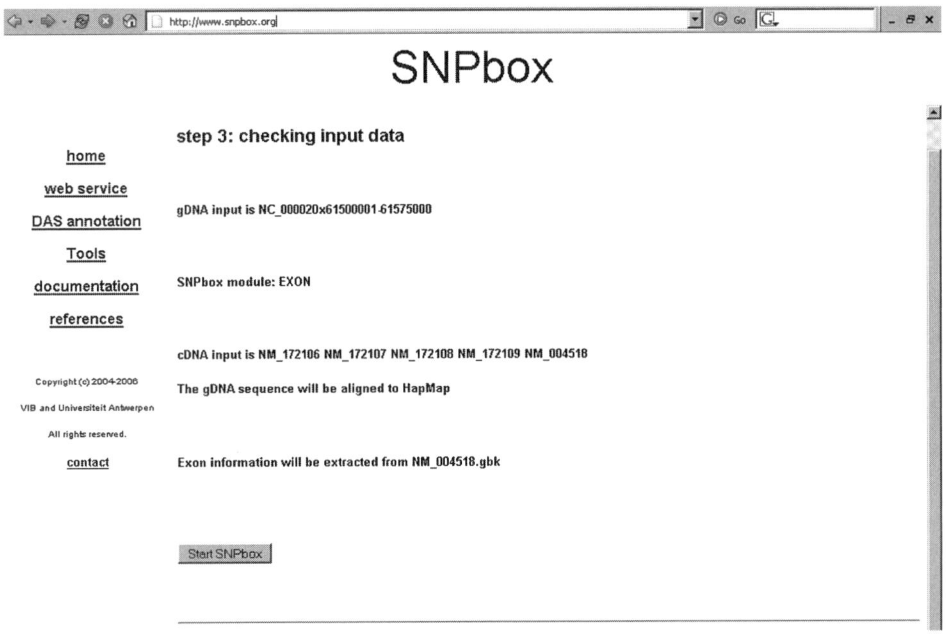

Fig. 10. Screenshot of input overview before SNPbox is started. If all input is according to the requirements, the "Start SNPbox" is available and has to be pressed to start the job.

(*see* **Fig. 12**). By hovering over these lines, information about the features appears, e.g., type of repeat and exon number. By clicking on these lines, a new window is opened showing the DNA sequence of the selected feature.

The lower frame provides a textual output, with a link to a tab-delimited file containing all primer information that is necessary for primer ordering (*see* **Fig. 13**). Also, FASTA files can be downloaded containing the masked and unmasked genomic sequences, primer sequences, PCR product sequences, SNPs with flanking sequence, and repetitive sequences (*see* **Fig. 14**).

4. Pre-Designed Primer Pairs

In 2002, the Distributed Annotation System (DAS, http://www.biodas.org) was described by Dowel et al. *(9)* as a system to cope with the increasing amount of annotation data for the human genome. This system makes it possible to physically separate those data: one reference server is needed hosting the reference sequence and annotation servers offer their information related to that reference sequence. Information from the different sources is brought

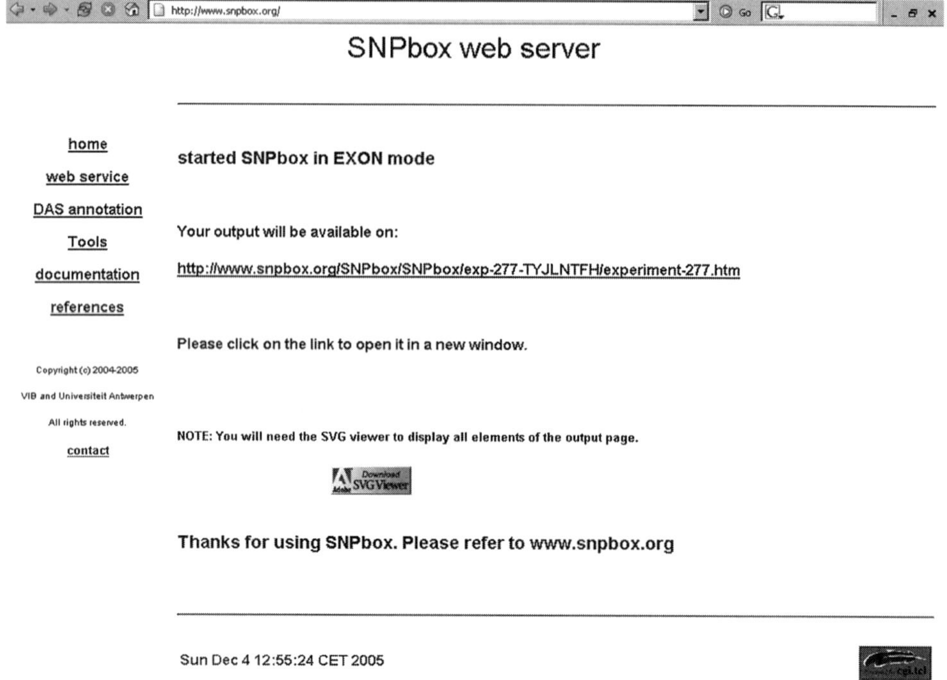

Fig. 11. Screenshot of page that is shown after a SNPbox job is started, with the link to the output page.

together in a DAS viewer. We used the gene annotation in Ensembl ***(10,11)*** (http://www.ensembl.org) to design primer pairs for all exons of the human and mouse genomes.

4.1. Setting Up the Ensembl Browser

To be able to view the pre-designed PCR products as an additional annotation in Ensembl's ContigView, following steps are to be followed. The information is stored locally in a cookie, so the Internet browser should allow cookies, preferably for all sessions.

- Go to Ensembl Human or Mouse and select a chromosome in the Karyotype overview, you will be forwarded to the MapView of the selected chromosome.
- Select a chromosomal region in MapView, you will be forwarded to ContigView containing the selected chromosomal region.

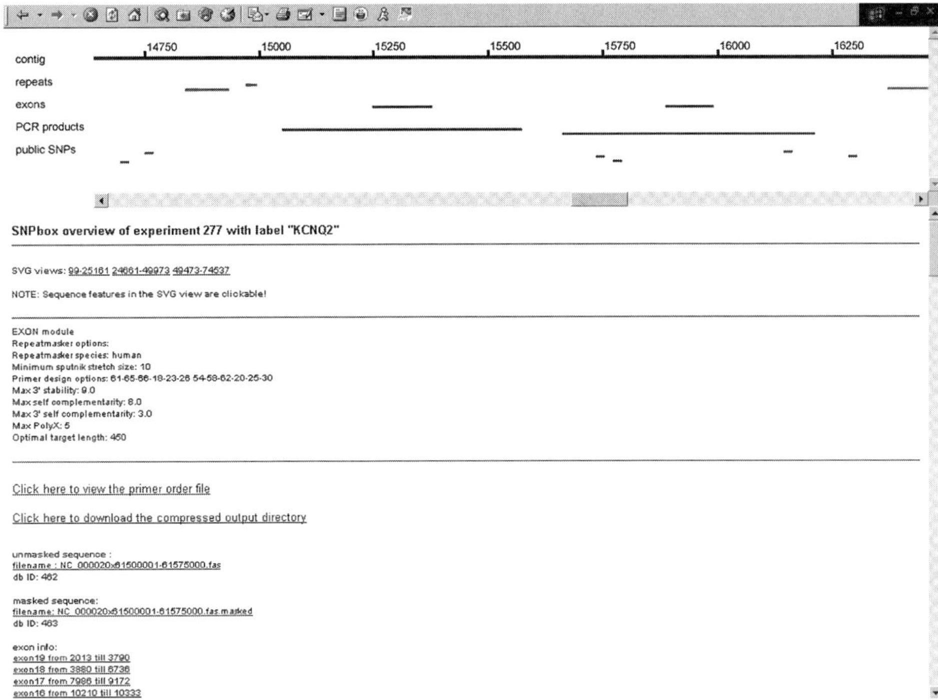

Fig. 12. Screenshot of output page. In the upper frame, the genomic sequence is represented together with the sequence features.

- Alternatively, you can search the Ensembl database, e.g., by gene name, and then choose to view the gene in the genomic location, which will also bring us to the ContigView.
- In the "Detailed View" of Ensembl's ContigView, go to "DAS Sources" and click on "Manage sources" (*see* **Fig. 15**). A new window "DasconfView" is opened, providing an overview of available DAS sources.
- Click on "Add Data Sources" on the left-hand side of the window.
- We enter step 1 of the DAS Wizard: "Data Location."
- Search for and select the appropriate SNPbox annotation from the list shown and press the "Next" button.
- This brings us to step 2 of the DAS Wizard "Data appearance."
- ContigView is selected by default, just press the "Next" button.
- We are forwarded to step 3 of the DAS Wizard: "Display Configuration." Leave all field unchanged and press the "Finish" button.
- We are redirected to the DASconfView overview, with SNPbox now added to the DAS source list.

	A	B	C	D	E	F	G	H
1	file generated on Sun Dec 4 15:05:20 CET 2005							
2								
3	contig sequence: NC_000020x61500001-61575000.fas							
4	user provided experiment name: KCNQ2							
5								
6	id	set	orient	sequence	position	temperature	amplicon_length	GC_stats(%)
7								
8	10109	1	F	GTTCGTGCCCTGAGGAGAGG	1861	64.543	557	min:33 mean:57 max:76 framelength:46
9	10110	1	R	TCTGCACTTTTGTTCATGCTCCA	2417	65.181	557	min:33 mean:57 max:76 framelength:46
10	10111	2	F	GACCTCAGGGAGTTCCTGTCCAC	2715	65.64	553	min:48 mean:64 max:76 framelength:50
11	10112	2	R	GCGGCTTCCTGAAGGATGTG	3267	65.36	553	min:48 mean:64 max:76 framelength:50
12	10113	3	F	GGGTGTGCAGCTTCTCTAGGTCT	3174	63.355	559	min:61 mean:67 max:72 framelength:46
13	10114	3	R	GTATGAGAGGGATGGCTCACGA	3732	63.701	559	min:61 mean:67 max:72 framelength:46
14	10115	4	F	CTGCTGGGCACCCTCCAT	3625	64.653	559	min:54 mean:65 max:72 framelength:46
15	10116	4	R	CCTCCTGAGCTATGAATGCCGTA	4183	64.699	559	min:54 mean:65 max:72 framelength:46
16	10117	5	F	GAGCTTTGTGAACCGTCCCTCTC	4058	65.674	558	min:48 mean:62 max:70 framelength:46
17	10118	5	R	AGGACACGCTGCCCCTCA	4615	65.472	558	min:48 mean:62 max:70 framelength:46
18	10119	6	F	CAGTGCCTGGGGTGATTTC	4527	61.484	560	min:50 mean:66 max:74 framelength:46
19	10120	6	R	TGTGGCTTTAGGGGTCGTG	5086	62.039	560	min:50 mean:66 max:74 framelength:46
20	10121	7	F	TTGCCATAAGGTGGCACTGGT	5618	64.854	560	min:59 mean:66 max:74 framelength:46
21	10122	7	R	AGTATGGCCCCACCTGCTCTTT	6177	65.3	560	min:59 mean:66 max:74 framelength:46
22	10123	8	F	CCCATGGGAGGAAAGAGCAG	6011	64.244	446	min:65 mean:68 max:71 framelength:49
23	10124	8	R	CTGGGGTCCATGTCCCTTTG	6456	64.853	446	min:65 mean:68 max:71 framelength:49
24	10125	9	F	TCCAGGTAGCTGAGACGGAACTG	6309	65.098	560	min:61 mean:68 max:78 framelength:46
25	10126	9	R	GAAAGCAGCATTTGAGGGTCGTA	6868	64.709	560	min:61 mean:68 max:78 framelength:46
26	10127	10	F	GAAGACGCCAACCAATAGCTCTGT	7842	64.968	556	min:54 mean:67 max:76 framelength:46
27	10128	10	R	CACGGTGCCTCCGAGGTTT	8397	65.753	556	min:54 mean:67 max:76 framelength:46
28	10129	11	F	GTCGCTGTCCCGCAGGTT	8703	64.807	549	min:57 mean:71 max:82 framelength:49
29	10130	11	R	CCCCATTTCCTGGGAGCAG	9251	65.084	549	min:57 mean:71 max:82 framelength:49
30	10131	12	F	TTGGGTACTGTCAGGAGGGAAGG	9985	65.72	545	min:61 mean:69 max:76 framelength:49
31	10132	12	R	GAACCTCACGGTGCCAGAAGAC	10529	65.515	545	min:61 mean:69 max:76 framelength:49
32	10133	13	F	GGAGAGATGGGAGAGACAGCAGA	15049	64.464	525	min:40 mean:65 max:79 framelength:47
33	10134	13	R	CCTGCAAGCCCCACCTGT	15573	64.649	525	min:40 mean:65 max:79 framelength:47

Fig. 13. Example of the tab-delimited output file.

Name	Type	Modified	Size	Ratio	Packed	Path
parameters.txt	Text Document	4/12/2005 15:05	281	0%	281	.\exp-277-TYJLNTFH\
error.log	Text Document	4/12/2005 15:05	0	0%	0	.\exp-277-TYJLNTFH\
legend.svg	SVG Document	4/12/2005 15:05	418	0%	418	.\exp-277-TYJLNTFH\
exp-277-3.svg	SVG Document	4/12/2005 15:05	83.768	0%	83.768	.\exp-277-TYJLNTFH\
exp-277-2.svg	SVG Document	4/12/2005 15:05	85.727	0%	85.727	.\exp-277-TYJLNTFH\
exp-277-1.svg	SVG Document	4/12/2005 15:05	92.150	0%	92.150	.\exp-277-TYJLNTFH\
NM_172109.fas.spidey.out	OUT File	4/12/2005 15:05	1.149	0%	1.149	.\exp-277-TYJLNTFH\
NM_172108.fas.spidey.out	OUT File	4/12/2005 15:05	1.909	0%	1.909	.\exp-277-TYJLNTFH\
NM_172107.fas.spidey.out	OUT File	4/12/2005 15:05	2.002	0%	2.002	.\exp-277-TYJLNTFH\
NM_172106.fas.spidey.out	OUT File	4/12/2005 15:05	1.892	0%	1.892	.\exp-277-TYJLNTFH\
NM_004518.fas.spidey.out	OUT File	4/12/2005 15:04	2.031	0%	2.031	.\exp-277-TYJLNTFH\
NC.blast.out	OUT File	4/12/2005 15:05	223.806	0%	223.806	.\exp-277-TYJLNTFH\
primers2order-exp277.xls	Microsoft Ex...	4/12/2005 15:05	4.709	0%	4.709	.\exp-277-TYJLNTFH\
NC_000020x61500001-61575000.fas.masked	MASKED File	4/12/2005 15:04	76.592	0%	76.592	.\exp-277-TYJLNTFH\
legendframe.htm	HTM File	4/12/2005 15:05	268	0%	268	.\exp-277-TYJLNTFH\
frame2-3-277.htm	HTM File	4/12/2005 15:05	393	0%	393	.\exp-277-TYJLNTFH\
frame2-2-277.htm	HTM File	4/12/2005 15:05	393	0%	393	.\exp-277-TYJLNTFH\
frame2-1-277.htm	HTM File	4/12/2005 15:05	390	0%	390	.\exp-277-TYJLNTFH\
frame1-277.htm	HTM File	4/12/2005 15:05	40.649	0%	40.649	.\exp-277-TYJLNTFH\
experiment-277.htm	HTM File	4/12/2005 15:05	655	0%	655	.\exp-277-TYJLNTFH\
repeat8934-8955.fas	FAS File	4/12/2005 15:05	66	0%	66	.\exp-277-TYJLNTFH\repeats\
repeat774-845.fas	FAS File	4/12/2005 15:05	114	0%	114	.\exp-277-TYJLNTFH\repeats\
repeat74445-74499.fas	FAS File	4/12/2005 15:05	102	0%	102	.\exp-277-TYJLNTFH\repeats\
repeat74301-74355.fas	FAS File	4/12/2005 15:05	102	0%	102	.\exp-277-TYJLNTFH\repeats\
repeat74071-74088.fas	FAS File	4/12/2005 15:05	63	0%	63	.\exp-277-TYJLNTFH\repeats\
repeat74066-74089.fas	FAS File	4/12/2005 15:05	70	0%	70	.\exp-277-TYJLNTFH\repeats\
repeat73895-73922.fas	FAS File	4/12/2005 15:05	73	0%	73	.\exp-277-TYJLNTFH\repeats\
repeat73337-73364.fas	FAS File	4/12/2005 15:05	73	0%	73	.\exp-277-TYJLNTFH\repeats\
repeat73065-73082.fas	FAS File	4/12/2005 15:05	63	0%	63	.\exp-277-TYJLNTFH\repeats\
repeat72875-72897.fas	FAS File	4/12/2005 15:05	68	0%	68	.\exp-277-TYJLNTFH\repeats\
repeat70657-70733.fas	FAS File	4/12/2005 15:05	118	0%	118	.\exp-277-TYJLNTFH\repeats\
repeat69146-69170.fas	FAS File	4/12/2005 15:05	70	0%	70	.\exp-277-TYJLNTFH\repeats\
repeat6904-6920.fas	FAS File	4/12/2005 15:05	60	0%	60	.\exp-277-TYJLNTFH\repeats\

Fig. 14. Overview of available files in the zipped downloadable file.

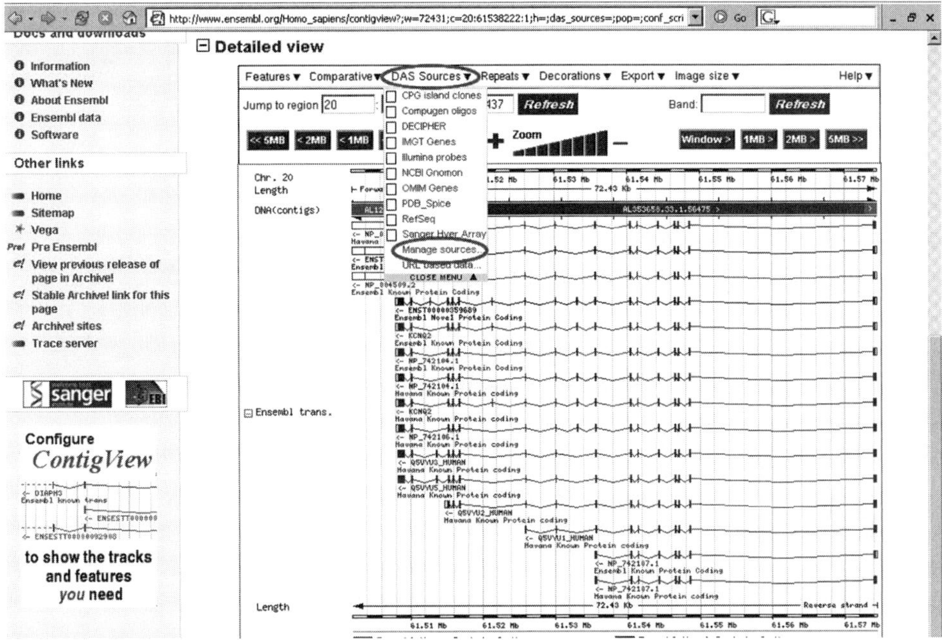

Fig. 15. Screenshot of Ensembl's ContigView. The "DAS Sources" tab is highlighted together with the "Manage sources" to indicate how a DAS source can be added to the Ensembl view.

- Close the window and select "DAS sources" in the detailed view. Now, SNPbox is available as selectable DAS source.
- Select SNPbox as DAS source and close the menu.
- The ContigView page is reloaded, and SNPbox features are now displayed (*see* **Fig. 16**).

When clicking on a blue SNPbox feature, a box appears with additional information on the feature. The last field "DAS link" with a SNPbox unique identifier is selectable and redirects the user to a page on the SNPbox Web server where detailed information is given on the SNPbox primers for all exons of the gene.

A second way to retrieve information on pre-designed primers for Ensembl genes is by using a tool available at the "Tools" subsection of the SNPbox Web site. Here, the user only needs to provide the Ensembl gene ID to obtain all primer information for that gene.

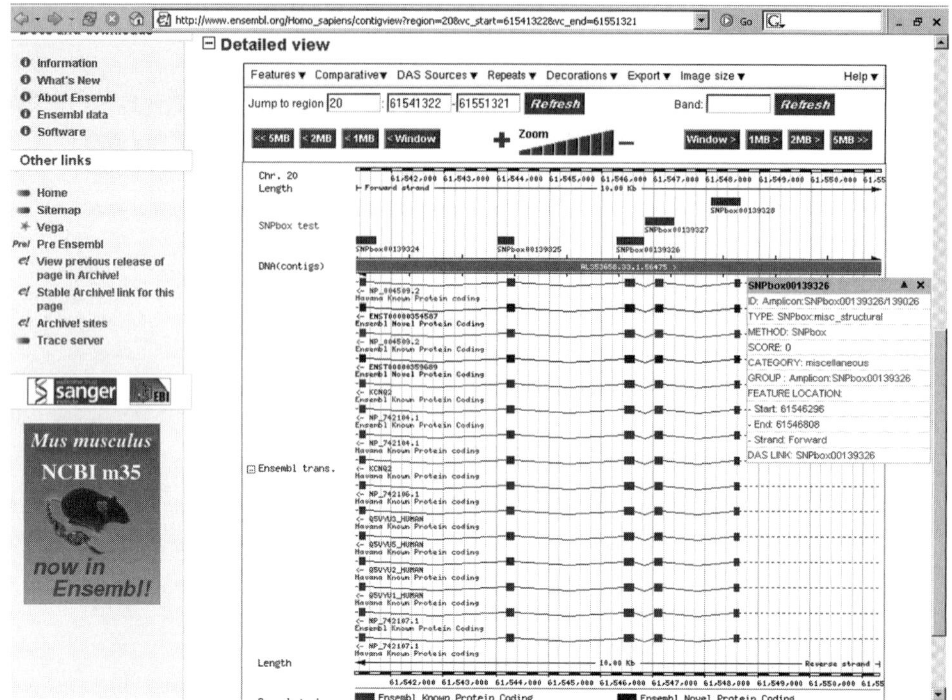

Fig. 16. Ensembl with SNPbox features added to ContigView. When a SNPbox feature is selected, a box appears with additional information about the feature. Clicking on the last line "DAS LINK," the user is forwarded to a page on the SNPbox Web server with primer details for all SNPbox PCR products for the selected gene.

Important notice: The above text describes how SNPbox annotation can be added to Ensembl. As the Ensembl team at the European Bioinformatics Institute and Sanger Institute is continuously improving Ensembl, we might face the moment that the procedure to add DAS sources is altered. Therefore, this procedure is also described in the "DAS annotation" section on the SNPbox Web server and will be updated in case the described procedure changes.

References

1. Lander, E.S., Linton, L.M., Birren, B., Nusbaum, C., Zody, M.C., Baldwin, J., Devon, K., Dewar, K., Doyle, M., FitzHugh, W. et al. (2001) Initial sequencing and analysis of the human genome. *Nature*, 409, 860–921.
2. Venter, J.C., Adams, M.D., Myers, E.W., Li, P.W., Mural, R.J., Sutton, G.G., Smith, H.O., Yandell, M., Evans, C.A., Holt, R.A. et al. (2001) The sequence of the human genome. *Science*, 291, 1304–1351.

3. Rozen, S. and Skaletsky, H.J. (2000) Primer3 on the WWW for general users and for biologist programmers. *Methods Mol. Biol.*, 132, 365–386.
4. Weckx, S., De Rijk, P., Van Broeckhoven, C. and Del Favero, J. (2005) SNPbox: a modular software package for large-scale primer design. *Bioinformatics*, 21, 385–387.
5. Smit, A.F. (1996) The origin of interspersed repeats in the human genome. *Curr. Opin. Genet. Dev.*, 6, 743–748.
6. Weckx, S., De Rijk, P., Van Broeckhoven, C. and Del Favero, J. (2004) SNPbox: web-based high-throughput primer design from gene to genome. *Nucleic Acids Res.*, 32, W170–W172.
7. Wheelan, S.J., Church, D.M. and Ostell, J.M. (2001) Spidey: a tool for mRNA-to-genomic alignments. *Genome Res.*, 11, 1952–1957.
8. HapMap Consortium. (2003) The International HapMap Project. *Nature*, 426, 789–796.
9. Dowel, R.D., Jokerst, R.M., Day, A., Eddy, S.R. and Stein, L.D. (2002) The distributed annotation system. *BMC Bioinformatics*, 2, 7.
10. Birney, E., Andrews, D., Caccamo, M., Chen, Y., Clarke, L., Coates, G., Cox, T., Cunningham, F., Curwen, V., Cutts, T. et al. (2006) Ensembl 2006. *Nucleic Acids Res.*, 34, D556–D561.
11. Hubbard, T., Andrews, D., Caccamo, M., Cameron, G., Chen, Y., Clamp, M., Clarke, L., Coates, G., Cox, T., Cunningham, F. et al. (2005) Ensembl 2005. *Nucleic Acids Res.*, 33, D447–D453.

10

Fast Masking of Repeated Primer Binding Sites in Eukaryotic Genomes

Reidar Andreson, Lauris Kaplinski, and Maido Remm

Summary

In this article, we describe the working principle and a list of practical applications for GenomeMasker—a program that finds and masks all repeated DNA motifs in fully sequenced genomes. The GenomeMasker exhaustively finds and masks all repeated DNA motifs in studied genomes. The software is optimized for polymerase chain reaction (PCR) primer design. The algorithm is designed for high-throughput work, allowing masking of large DNA regions, even entire eukaryotic genomes. Additionally, the software is able to predict all alternative PCR products from studied genomes for thousands of candidate PCR primer pairs. Practical applications of the GenomeMasker are shown for command-line version of the GenomeMasker, which can be downloaded from http://bioinfo.ut.ee/download/. Graphical Web interfaces with limited options are available at http://bioinfo.ut.ee/genometester/ and http://bioinfo.ut.ee/snpmasker/.

Key Words: PCR; DNA repeats; primer design; microarrays; DNA masking.

1. Introduction

Modern genomic technologies allow studying thousands of genomic regions from each DNA sample. Most of these technologies require polymerase chain reaction (PCR) amplification to achieve sufficiently strong signals. Although many currently available high-throughput technologies use one single pair of universal PCR primers, most applications still need a large number of custom-designed PCR primers. Therefore, there is an urgent need for automatic PCR primer design methods. The automatic PCR primer design is trusted only if it generates primers with a success rate equal to or higher than the manual

From: *Methods in Molecular Biology, vol. 402: PCR Primer Design*
Edited by: A. Yuryev © Humana Press, Totowa, NJ

primer design. Although not all aspects of the PCR process are thoroughly described and modeled, there are certain factors that are known to affect the PCR success rate. One of such factors is primer overlap with repeated regions on the template DNA. The primers that bind to the repeats are more likely to fail or generate false products.

We started studying that problem in early 2001 when we got involved in a large-scale genotyping experiment covering the whole human chromosome 22 *(1)*. The study analyzed 1278 single-nucleotide polymorphisms (SNPs). For each SNP, a separate PCR primer pair was designed. We planned to find the correlations between the PCR primer pair success rate and the sequence properties of the primers. One of the factors studied was the number of predicted primer binding sites. However, it quickly turned out that finding the number of binding sites for thousands of primer pairs from the entire human genome is not an easy task. One of the difficulties was that we did not know exactly what to search for. The binding site of the primer can be defined in various ways. For example, the binding site can be defined as eight nucleotides from the primer 3′-end, or 12 nucleotides from the primer 3′-end, or 16 nucleotides from the primer 3′-end, or any other number of nucleotides from the primer 3′-end. Also, it can be defined as the exact match, 100% identical to the primer or similarity with one or two mismatches. Furthermore, the binding site can be defined as variable length string from the 3′-end of the primer. In this case, the length of the primer varies, so that the binding energy exceeds certain ΔG level for many different ΔG values. If we want to study all these potential models of binding sites for their effect on prediction of PCR success rate, we first have to calculate the number of binding sites for all the primer pairs with each binding site model. In our case, this meant finding the binding sites for 1278 primer pairs in hundreds of different ways from the human genome.

We tried to use existing programs for finding and counting primer binding sites from the human genome. The BLAST program is most frequently used for this purpose in multiple applications *(2–4)*. Unfortunately, the speed of BLAST is not sufficient for counting primer binding sites in large eukaryotic genomes with large number of primers. The speed can be increased by using MEGABLAST *(5)*, SSAHA *(6)*, or BLAT *(7)* that are specifically designed for homology search from large genomes. Unfortunately, these faster programs are not optimized for primer design tasks, and thus recording the number and the location of the predicted primer binding sites requires additional efforts.

Thus, we decided to create software that counts all primer binding sites in the human genome within seconds and reports potential PCR products for thousands of primer pairs. The program, named GenomeTester, is based on exhaustive counting and recording of the locations of all potential binding sites from the human genome. The locations are stored in a binary hash data structure, which allows extremely fast retrieval of the number and the location of all binding sites for any given primer.

To simplify the design of PCR primers in future, we have also added a possibility to mask repeated regions on a template DNA. This helps to avoid the design of PCR primers with extensive number of binding sites and thus increase the success rate of the designed primers. Masking of repeats on the template DNA is a common approach that is used to mark specific regions in DNA. DUST (ftp://ftp.ncbi.nlm.nih.gov/pub/tatusov/dust/) and TandemRepeatsFinder *(8)* are commonly used for masking simple (short) repeat motifs. RepeatMasker is a universal program that is used for masking out several kinds of repeats and is therefore mostly used for this kind of sequence analysis (Smit, A.F.A., Hubley, R. and Green, P., http://www.repeatmasker.org/). Similarly, BLAST *(9)* can be used to mask the non-unique regions of the genome ***(10,11)***. Our masking software, named GenomeMasker, is dedicated to masking of repeated primer binding sites in large genomes. The details of the algorithm are briefly described in Chapter 2. Examples of practical use of our programs are shown in Chapter 3.

2. Working Principle and Data Structures

The efficiency of both GenomeTester and GenomeMasker software is based on pre-processing of genomic sequence into specific data structure—the hash structure.

2.1. GenomeMasker

In GenomeMasker, the hash structure contains list of all repeated sequence motifs from a given genome, encoded into 32-bit integers. The encoding is done by allocating two consecutive bits for each nucleotide in a word. Thus, the length of repeated sequence motif X in current implementation is in range between 8 and 16 nucleotides. Our group is using 16 nucleotides as default value of X, because it seems to give the best separation between high and low success rate PCR primers. Most of the following examples in this article are also based on repeat length $X = 16$. The sequence motif is defined as "repeated" if given nucleotide sequence occurs in the given genome more than Y times, where Y is an integer chosen by user (1, 2, 3, etc.). For example, if motif length $X = 16$ nucleotides and tolerated repeat number $Y = 1$, then all 16 nucleotide long sequence motifs that occur more than once in the given genome are put into list of repeats and stored in the hash structure. The entire hash structure of repeated motifs is sorted for faster access and written into blacklist file.

The program GenomeMasker uses this blacklist file as a reference to quickly mask the template sequence for PCR primer design. The GenomeMasker iterates over the whole template sequence (which can be the entire genome) with window length X nucleotides and with step 1 nucleotide.

For each window, it checks whether the sequence motif within the window or its reverse complement is recorded in the blacklist file. If the given sequence motif is in the blacklist, the corresponding window in the template sequence is masked (*see* **Fig. 1A**). Theoretically, primers with partial overlap with repeats can work in PCR. Therefore, we can design PCR primers that partly overlap with repeated motif as long as the last 16 nucleotides from the 3′-end of the primer are not repeated (*see* **Fig. 1A**). This will lead to asymmetric masking because primers are single stranded and the 3′-end of repeat is different on upper and lower strand. For example, the primer A from upper strand can work fine, whereas its reverse complement from lower strand (primer B) may fail because its 3′-end is overlapping with repeat (*see* **Fig. 1A**). Furthermore, small overlap with the repeated region can be tolerated even if it happens in primers 3′-end. Therefore, we do not have to mask the entire repeat but just a couple of nucleotides from the repeats 3′-end (*see* **Fig. 1B**). We have initiated a comprehensive experimental analysis of factors influencing PCR primer success. Our preliminary results indicate that overlaps with repeated regions can be tolerated as long as the last 16 nucleotides of a primer are not repeated. Thus, masking of single nucleotide at the 3′-end of repeat may be sufficient to avoid PCR primers with low success rate (*see* **Fig. 1C**). Nevertheless, the length of masked region Z can be changed with a special option in GenomeMasker software, allowing each user to select his/her own settings for Z.

2.2. GM_PRIMER3

With GenomeMasker, the sequence can be masked by any user-defined character; however, the most useful masking style is with lowercase letters. The lowercase masking maintains the sequence information in masked regions and allows subsequent primer design from the masked sequence even if some primers overlap with masked nucleotides. To take advantage of lowercase-masked sequence, we have modified the well-known program PRIMER3 ***(12)***. The overall functionality and algorithm of the program is the same as in the original PRIMER3, but we have added a new filtering feature that rejects the primer candidates with lowercase letters at their 3′-end and new parameters for the calculation of melting temperature of primers. Although standard version of

A. 16-nucleotide masking (allowed 3'-end overlap with repeat = 0 nt.)

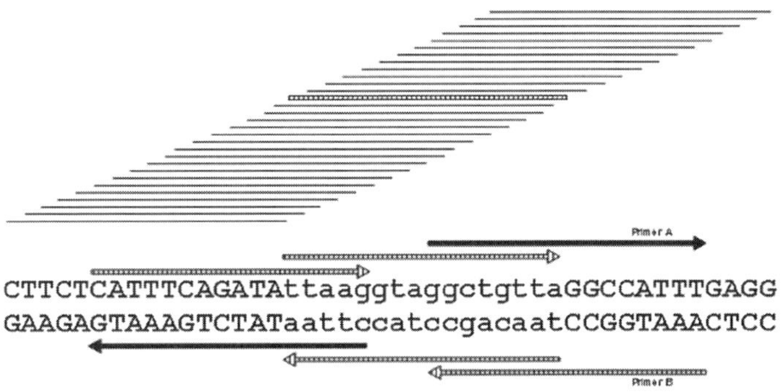

B. 10- nucleotide masking (allowed 3'-end overlap with repeat = 6 nt.)

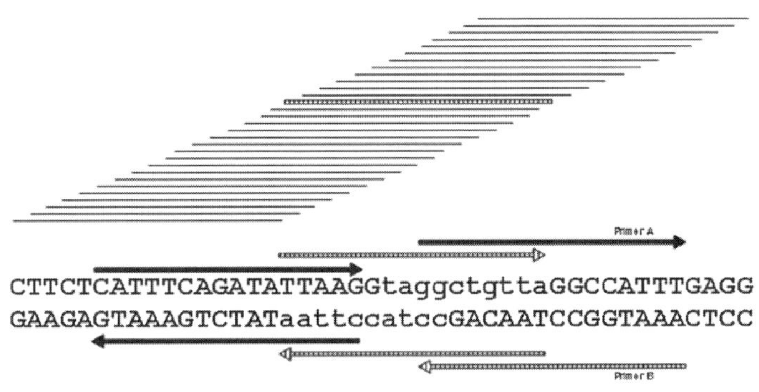

C. 1- nucleotide masking (allowed 3'-end overlap with repeat = 15 nt.)

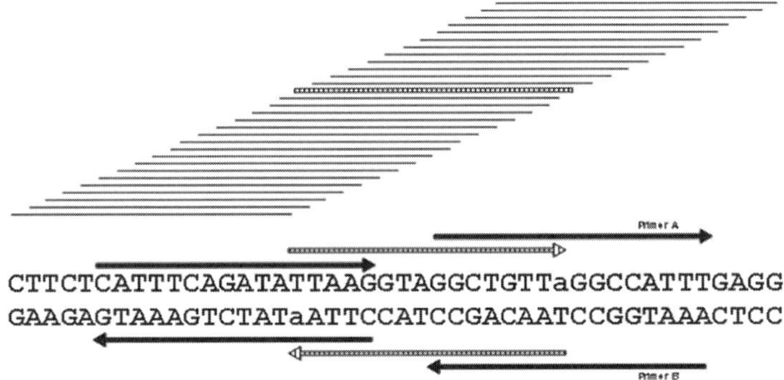

Fig. 1. Principles of masking and subsequent primer design with GenomeMasker software package. The repeated 16-nucleotide motif is striped; non-repeated motifs are

the PRIMER3 uses nearest-neighbor model to calculate melting temperature, the parameters used for that are rather old *(13)*. We replaced the nearest-neighbor parameters with a newer set *(14)*, added sequence-dependent salt correction formula *(15)*, and correction for concentration of divalent cations *(16)*.

The modified version of PRIMER3 called *GM_PRIMER3* is available at http://bioinfo.ut.ee/download/, and an online version of the program can be used at http://bioinfo.ut.ee/mprimer3/.

2.3. GenomeTester

GenomeTester is a program for counting potential binding sites and potential products for each tested PCR primer. During the pre-processing of the genome sequence, the GenomeTester exhaustively counts all locations of all possible PCR primers. The data structure for pre-processed genomic data is hash structure, similar to GenomeMasker. The main difference from abovementioned GenomeMasker program is that the GenomeMasker stores only the number of binding sites for each potential primer, whereas GenomeTester also stores all locations (genome coordinates) for each potential primer. The hash structure created during pre-processing is sorted and saved into file(s). The hash structure allows fast identification of PCR primer binding sites and PCR products for large number of primers.

Similar hash structure is used by the program SSAHA that is designed to run fast-sequence searches from genomic sequences. However, the GenomeTester uses slightly different data structure that allocates equal amount of memory for each location and thus makes it faster in subsequent searches.

3. Practical Applications

3.1. GenomeMasker Application

GenomeMasker is suitable for users who need to design PCR primers that are unique in given genomic DNA. It can be described as an additional filtering stage to avoid primer candidates with low success rate. All this can be done

Fig. 1. shown in black. The rejected primers are striped, and accepted primers are in black. The number of masked nucleotides can be selected by the user. Full-length masking of repeated motifs is most conservative, allowing design of only two of six primers on this example (**A**), whereas 10-nucleotide masking allows design of three (**B**) and single-nucleotide masking four of six primers (**C**). Please note that the method generates asymmetric masking only if less than 16 nucleotides of repeated 16-nucleotide motif are masked.

before the actual primer design starts, and the output of the GenomeMasker can be used for selecting the successful primer candidates.

GenomeMasker application contains the following executables:

- *glistmaker* creates a list of repeated (occurring more frequently than user-defined threshold) words in a given genome. This step has to be performed only once for a given set of genomic data and chosen word length.
- *gmasker* performs a masking procedure for each studied FASTA sequence.

3.1.1. *The Usage of* glistmaker

The GenomeMasker requires pre-processed data structure (blacklist file) as a reference to mask DNA sequences. The executable *glistmaker* creates binary blacklist file for GenomeMasker application from the genomic data in the FASTA format. Here are the options for *glistmaker*:

```
prompt> ./glistmaker -h
Usage: glistmaker OPTIONS inputfilelist outputfile wordsize
-v - Print version and exit
-h - Print this usage screen and exit
-d - Turn on debugging output
-overreplimit NUMBER - Specify overrepresentation cutoff
(default 10)
```

See **Note 1** for detailed explanation of input parameters and files. The following is a command-line example of running *glistmaker*:

```
prompt> ./glistmaker -overreplimit 100
human_chr_list.txt human_repeats.list 12
```

Command shown above will create a blacklist file ("human_repeats.list") that contains all 12-mer words having more than 100 different locations in all chromosomes included to "human_chr_list.txt" file.

3.1.2. *The Usage of* gmasker

After creating the blacklist file, the user can start masking sequence files containing template DNA regions (in FASTA format) with the second executable in GenomeMasker application called *gmasker*. Options for the *gmasker* are shown below:

```
prompt> ./gmasker -h
Usage: gmasker OPTIONS blacklistfile maskingletter
maskingtype [start end]
-v - Print version and exit
-h - Print this usage screen and exit
-d - Turn on debugging output
-u - Convert sequence to uppercase before processing
-nbases NUMBER - How many bases from 3' end to mask
(default 1)
```

See **Note 2** for detailed explanation of input parameters and files.

User can define the sequence file that should be masked by *gmasker* and manage output with different ways using Unix pipes (“|,” “<,” and “>”). Here are some examples:

```
prompt> cat sequences.fas | ./gmasker
human_repeats.list N both> masked_sequences.fas
prompt> ./gmasker human_repeats.list N both
<sequences.fas> masked_sequences.fas
```

In both cases shown above, “sequences.fas” will be used as an input file (*see* example sequence in **Fig. 2**) and the results are written into “masked_sequences.fas”.

The final output of GenomeMasker application is a masked FASTA file, where 3′-end of repeated words are replaced with a user-defined character. **Figure 3** shows additional masking possibilities and command-line examples to run *gmasker*.

3.2. GM_PRIMER3 Application

To design primers with *GM_PRIMER3*, user needs several supporting programs. Current application includes two Perl scripts, which help to automate the primer design with the *GM_PRIMER3*:

```
>user sequence
TTAACGTTTTCAAAAAGTTAACAGTACCTATGTCTTCATAGTATTTATTACATTCAGAAATTTTAATTTAC
CTACCTTTCTTATCCATAGTTTCTAATCTTATGAAATAAGCATGCATTTGTTCTGTAGTAAGAAAAATAAT
TTTTATTCTGTTATTTAAATAAAATGTATATATACTCATAGTTGAGTAATAAAACTTTTCATCTAGATAAG
TTTTTTAAAAMTGTGTAATACATTAAATTAAATTTATTGTGAGTAGGAATTACCTACATTTATATATTTAA
ATAAATAGGTTTTTAATAAAATAACATTTACAGAGTTGCAGTCATTGTTGTATTAAGAGTTCTAAATACTT
TTATGTATGTTAGGCTGACCTGAAAGATCCAGAATCGATCTGATTTTCCTTGATCTGTC
```

Fig. 2. Example input file for *gmasker*.

A.

```
prompt> cat sequences.fas | ./gmasker human_repeats.list l target
220 230 > masked_sequences.fas

>user sequence (masked_sequences.fas)
TTAACGTTTTCAAAAAGTTAacagtACCTATGTCTTCATAGTAtttAttaCattcAGaaaTTttaatttac
CtaCCtTTCTtATCCATAGTTTCTAATCttATGAAaTAagCATGCaTTtGTTcTGTAGTAAgaaaaataat
ttttattctgttatttaaataaaatgtatatatactcataGTTGAGTAATAAAACTtTtcatctAGATAAG
TTTTTTAAAAMTGTGTAAtacattaaaTTAAATTTATTGTGAGTAGGaAttaCctacatttatatatttaa
ataaataggttttaataaaATaACATTTACAGAGTTGCAGTCATTGTTGTATTAAGagtTCtaAataCTT
TTATGTATGTTAGGCTGACCTGAAAGATCCAGAATCGATCTGATTTTCCTTGATCTGTC
```

B.

```
prompt> cat sequences.fas | ./gmasker human_repeats.list N both >
masked_sequences.fas

>user sequence
TTAACNNNNNCAAAAAGTTANNNNNACCNNNGNNNTNNNNGTNNNNANNNNNNNNNGNNNTNNNNNNNNNN
CNNCCNTTCTNATNNATAGTNTCNNATCNNNTGNANTNNNCATGCNNNNNNNNNNNNNNNNNNNNNNNNNN
NNNNNNNNNNNNNNNNNNNNNNNNNNNNNNNNNNNNNNNNGNTNNNNNNTAAAACTNTNNNNNNNNATAAG
TTTTTTANNNMNGNNNNNNNNNNNNNNTNNNNNNNNNNNNNNGTAGGNANNNCNNNNNNNNNNNNNNNNNN
NNNNNNNNNNNNNNNNNNNNNNNNNNNNNNNNNNTTNCAGTCATTGTTGTATTAAGNNNTCNNANNNCTT
TNNNGTNNGNNNGGCTGACCTGAAAGATCCAGAATCGATCTGATTTTCCTTGATCTGTC
```

C.

```
prompt> ./gmasker -nbases 10 human_repeats.list # backward <
sequences.fas > masked_sequences.fas

>user sequence
```

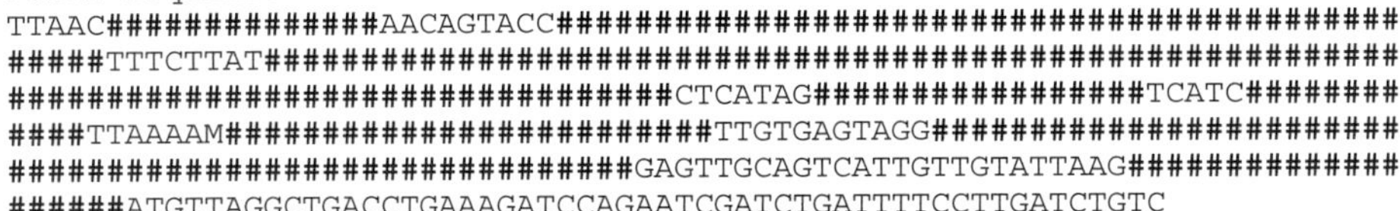

```
TTAAC##############AACAGTACC###########################################
#####TTTCTTAT##########################################################
##################################CTCATAG#################TCATC########
####TTAAAAM#########################TTGTGAGTAGG########################
###############################GAGTTGCAGTCATTGTTGTATTAAG##############
######ATGTTAGGCTGACCTGAAAGATCCAGAATCGATCTGATTTTCCTTGATCTGTC
```

Fig. 3. Different masking possibilities with GenomeMasker application. The first section (**A**) illustrates a result of the *gmasker* using a "target" masking type. For the PCR primer design, only the upper strand should be masked on the left side of the target region and only the lower strand should be masked on the right side of the target region. The middle part (bases from 220 to 230) is the target region, which is chosen to be amplified. The character "l" defines that 3′-end of the repeated words are masked with lower case letters. Next two sections demonstrate alternative masking possibilities: (**B**) upper and lower strand ("both") masking and (**C**) lower strand ("backward") masking. With special option "-nbases," the user can define how many nucleotides will be masked by *gmasker*. The masking was done by using NCBI build 35.1 human genome assembly. GenomeMasker blacklist was created by using wordsize 16, and the maximum number of allowed locations for each word was set to 10.

- *fasta_to_p3.pl* converts FASTA format sequences to *GM_PRIMER3* input format.
- *p3_to_table.pl* converts *GM_PRIMER3* output to tab-delimited table format with the following columns: name, sense_primer, antisense_primer, and product sequence. This table format can be used as an input for the *gtester* executable (*see* **Subheading 3.3.**) and for MultiPLX program ***(17)***.

3.2.1. *The Usage of* GM_PRIMER3

The following parameters need to be changed in *fasta_to_p3.pl* script before primer design can begin:

```
PRIMER_PRODUCT_SIZE_RANGE=100-600
PRIMER_PRODUCT_OPT_SIZE=200
PRIMER_OPT_SIZE=21
PRIMER_MIN_SIZE=18
PRIMER_MAX_SIZE=26
PRIMER_OPT_TM=62
PRIMER_MIN_TM=59
PRIMER_MAX_TM=65
PRIMER_MAX_DIFF_TM=4
PRIMER_OPT_GC_PERCENT=35
PRIMER_MIN_GC=20
PRIMER_MAX_GC=70
PRIMER_SALT_CONC=20
PRIMER_FILE_FLAG=0
PRIMER_EXPLAIN_FLAG=1
PRIMER_MAX_POLY_X=4
PRIMER_NUM_RETURN=1
TARGET=475,51
```

Of course, the user can add additional parameters here; the complete list of possible options and their explanation is available at *GM_PRIMER3* Web site (http://bioinfo.ut.ee/mprimer3/). Otherwise, the default values will be used for other parameters.

Here is a command-line example to run *GM_PRIMER3* for primer design:

```
prompt> cat masked_sequence.fas | ./fasta_to_p3.pl |
./gm_primer3 | ./p3_to_table.pl> primers.txt
```

The result of successful primer design is a tab-delimited "primers.txt" file that contains primer id, primer, and product sequences. If *GM_PRIMER3* is unable to find good primer candidates, these columns are filled with dashes. What to do with the regions, where *GM_PRIMER3* could not design primers?

The first thing is to try relaxing various *GM_PRIMER3* parameters, because too strict parameters may force the *GM_PRIMER3* to reject many good primer candidates (*see* **Note 3** for more detailed parameter explanations). Second, make sure that there are adequate stretches of non-Ns in the regions in which you wish to pick primers. Finally, in some cases, the region of interest is full of repeats, and therefore, *GM_PRIMER3* cannot design primers for a given sequence.

3.3. GenomeTester Application

Having too many binding sites will typically result in failed PCR, but that can be eliminated with GenomeMasker application. A different problem emerges if two primers give several alternative products in PCR. Amplifying more than one product is undesirable because alternative PCR products could cause false-positive signals in genotyping. The GenomeTester programs can be used to make sure if designed primers produce single PCR product or not.

GenomeTester application contains the following executables:

- *gindexer* creates index files containing locations of all the predicted binding sites in a given genome.
- *gtester* counts all locations of all possible PCR primers and predicts PCR products.

3.3.1. The Usage of gindexer

The program *gindexer* is needed to create index files for *gtester* to work. The executable *gindexer* creates binary index files for GenomeTester application from the genomic data (in FASTA format). Here are the options for *gindexer*:

```
prompt> ./gindexer -h
Usage: gindexer OPTIONS inputfile outputfile
-v - Print version and exit
-h - Print this usage screen and exit
-d - Turn on debugging output
-wordsize LENGTH - Specify word size for index (default 16)
```

Please *see* **Note 4** for detailed explanation of input parameters and files.

Different possibilities to execute *gindexer* are shown below:

```
prompt> ./gindexer ecoli_genome.fas ecoli
prompt> ./gindexer -wordsize 12 chr22.fas chr22
prompt> ./gindexer /home/db/human_DNA/chr/1.fa indexes/1
prompt> ./gindexer /home/db/human_DNA/chr/2.fa indexes/2
...
prompt> ./gindexer /home/db/human_DNA/chr/Y.fa indexes/Y
```

The first example creates index files (ecolia.location, ecolit.location, ecolic.location, and ecolig.location) for *Escherichia coli* complete genome. The second example shows the possibility to use smaller word length (-wordsize 12) to create indexes for given chromosome sequence. The other three examples illustrate the series of executions to create indexes for human chromosomes.

3.3.2. The Usage of gtester

After creating index files for chromosomes or genomic sequences, the user can start using *gtester* program. Here are the command-line options for *gtester*:

```
prompt> ./gtester -h
Usage: gtester OPTIONS primerfile locationsfile
-v - Print version and exit
-h - Print this usage screen and exit
-d - Turn on debugging output
-maxprodlen LENGTH - Specify maximum product length
(default 1000bp)
-limit NUMBER - Maximum number of binding sites to track
(default 1000)
```

See **Note 5** for detailed explanation of input parameters and files.

The basic execution of *gtester* is simple:

```
prompt> ./gtester primers.txt
human_indexes_list.txt
```

The file "primers.txt" is the output of *GM_PRIMER3* application, and "human_indexes_list.txt" is the list of filenames for all indexes of human chromosomes. The command above will generate three result files ending with specific suffixes (.gt1, .gt2, and .gt3): primers.txt.gt1, primers.txt.gt2, and primers.txt.gt3. **Figure 4** illustrates the columns of these three files.

There are also alternative possibilities to use *gtester*:

```
prompt> ./gtester -maxprodlen 10000 -limit 5000
primers.txt human_indexes_list.txt
prompt> ./gtester primers.txt human_indexes_list.txt
human_repeats.list
```

The first example illustrates the possibility to define the maximum length of products and the number of binding sites. The second example shows how to use a pre-processed blacklist file created by *glistmaker* to reduce the calculation time *gtester*.

A. primers.txt.gt1

NAME
NUMBER OF BINDING SITES FOR PRIMER A (left primer)
NUMBER OF BINDING SITES FOR PRIMER B (right primer)
NUMBER OF PRODUCTS

B. primers.txt.gt2

NAME
PRODUCT NUMBER
CHR
LOCATION (start nucleotide)
LENGTH (bp)
TYPE OF PRODUCT
1: PrimerA-PrimerB (sense strand product)
-1: PrimerB-PrimerA (antisense strand product)
2: PrimerA-PrimerA
-2: PrimerB-PrimerB

C. primers .txt .gt3

NAME OF THE PRIMER PAIR
PRIMER (A or B)
STRAND (1=sense, -1=antisense)
CHR
LOCATION OF THE 5' END OF THE PRIMER

Fig. 4. Description of three output files of *gtester*. The *gtester* executable produces three result files: (**A**) a file with the number of primer binding sites and with the number of products, (**B**) a file with the description of all PCR products, and (**C**) a file with the description of all primer binding sites.

4. Notes

Note 1: Input Parameters and Files for glistmaker

The first required input file—"inputfilelist"—is a text file containing locations of the chromosome or contig files (in FASTA format) in the user system, one file name per line:

```
/home/db/human_DNA/chr/1.fa
/home/db/human_DNA/chr/2.fa
/home/db/human_DNA/chr/3.fa
/home/db/human_DNA/chr/4.fa
...
/home/db/human_DNA/chr/Y.fa
```

The second file—"outputfile"—is the name of the blacklist file (in binary format) that the *glistmaker* creates.

The third parameter is called "-wordsize" and must be defined by the user. It represents the length of the word (sequence window) that program will use to find and store repeats to the blacklist file. The maximum length of the word can be 16 and minimum 8 nucleotides.

Optional parameter—"–overreplimit"—should be changed if the user wishes to use smaller word size than 16 nucleotides (by default, it is 10). Otherwise, sequences might be masked with 100%, because short words are more common in genomes. That cutoff defines the maximum number of different locations any word can have in a given DNA sequence before it is stored to the blacklist file.

Note 2: Input Parameters and Files for gmasker

The first required input file for *gmasker* is "human_repeats.list"—a preprocessed blacklist file created by *glistmaker.*

The next parameter is called a "maskingletter." This can be almost any letter, typical examples are "*N*" or "*X*." The only exception is "*l*" (or "*L*"), which triggers the lower case masking (3′-ends of over-represented words are in lower case, and 3′-ends of words with acceptable frequency are in upper case).

The third parameter, which the user must define, is "maskingtype." There are four possible options for defining the type of masking repeats in a sequence: forward, backward, both, and target. As the PCR primer is single stranded and thus can bind to only one strand, we can mask nucleotides strand specifically, depending whether we need primers for upper or lower (forward or backward) strand. It can be thought as masking repeat's 3′-end only and leaving some nucleotides from its 5′-end unmasked. Which end of repeat is 3′-end depends on whether we design primers for upper or lower strand (*see* **Fig. 1**). The third masking type "both" mask repeats using both strands. The most useful option for primer design is "target." This type masks upper strand in front of a target region and lower strand behind the target region. The target in this case is the region that should be amplified with PRIMER3. START and END define start and end positions of the target region in a given input sequence.

Parameter "-nbases" defines the number of bases from 3′-end of the primer candidate that *gmasker* should mask (default 1). The value of this parameter can vary from 1 to 16.

Note 3: Changing Parameters in GM_PRIMER3

Most of the *GM_PRIMER3* default options are tuned to suit the situation where the user wants only good primers for very similar PCR conditions and would rather discard regions with no good primers than work with suspicious primer candidates. However, when the user is in a situation, where primers must be designed for very narrow region, parameter relaxation is inevitable. The user should relax the constraints that are least important in particular case and that are most likely preventing primers from being acceptable. Here is the list of options we have commonly used to relax in our practice:

```
PRIMER_PRODUCT_SIZE_RANGE to 1000
PRIMER_MAX_SIZE to 30
PRIMER_MIN_TM to 54
PRIMER_MAX_GC to 80
PRIMER_MAX_POLY_X to 5
PRIMER_MAX_DIFF_TM to 5
```

Note 4: Input Parameters and Files for gindexer

The first required input file—"inputfile"—is a FASTA file with one single sequence (e.g., the assembled chromosome sequence). Multiple FASTA files (e.g., multiple contig sequences) are not supported at the moment. One possible workaround is that we save each sequence to a separate file and create indexes for all these files. In any case, if we have to index more than one file, it would be practical to write a shell script that does it for each file.

The second parameter "outputfile" is the prefix for the output file. We can add directory names in front of the output file name. For each input file, four different binary index files will be created—all words starting with A, C, T, and G nucleotide. For example, for 24 human chromosomes, *gindexer* creates a total number of 96 files.

The parameter called "-wordsize" defines the length of the words that *gindexer* is using for creating indexes. Default word length for the indexes is 16, but it can be changed within the range between 8 and 16 nucleotides.

Note 5: Input Parameters and Files for gtester

The first input file—"primerfile"—is a tab-delimited text file containing the following data columns: primer id, left PCR primer sequence, and right PCR primer sequence. Additional columns will be ignored.

The second input file is "locationsfile." This is a text file containing file names that were indexed with *gindexer*. Each file name should be on separate line. An example of "locationsfile":

```
/home/db/human_DNA/genometester/1
/home/db/human_DNA/genometester/2
/home/db/human_DNA/genometester/3
/home/db/human_DNA/genometester/4
...
/home/db/human_DNA/genometester/Y
```

Note that file names do not include extensions; *gtester* requires only chromosomal or genomic file names and adds extensions (a.location, t.location, c.location, and g.location) for each file by itself.

The third input file—"blacklistfile"—is a pre-processed index file created by *glistmaker*. This is an optional parameter that might speed up *gtester*. If this file name is defined, the *gtester* will not test and record locations of the words that are already listed in the blacklist. Please note that this index can only be used if the word length in *gindexer* location indexes is the same as in *glistmaker* blacklist.

The user can define the maximum product length of two primers that *gtester* is searching with a parameter called "-maxprodlen" (default 1000 bp).

Additionally, the user can define the maximum number of primer binding sites that *gtester* tracks (default 1000)—"-limit". For shorter word lengths (less than 16 bp), the user should use larger cutoff value for the maximum number of binding sites.

Acknowledgments

The development of GenomeMasker package was supported by the Estonian Ministry of Education and Research grant 0182649s04, grant 6041 from Estonian Science Foundation, and grant EU19730 from Enterprise Estonia. The authors thank Katre Palm for a valuable help with English grammar.

References

1. Dawson, E., Abecasis, G.R., Bumpstead, S., Chen, Y., Hunt, S., Beare, D.M., Pabial, J., Dibling, T., Tinsley, E., Kirby, S. et al. (2002) A first-generation linkage disequilibrium map of human chromosome 22. *Nature*, **418**, 544–548.
2. Rouchka, E.C., Khalyfa, A. and Cooper, N.G. (2005) MPrime: efficient large scale multiple primer and oligonucleotide design for customized gene microarrays. *BMC Bioinformatics*, **6**, 175.
3. Xu, D., Li, G., Wu, L., Zhou, J. and Xu, Y. (2002) PRIMEGENS: robust and efficient design of gene-specific probes for microarray analysis. *Bioinformatics*, **18**, 1432–1437.

4. Weckx, S., De Rijk, P., Van Broeckhoven, C. and Del-Favero, J. (2005) SNPbox: a modular software package for large-scale primer design. *Bioinformatics*, **21**, 385–387.
5. Zhang, Z., Schwartz, S., Wagner, L. and Miller, W. (2000) A greedy algorithm for aligning DNA sequences. *J Comput Biol*, **7**, 203–214.
6. Ning, Z., Cox, A.J. and Mullikin, J.C. (2001) SSAHA: a fast search method for large DNA databases. *Genome Res*, **11**, 1725–1729.
7. Kent, W.J. (2002) BLAT–the BLAST-like alignment tool. *Genome Res*, **12**, 656–664.
8. Benson, G. (1999) Tandem repeats finder: a program to analyze DNA sequences. *Nucleic Acids Res*, **27**, 573–580.
9. Altschul, S.F., Madden, T.L., Schaffer, A.A., Zhang, J., Zhang, Z., Miller, W. and Lipman, D.J. (1997) Gapped BLAST and PSI-BLAST: a new generation of protein database search programs. *Nucleic Acids Res*, **25**, 3389–3402.
10. Rouillard, J.M., Zuker, M. and Gulari, E. (2003) OligoArray 2.0: design of oligonucleotide probes for DNA microarrays using a thermodynamic approach. *Nucleic Acids Res*, **31**, 3057–3062.
11. van Hijum, S.A., de Jong, A., Buist, G., Kok, J. and Kuipers, O.P. (2003) UniFrag and GenomePrimer: selection of primers for genome-wide production of unique amplicons. *Bioinformatics*, **19**, 1580–1582.
12. Rozen, S. and Skaletsky, H. (2000) Primer3 on the WWW for general users and for biologist programmers. *Methods Mol Biol*, **132**, 365–386.
13. Breslauer, K.J., Frank, R., Blocker, H. and Marky, L.A. (1986) Predicting DNA duplex stability from the base sequence. *Proc Natl Acad Sci USA*, **83**, 3746–3750.
14. SantaLucia, J., Jr. (1998) A unified view of polymer, dumbbell, and oligonucleotide DNA nearest-neighbor thermodynamics. *Proc Natl Acad Sci USA*, **95**, 1460–1465.
15. Owczarzy, R., You, Y., Moreira, B.G., Manthey, J.A., Huang, L., Behlke, M.A. and Walder, J.A. (2004) Effects of sodium ions on DNA duplex oligomers: improved predictions of melting temperatures. *Biochemistry*, **43**, 3537–3554.
16. von Ahsen, N., Wittwer, C.T. and Schutz, E. (2001) Oligonucleotide melting temperatures under PCR conditions: nearest-neighbor corrections for Mg(2+), deoxynucleotide triphosphate, and dimethyl sulfoxide concentrations with comparison to alternative empirical formulas. *Clin Chem*, **47**, 1956–1961.
17. Kaplinski, L., Andreson, R., Puurand, T. and Remm, M. (2005) MultiPLX: automatic grouping and evaluation of PCR primers. *Bioinformatics*, **21**, 1701–1702.

B

Multiplex PCR Primer Design

11

Degenerate Primer Design

Theoretical Analysis and the HYDEN Program

Chaim Linhart and Ron Shamir

Summary

A polymerase chain reaction (PCR) primer sequence is called *degenerate* if some of its positions have several possible bases. The *degeneracy* of the primer is the number of unique sequence combinations it contains. We study the problem of designing a pair of primers with prescribed degeneracy that match a maximum number of given input sequences. Such problems occur, for example, when studying a family of genes that is known only in part or is known in a related species. We discuss the complexity of several versions of the problem and give approximation algorithms for one simplified variant. On the basis of these algorithms, we developed a program called HYDEN for designing highly degenerate primers for a set of genomic sequences. We describe HYDEN, and report on its success in several applications for identifying olfactory receptor genes in mammals.

Key Words: Degenerate primers for PCR; DPD; HYDEN; olfactory receptor genes.

1. Introduction

A *degenerate* polymerase chain reaction (PCR) primer is a primer sequence that contains several possible bases in one or more positions *(1)*. For example, in the primer GG{C,G}A{C,G,T}A, the third position is C or G, and the fifth is C, G, or T. The *degeneracy* of the primer is the total number of sequence combinations it contains. For example, the degeneracy of the above primer is 6. Degenerate primers are as easy and cheap to produce as regular unique primers, are useful for amplifying several related genomic sequences, and have been used in various applications. Most extant applications use low degeneracy

From: *Methods in Molecular Biology, vol. 402: PCR Primer Design*
Edited by: A. Yuryev © Humana Press, Totowa, NJ

of up to hundreds. In this chapter, we study the problem of designing primers of high degeneracy from the theoretical and practical perspectives.

Suppose one has a collection of related target sequences, for example, DNA sequences of homologous genes, and the goal is to design primers that will match as many of them as possible, as well as perhaps additional related sequences that are unknown yet. A naïve solution would be to align the sequences without gaps, count the number of different nucleotides in each position along the alignment, and seek a primer-length window (typically 20–30) where the product of the counts is low. Such solution is insufficient because of gaps, the inappropriate objective function of the alignment, and, most notably, the exceedingly high degeneracy; when degeneracy is too high, unrelated sequences may be amplified as well, and specificity will decrease. We may have to compromise by aiming to match many but not necessarily all the sequences. We describe here an ad hoc method for designing primers that will allow tradeoff between the degeneracy and the coverage (the number of matched input sequences). We call this problem degenerate primer design (DPD). In the next sections, we define and analyze several variants of DPD and describe a program we developed, called HYDEN, for producing high-degeneracy primers. Finally, we report results of several projects that used degenerate primers for amplifying olfactory receptor (OR) genes in various mammals. The theoretical results have been described in detail in **ref.** *2* and are reprinted with permission. The experimental results are based on **ref.** *3*.

1.1. Related Problems

DPD is related to the primer selection problem (PSP) *(4)*, in which the goal is to minimize the number of (non-degenerate) primers required to amplify a set of DNA sequences. Several algorithms have been developed to solve this problem, and some take into account various biological considerations and technical constraints *(5)*. However, for large gene families, the number of primers needed to cover a sufficient portion of the genes without losing specificity is rather large. Furthermore, as the primers are not degenerate, they do not amplify many of the unknown related genes. Also, in contrast to a single pair of degenerate primers for DPD, here the cost of generating the primers depends on the set size.

As a degenerate primer can be viewed as a motif, DPD is also related to motif finding. However, there are marked differences: Motif finding algorithms (e.g., MEME *(6)*, Gibbs Sampler *(7)*) usually produce a profile matrix or a hidden markov model (HMM), with no constraint on the maximum degeneracy. Some combinatorial motif finding algorithms do use consensus with degenerate positions, but their goal is to find a "surprising" motif, that is, a pattern that is unlikely given the background sequence probabilities. In DPD, on the contrary the "surprise" in a primer is irrelevant, and we care about degeneracy and coverage instead.

2. Theoretical Analysis

In this section, we formally define several versions of DPD, report on hardness results, and briefly describe approximation algorithms for one key variant of the problem. The full proofs are given in **ref.** *2* and *8*. For basic background on algorithms and complexity, we refer the reader to **ref.** *9*.

2.1. Problem Definition

Given a set of DNA sequences, our goal is to design a pair of degenerate primers, so that the primers match and amplify (in the PCR sense) as many of the input sequences as possible. To obtain primers that match a large number of genes, one should obviously use highly degenerate primers. On the contrary, to reduce the chance of amplifying non-related sequences, the degeneracy must be bounded.

The following notation will help us formally define the problems. Let Σ denote a finite fixed alphabet. In the case of DNA sequences, $\Sigma = \{A, C, G, T\}$. A *degenerate string*, or *primer*, is a string P with several possible characters at each position, that is, $P = p_1 p_2 \ldots p_k$, where $p_i \subseteq \Sigma,\ p_i \neq \emptyset$. k is the *length* of the primer. The number of possible character sets at a single position is $\sigma = 2^{|\Sigma|} - 1$. The *degeneracy* of P is $d(P) = \prod_{i=1}^{k} |p_i|$. For example, the primer $P^* = \{A\}\{C, G\}\{A, C, G, T\}\{G\}\{T\}$ is of length 5 and degeneracy 8. At *non-degenerate positions*, that is, positions that contain a single character, we shall often omit the brackets. We will sometimes use an asterisk to denote a fully degenerate position, that is, a position that includes all possible characters. Hence, $P^* = A\{C, G\} * GT$. An alternative way to describe a primer is using the IUPAC nucleotide code: $P^* =$ ASNGT. Let $\delta(P)$ be the number of degenerate positions in P. Clearly, $\lceil \log_{|\Sigma|} d(P) \rceil \leq \delta(P) \leq \lfloor \log_2 d(P) \rfloor$. A primer $P^1 = p_1^1 p_2^1 \ldots p_k^1$ is a *sub-primer* of a primer $P^2 = p_1^2 p_2^2 \ldots p_k^2$ of the same length, if $\forall i,\ 1 \leq i \leq k,\ p_i^1 \subseteq p_i^2$. The *union* of the primers P^1 and P^2, denoted by $P^1 \cup P^2$, is P^{12} where $p_i^{12} = p_i^1 \cup p_i^2$. A primer $P = p_1 p_2 \ldots p_k$ *matches* a string $S = s_1 s_2 \ldots s_l,\ s_i \in \Sigma$, if S contains a substring that can be extracted from P by selecting a single character at each position, that is, $\exists j,\ 0 \leq j \leq l-k$ s.t. $\forall i,\ 1 \leq i \leq k,\ s_{j+i} \in p_i$. For example, the primer P^* matches the string TGAGAGTC starting from the third position. A *mismatch* is a position i at which $s_{j+i} \notin p_i$. In actual PCR, a few mismatches usually do not prevent hybridization. Unless stated otherwise, we will not allow mismatches. We are now ready to define several problem variants.

Problem 1 *DPD: Given a set of n strings and integers k, d, and m, is there a primer of length k and degeneracy at most d that matches at least m input strings?*

We defined DPD as a decision problem rather than an optimization problem. Ideally, one wishes to optimize each of the parameters k, m, and d. As the value of k is usually predetermined by biological or technical constraints (e.g., in PCR experiments, k is usually between 20 and 30), we shall focus on optimizing either m, the *coverage* of the primer, or d, the primer's degeneracy. As we will explain later on, these two optimization problems remain difficult to solve even if simplified further. Specifically, when designing a primer that matches as many strings as possible, we shall assume that all input strings are of the same length as the primer. When minimizing the degeneracy of the primer, on the contrary, we will seek a full coverage of the input strings.

Problem 2 *Maximum Coverage DPD (MC-DPD): Given a set of strings of length k and an integer d, find a primer of length k and degeneracy at most d that matches a maximum number of input strings.*

Problem 3 *Minimum Degeneracy DPD (MD-DPD): Given a set of strings and an integer k, find a primer of length k and minimum degeneracy that matches all the input strings.*

We shall now define several generalizations of MC-DPD and MD-DPD. As mentioned earlier, a gene is usually amplified even if there are a few mismatches between the primer and the gene. In fact, mismatches near the 3′ extension site, that is close to the part of the gene that undergoes amplification, are typically more disruptive than internal mismatches ***(1)***. The following problem takes into account errors (mismatches) between the primer and the strings but ignores their position (i.e., we assume that all mismatches are equally disruptive).

Problem 4 *Minimum Degeneracy DPD with Errors (MD-EDPD): Given a set of n strings and integers k and e, find a primer of length k and minimum degeneracy that matches all the input strings with up to e errors (mismatches) per string.*

Under many circumstances, a single primer might not suffice, that is, provide satisfactory coverage, because of its limited degeneracy and the divergence of the input strings. A natural question is whether one could design several primers that, together, would match all the strings.

Problem 5 *Minimum Primers DPD (MP-DPD): Given a set of n strings of length k and an integer d, find a minimum number of primers of length k and degeneracy at most d, so that each input string is matched by at least one primer.*

In MP-DPD, we assume that all the input strings are of the same length as the primers. If we remove this constraint, that is, allow the strings to have arbitrary length, we get a more general problem. This variant of DPD, called Multiple DPD (MDPD), is studied in **ref.** ***10***.

The real problem of designing degenerate primers requires the construction of one or more *pairs* of primers, so that each of the given genes matches at least one of the primer pairs with only a few mismatches. For an effective PCR, we should require that the distance between the 5′- and the 3′-primer match site is large enough (i.e., the amplified region is sufficiently long for biological study). Other factors that influence PCR may also be incorporated, such as the positions of the mismatches and the GC content *(1)*. Our theoretical results focus on the simple, restricted DPD variants. As we shall now see, even those are hard.

2.2. Complexity Results

Using exhaustive search algorithms, it is possible to solve restricted cases of DPD in polynomial time. For example, if $d = O(1)^*$, we could consider all $< L$ substrings, where L is the sum of the lengths of the input strings, and continue in one of two ways. First, we could try to increase the degeneracy of each candidate substring by adding new characters at various positions. There are no more than $\delta = \lfloor \log_2 d \rfloor$ degenerate positions in a primer whose degeneracy is d or less, as each such position at least doubles the total degeneracy. At each degenerate position, we could try all σ possible character sets. Thus, there are a total of less than $L \binom{k}{\delta} \sigma^{\delta}$ degenerate primers to check, and the total running time is $O(kL^2 \binom{k}{\delta} \sigma^{\delta})$. We shall later introduce an efficient approximation algorithm that is a variant of this exhaustive search.

*(See **ref.** ***9*** for the definition of the Oh notation)

Second, we could take each suppose P^1 is a substring of the input string S^1. P^1 can be viewed as a non-degenerate primer that matches S^1. Let S^2 be an input string that P^1 does not match, and let P^2 be a substring of S^2. Obviously, $P^1 \neq P^2$. Let $P^{12} = P^1 \cup P^2$. P^{12} is a degenerate primer that matches both S^1 and S^2, and its degeneracy is larger than that of P^1 and P^2, because it strictly contains them. Now, P^{12} can be expanded using a third primer, P^3, which is a substring of an input string that is not matched by P^{12}, and so on. We continue to expand the primer as long as its degeneracy does not exceed d. In each step, we consider all substrings of the yet unmatched input strings and add (in terms of the union operation) each substring to the primer, in its turn. As the degeneracy of the primer increases in each step by at least 1 (more accurately, by a factor of at least $|\Sigma|/(|\Sigma|-1)$, the number of steps is no more than d. Therefore, the running time of the algorithm is $O(kLL^d)$. Theorem 6 summarizes restricted cases of DPD that can be solved in polynomial time *(2)*.

Theorem 6 *DPD is polynomial when* $d = O(1)$, *or* $m = O(1)$, *or* $k = O(\log L)$.

Unfortunately, all the versions of DPD we defined are, in the general case, difficult problems *(8)*.

Theorem 7 *The following problems are NP-Complete: MC-DPD (for* $|\Sigma| \geq 2$*), MD-DPD (for* $|\Sigma| \geq 3$*), MD-EDPD (for* $|\Sigma| \geq 2$*, even if* $e = 1$ *and all input strings are of length k), and MP-DPD (for* $|\Sigma| \geq 2$*).*

Furthermore, in MD-DPD and MD-EDPD, it is difficult to approximate the *number* of degenerate positions in an optimal primer *(8)*.

Theorem 8 *Assuming* $P \neq NP$*, there is no polynomial time algorithm that approximates the number of degenerate positions in: (a) MD-DPD, within a factor of* $c \cdot \log n$*, for some constant* $c > 0$*; (b) MD-EDPD, within a factor of 1.36, even when* $e = 1$ *and all strings are of length k.*

3. Approximation Algorithms

In this section, we describe several polynomial approximation algorithms for MC-DPD over the binary alphabet—$\Sigma = \{0, 1\}$. In this case, the number of degenerate positions in a primer is always $\delta(P) = \log_2 d(P)$.

3.1. Simple Approximations

Denote by $M(P)$ the set of input strings matched by a primer P. Let P^o be an optimal solution with degeneracy d to an instance of MC-DPD. Like any other primer with degeneracy d, P^o is a union of d non-degenerate primers (strings of length k): $P^o = \bigcup_{i=1}^{d} P^i$, where $P^1, \ldots, P^d$ constitute *all* the non-degenerate sub-primers of P^o, and $M(P^o) = \bigcup_{i=1}^{d} M(P^i)$. Let P^m be a sub-primer with the largest coverage, that is, $|M(P^m)| = \max_{i=1}^{d} \{|M(P^i)|\}$. Then, obviously, $|M(P^o)| \leq d \cdot |M(P^m)|$. It is now clear how one can obtain a d-approximation to P. Simply traverse all k-long substrings of the input strings and choose a substring P_0 that matches a maximum number of input strings. As $|M(P^m)| \leq |M(P_0)|$, we get $|M(P_0)| \geq |M(P^o)|/d$. The algorithm runs in time $O(kL^2)(= O(k^3n^2)$, because in MC-DPD, $L = nk$). The running time can be reduced to $O(kL)$ using a hash table to store the number of strings matched by each substring. Notice that the output of the above algorithm is an optimal non-degenerate primer P_0, and its approximation ratio is d.

We now describe another algorithm, which starts with a completely degenerate primer, and gradually refines, or "contracts," it. Let P^k be a completely degenerate primer of length k and degeneracy 2^k. P^k covers all the input strings: $|M(P^k)| = n$. We shall now reduce the degeneracy of P^k to d, by replacing $k - \delta$ ($\delta = \log_2 d$) degenerate positions with simple characters. Denote by $P_i^k (i \in \{0, 1\})$ the primer that begins with the character i, followed by $k - 1$ degeneracies. For example, if $k = 3$, then $P_0^k = 0**$ and $P_1^k = 1**$. Clearly, $M(P^k) = M(P_0^k) \cup M(P_1^k)$; hence, by choosing either P_0^k or P_1^k, we get a primer whose coverage is at least $n/2$. Similarly, we can de-degenerate, or *refine*, the second position in the primer, that is, replace it with "0" or "1," whichever is better, and obtain a primer with degeneracy 2^{k-2} that matches at least $n/4$ input strings, and so on. After $k - \delta$ steps, we have a primer with the required degeneracy d, whose coverage is at least $n/2^{k-\delta}$ and therefore at least $m_o/2^{k-\delta}$. The total running time of the algorithm is $O((k - \delta)n)$, as it suffices to examine the first $(k - \delta)$ characters in each input string.

Combining the two approximation algorithms we have just described, we can approximate MC-DPD within a factor of $2^{k/2}$: if $\delta < \frac{k}{2}$, we run the first algorithm; otherwise, we execute the second algorithm. In summary, we have the following proposition.

Proposition 9 *MC-DPD can be approximated within a factor of* $2^{k/2}$ *in time O(kL).*

3.2. Approximating the Number of Unmatched Strings

Unlike the previous algorithms we studied, we shall now describe several algorithms that approximate the number of *unmatched* strings. In other words, we now treat MC-DPD as a minimization problem, designated MC-DPD*, in which the goal is to minimize the number of input strings that the primer does not match. This does not alter the optimization problem, only the way in which we measure the quality of the approximation. We say that an algorithm approximates MC-DPD* within ratio $r\,(r > 1)$ if the number of strings not covered by the primer it designs is no more than ru_o, where u_o is the optimal solution value.

The first two algorithms construct the *column distribution matrix D(b,i)* that holds the number of appearances, or *count*, of each character at each position. Formally, denote by $S^j = s_1^j s_2^j \ldots s_k^j$ the jth input string, $1 \leq j \leq n$, then $\forall\ b \in \Sigma,\ 1 \leq i \leq k,\ D(b, i) = |\{j | s_i^j = b\}|$. Let $P^o = p_1^o p_2^o \ldots p_k^o$ be an optimal primer of degeneracy d, with $\delta = \log_2 d$ degenerate positions. Suppose P^o covers m_o input strings, that is, $u_o = n - m_o$. Clearly, $\forall b \notin p_i^o,\ D(b, i) \leq u_o$, and for each non-degenerate position i in P^o, $D(p_i^o,\ i) \geq m_o$. As P^o contains $k - \delta$ non-degenerate positions, it follows that there are $k - \delta$ (or more) columns in D with a value at least m_o. Given a column distribution matrix D, we define the *leading value* of column i, denoted $v(i)$, as the largest value in that column $v(i) = \max\{D(b, i) | b \in \Sigma\}$. Similarly, the *leading character* of column i is a character $c(i)$, whose count is the leading value $D(c(i), i) = v(i)$. Let $v(i_1) \geq v(i_2) \geq \ldots \geq v(i_k)$ be the leading values in D, sorted from largest to smallest. The following lemma follows from the discussion above.

Lemma 10 *If P^o covers m_o strings, then $v(i_{k-\delta}) \geq m_o$.*

The CONTRACTION Algorithm

The CONTRACTION algorithm selects the $k - \delta$ largest leading values in D and sets the output primer P^c to contain the $k - \delta$ corresponding leading characters and degeneracies at the rest of the positions, that is

$$\forall 1 \leq i \leq k,\ p_i^c = \begin{cases} c(i) & i \in \{i_1, \ldots, i_{k-\delta}\} \\ \{0, 1\} & \textit{otherwise} \end{cases}$$

An alternative way to describe CONTRACTION is as follows. The algorithm starts with a fully degenerate primer and contracts it iteratively. In each iteration, the algorithm discards the character with the smallest count. In other words, it examines all the remaining degenerate positions, chooses a position i that contains a character b, whose count $D(b, i)$ is the smallest, and removes b from position i in the primer. The algorithm stops once the degeneracy of the primer reaches d. In a sense, this is a smart variation of the simple $2^{k-\delta}$ approximation algorithm we saw earlier—CONTRACTION uses the column distribution matrix to guide it in selecting good positions to refine, instead of choosing them arbitrarily. **Figure 1** illustrates an execution of CONTRACTION.

The running time of CONTRACTION is linear in the length of the input, $O(nk)$, because this is the time it takes to compute the column distribution matrix D, and the $k - \delta$ largest leading values can be found in time $O(k)$ ***(11)***. At each degenerate position, the primer P^c has no mismatches with the input strings. According to Lemma 10, at each non-degenerate position, P^c has a mismatch with at most u_o input strings. The total number of strings P^c does not match cannot exceed the sum of the number of mismatches at each position, which is bounded by $(k - \delta)u_o$. In conclusion, we have the following theorem.

Theorem 11 *CONTRACTION approximates MC-DPD* within a factor of $(k - \delta)$ in time $O(nk)$.*

The EXPANSION Algorithm

The second algorithm, called EXPANSION, performs n iterations. In each iteration, it expands (degenerates) an input string. In the jth iteration, EXPANSION computes the matrix D'_j:

$$\forall b \in \{0, 1\},\ 1 \le i \le k,\ D'_j(b, i) = \begin{cases} 0 & s_i^j = b \\ d(b, i) & otherwise \end{cases}$$

Intuitively, $D'_j(b, i)$ is the number of strings that will be mismatched because of setting the ith position in the primer to s_i^j while their ith position is b. EXPANSION then selects the δ largest leading values in D'_j: $v'_j(i_1), \ldots, v'_j(i_\delta)$ and uses them to expand S^j and create a primer $P^j = p_1^j \ldots p_k^j$, as follows:

Input: $n = 8$, $k = 9$, $d = 2^4$

S^1: 011010101
S^2: 010010000
S^3: 111010100
S^4: 011111001
S^5: 111010101
S^6: 001111100
S^7: 101011110
S^8: 111010001

⟹

Column distribution matrix D:

4 2 1 **6** 0 5 3 **7** 4
4 **6** **7** 2 **8** 3 5 1 4

⇓

Output:

P^c: * 1 1 0 1 * * 0 *

Fig. 1. Example of an execution of CONTRACTION on eight strings. The five $(k - \delta)$ largest leading values in D are marked in bold face. The primer P^c covers four input strings — S^1, S^3, S^5, and S^8.

$$\forall 1 \leq i \leq k,\ p_i^j = \begin{cases} \{0,1\} & i \in \{i_1, \ldots, i_\delta\} \\ s_i^j & otherwise \end{cases}$$

The output of the algorithm, P^c, is the best primer P^j it found in the n iterations.

Denote by m_c and m_e the number of strings covered by the primers P^c and P^e, respectively. It is possible to show that $m_e \geq m_c$ *(2)*, which implies that EXPANSION also guarantees a $(k - \delta)$ approximation to MC-DPD*. In fact, in some cases, EXPANSION may find a better primer than CONTRACTION, as demonstrated in **Fig. 2**. On the down side, EXPANSION is slower—its running time is $O(n^2k)$, dominated by the coverage computation of the n primers it constructs.

Corollary 12 *EXPANSION approximates MC-DPD* within a factor of* $(k - \delta)$ *in time* $O(n^2k)$.

Input: $n = 8$, $k = 9$, $d = 2^4$

S^1: 011010101		**Column distribution matrix D:**
S^2: 010010000	⟹	4 2 1 6 0 5 3 7 4
S^3: 111010100		4 6 7 2 8 3 5 1 4
...(as in Figure 1)		⇓
Starting string: S^1	⇒	D': 0 2 1 0 0 0 3 0 4
		4 0 0 2 0 3 0 1 0
		⇓
		P^1: * 1 1 0 1 * * 0 *
Starting string: S^2	⇒	D': 0 2 0 0 0 0 0 0 0
		4 0 7 2 0 3 5 1 4
		⇓
		P^2: * 1 * 0 1 0 * 0 *

Fig. 2. Illustration of the first two iterations of EXPANSION on the eight strings from Fig. 1. The four (δ) largest leading values in D' are marked in bold face. The expansion of S^1 (P^1) covers four strings and is identical to the primer constructed by CONTRACTION. The expansion of S^2 (P^2) covers five input strings — S^1, S^2, S^3, S^5, and S^8.

The CONTRACTION-X Algorithm

We now present an improved version of CONTRACTION, called CONTRACTION-X, that yields better approximations at the expense of longer running times. A similar improvement could be developed for the EXPANSION algorithm, as well. The main idea we employ is to examine several positions simultaneously and decide which are best to refine (i.e., de-degenerate), instead of checking the distribution at each position separately. Formally, let x be a pre-defined integer, $1 \leq x \leq k-\delta$. For simplicity, assume $x|(k-\delta)$. Denote by $\bar{b}=(b_1,\ldots,b_x)$ a binary vector of length x, or x-tuple, and denote by $\bar{i}=(i_1,\ldots,i_x)$, $1 \leq i_j \leq k$, a set of x distinct positions. Define the *multi-column distribution matrix* $MD(\bar{b},\bar{i})$ as the count of the x bits of $\bar{b}$ at positions $i_1,\ldots,i_x$ in the input strings, that is

$$MD((b_1,\ldots,b_x),(i_1,\ldots,i_x)) = |\{j|s_{i_1} = b_1,\ldots,s^j_{i_x} = b_x\}|$$

Let P^o be an optimal primer and denote by u_o the number of input strings it does not match. CONTRACTION-X starts with a completely degenerate primer, $P^x = p^x_1 \ldots p^x_k$, $p^x_j = \{0, 1\}$, and iteratively refines it. In the first iteration, it selects an x-tuple with the largest count and sets the x corresponding positions in the primer to contain the bits of the x-tuple. In other words, if $MD(\bar{b}',\bar{i}') = \max\{MD(\bar{b},\bar{i})\}$, then $\forall 1 \leq j \leq x$, $p^x_{i'_j} = b'_j$. In the next iteration, CONTRACTION-X continues to refine P^x in a similar fashion. It examines all x-tuples in positions that are still degenerate, that is, that were not refined in the first iteration, selects an x-tuple with the largest count, and sets the corresponding positions in P^x accordingly. The algorithm performs $(k-\delta)/x$ iterations, as above, and reports the obtained primer P^x. Since in each iteration, it refines x new positions, the output primer contains exactly δ degeneracies as required. If $x \nmid (k-\delta)$ and denote $r = (k-\delta)$ *mod* x, then CONTRACTION-X performs $[(k-\delta)/x]$ iterations as above, and an additional iteration, in which it refines only r positions, that is, it computes the count of every r-tuple at each subset of r positions that are still degenerate, selects the largest one, and refines those positions accordingly. The performance of CONTRACTION-X is summarized in the following theorem *(2)*.

Theorem 13 *CONTRACTION-X approximates MC-DPD* within a factor of* $[(k-\delta)/x]$ *in time* $O(\binom{k}{x} n(k-\delta))$ *and space* $O(\binom{k}{x} nx)$.

Notice that for $x = 1$, CONTRACTION-X is identical to CONTRACTION. In the other extreme case, when $x = k-\delta$, CONTRACTION-X effectively considers all k-long primers with δ degeneracies, and it therefore always yields an optimal primer. The multi-column distribution matrix is also utilized in Multiprofiler, a motif finding algorithm that has recently been reported to detect particularly subtle motifs ***(12)***.

3.3. Non-Binary Alphabets

So far, we have discussed several approximation algorithms for MC-DPD when $|\Sigma| = 2$. However, in many real-life applications, the alphabet is not binary, as is the case when designing primers for genomic sequences ($|\Sigma| = 4$). The simple approximations described in **Subheading 3.1** are easily generalized to larger alphabets, as follows. Let P^o be an optimal primer of length k and degeneracy d for a given set of n strings over Σ. Let m_o be the coverage of P^o. The primer P^o is a union of d non-degenerate primers, and the number of strings covered by P_o is at most the sum of the coverage of these non-degenerate primers. Hence, an optimal non-degenerate primer, which is simply a k-long substring that appears in the largest number of input strings, covers at least m_o/d strings.

As in the binary case, we can also devise a simple contraction algorithm for non-binary alphabets. For convenience, denote $\alpha = |\Sigma|$, and $\delta' = \lfloor \log_\alpha d \rfloor$. A completely degenerate primer of length k has degeneracy α^k and coverage n. By replacing the first degeneracy in the primer with a simple character (one that gives the largest coverage), we get a primer with degeneracy α^{k-1} that covers at least n/α strings. We similarly refine positions $2, \ldots, k - \delta'$ and obtain a primer with degeneracy at most d and whose coverage is at least $n/\alpha^{k-\delta'}$ and therefore at least $m_o/\alpha^{k-\delta'}$.

Both algorithms we have just outlined run in time $O(kL)$. Combining them, we get a $|\Sigma|^{\lceil k/2 \rceil}$ approximation algorithm for MC-DPD: if $d \geq |\Sigma|^{\lceil k/2 \rceil}$, then $\alpha^{k-\delta'} \leq |\Sigma|^{\lceil k/2 \rceil}$, so we run the second algorithm; otherwise, we run the first algorithm (compare to Proposition 9).

Proposition 14 *When $|\Sigma| > 2$, MC-DPD can be approximated within a factor of $|\Sigma|^{\lceil k/2 \rceil}$ in time O(kL).*

Unfortunately, the results we obtained in **Subheading 3.2** for the CONTRACTION and EXPANSION algorithms do not hold for non-binary alphabets. There are two complications in large alphabets. First, there is more than one possibility for a degenerate position. When $|\Sigma| = 2$, every degenerate position in the primer is $\{0, 1\}$, whereas when $|\Sigma| > 2$, we need to choose one among several possible degeneracies (subsets of Σ with more than one character) at each degenerate position. Second, there is the additional complexity in deciding how to partition the degeneracy among the positions. In the binary case, the degeneracy is always of the form 2^{δ}, where δ is the number of degenerate positions. However, when $|\Sigma| > 2$, the number of degenerate positions could be any one of many values. For example, if $d = 16$ and $|\Sigma| = 4$, there may be four degenerate positions (each one with degeneracy 2), three (4,2,2), or only two (4,4). In **Subheading 4** we describe heuristics for MC-DPD with non-binary alphabets that are based on CONTRACTION and EXPANSION and perform well in practice.

4. The HYDEN Program

We developed and implemented an efficient heuristic, called HYDEN ***(13)***, for designing highly degenerate primers. The input to HYDEN is a list of DNA sequences and a set of integers that control the number of primer pairs to design, the length and maximum degeneracy of the primers, and other parameters of the algorithm. HYDEN constructs primer pairs with the specified length and degeneracy that together cover many of the given sequences. Each primer is designed by running a 3-phase algorithm, outlined in **Fig. 3**. In the first phase, HYDEN locates conserved regions in the DNA sequences by finding ungapped local alignments with a low entropy score. In the second phase, it designs primers using variants of the CONTRACTION and EXPANSION algorithms. Finally, it uses a greedy hillclimbing procedure to improve the primers and selects the one with the largest coverage as the output. This procedure is repeated to design several pairs of 5′ and 3′ primers, as explained in **Subheading 4.2**. HYDEN is written in C++ and runs under Windows and Linux. HYDEN is freely available for academic use (http://www.cs.tau.ac.il/~rshamir/hyden).

HYDEN ($I = \{S^1, \ldots, S^n; k; d; e\}$):
Phase 1: $A_1, \ldots, A_{N_a} \leftarrow$ H-Align(I).
Phase 2: Foreach alignment A_i, $i = 1, \ldots, N_a$ do:
    $P_i^c \leftarrow$ H-Contraction(I; A_i).
    $P_i^e \leftarrow$ H-Expansion(I; A_i).
  Sort primers $\{P_i^c, P_i^e \mid i = 1, \ldots, N_a\}$ acc. to coverage.
Phase 3: Foreach primer $P \in$ {best N_g primers} do:
    $P \leftarrow$ H-Greedy(I; P).
Output the primer with the largest coverage found in Phase 3.

Fig. 3. The HYDEN algorithm for designing a single primer.

4.1. The HYDEN Algorithm

Let $I = \{S^1, \ldots, S^n; k; d; e\}$ be the input to HYDEN, where $S^1, \ldots, S^n$ are n strings over $\Sigma = \{A, C, G, T\}$ with a total length of L characters, and k, d, and e are the length, degeneracy, and mismatches parameters, respectively. Let N_a, $N_{a'}$, N_g, and N_h be additional integer parameters, whose roles will be explained soon. Denote by A an ungapped local alignment (alignment, in short) of the input strings, that is, a set of n substrings of length k (actually, A is a multi-set, because it may contain several copies of a substring). Denote by D_A the column distribution matrix of the substrings in A. To determine how well-conserved the alignment is, and thereby estimate how likely we are to construct a good primer from it, we compute its entropy score, H_A.

$$H_A = -\sum_{i=1}^{k}\sum_{b\in\Sigma} \frac{D_A(b,i)}{n} \cdot \log_2 \frac{D_A(b,i)}{n}$$

The lower the entropy score is, the less variable are the columns of A, and, intuitively, the greater the chances are for finding a primer that covers many of the substrings in A. The first phase of HYDEN, called H-ALIGN, exhaustively enumerates all substrings of length k in the input strings and generates an alignment for each one, as follows (*see* **Fig. 4**). Let $T = t_1t_2 \ldots t_k$ be a substring of length k. In each input string S^j, H-ALIGN finds the best match to T in terms of Hamming distance, that is, the k-long substring T^j of S^j that has the smallest number of mismatched characters with T. The substrings $T^1, \ldots, T^n$ (one of which is T itself) form the alignment A_T. After considering all $O(L)$ different substrings in the input, H-ALIGN obtains $O(L)$ alignments. The N_a alignments with the lowest entropy score are passed to the second phase. H-ALIGN runs in time $O(kL^2)$. Fortunately, a few simple heuristics, which we describe below, reduce the running time considerably with marginal impact on the quality of the results.

```
H-Align(I):
Foreach k-long substring T of S^1, ..., S^n do:
    A_T ← φ.
    Foreach string S^j, j = 1, ..., n do:
        Add to A_T the best match in S^j to T.
    D_{A_T} ← Column distribution matrix of A_T.
    H_{A_T} ← Entropy score of D_{A_T}.
Output N_a alignments with lowest entropy score.
```

Fig. 4. The basic alignment phase in HYDEN.

Let $A_h \subset A$ be an arbitrary subset of an alignment A, $|A_h| = N_h$. Provided that N_h is not too small, we can use A_h to estimate how well-conserved A is, or, in other words, we may assume that $H_{A_h} \approx H_A$. Thus, a more efficient version of H-ALIGN iterates all k-long substrings and aligns only N_h input strings to each one. Then, the $N_{a'}$ substrings, whose alignments received the lowest (partial) entropy scores, are re-aligned against all n input strings, their full entropy score, H_A, is computed, and the best $N_a(\leq N_{a'})$ alignments are passed to the next stage. If all input strings have approximately the same length, then this efficient version of H-ALIGN runs in time $O[kL((N_h/n)L + N_{a'})]$. Another improvement we applied exploits the fact that alignments obtained from highly overlapping substrings are very similar. Therefore, if the alignment we get from a substring $s_i \ldots s_{i+k-1}$ has a high entropy score, there is no point in checking the next substring $s_{i+1} \ldots s_{i+k}$, as it is highly unlikely to yield good results, too. In fact, if the entropy score is very poor, we may decide to skip more than one substring. In practice, this simple idea reduced the running time of H-ALIGN by another factor of 2–4.

The second phase constructs two primers from each of the N_a alignments. Given an alignment A with a column distribution matrix D_A, HYDEN runs two heuristics—H-CONTRACTION and H-EXPANSION. These algorithms are generalizations of the CONTRACTION and EXPANSION approximation algorithms, respectively, to non-binary alphabets. H-CONTRACTION starts with a fully degenerate primer and discards characters at degenerate positions with the smallest count in D_A until the primer reaches the required degeneracy, as shown in **Fig. 5**. H-EXPANSION employs an opposite approach. It uses the substring $T \in A$, from which A was constructed, as an initial non-degenerate primer, and repeatedly adds to it a character with the largest count as long as its degeneracy does not exceed the threshold d, as detailed in **Fig. 6**. Notice that the original EXPANSION algorithm repeats this procedure for each substring

H-Contraction (I; A):

Sort the counts: $D_A(b_1, i_1) \leq D_A(b_2, i_2) \leq \ldots \leq D_A(b_{4k}, i_{4k})$.
$P \leftarrow$ Fully degenerate primer; $j \leftarrow 1$.
While $d(P) > d$ and $j \leq 4k$ **do:**
  $P' \leftarrow P$ without character b_j at position i_j.
  If $d(P') \neq 0$ **then** $P \leftarrow P'$.
  $j \leftarrow j + 1$.
Output P.

Fig. 5. The H-CONTRACTION algorithm used by HYDEN.

```
H-Expansion (I; A):
Sort the counts: D_A(b_1, i_1) ≥ D_A(b_2, i_2) ≥ ... ≥ D_A(b_4k, i_4k).
Let T be the substring from which A was constructed.
P←T ; j←1.
While j ≤ 4k do:
    P'←P with character b_j added at position i_j.
    If d(P') ≤ d then P←P'.
    j←j + 1.
Output P.
```

Fig. 6. The H-EXPANSION algorithm used by HYDEN.

in A. However, early experiments demonstrated that in many cases, there is little difference between primers obtained by expanding different substrings in A. Therefore, in H-EXPANSION, we chose to expand only one substring from each alignment. Finally, the second phase of HYDEN computes the coverage of the $2N_a$ primers it constructed and selects the $N_g(\leq 2N_a)$ primers that match the largest number of input strings (with up to e mismatches). The running time of the second phase of HYDEN is $O(N_a kL)$.

The final phase of HYDEN tries to improve the N_g primers found in the previous phase using a simple hill-climbing procedure, called H-GREEDY. Given a primer P, H-GREEDY checks whether it can remove a character in a degenerate position in P and add a different character in any position instead, so that the coverage of the primer increases. This process is repeated as long as coverage is improving (*see* **Fig. 7**). Denote by r the number of iterations performed until a local maximum is reached. Then, the running time of H-GREEDY is $O(rk^3L)$. In our experiments, r was almost always below 5. To limit the running time in the general case, one could fix an upper bound $\bar{r}$ on the number of improvement iterations the algorithm performs, thereby setting the total running time of the third phase of HYDEN to $O(N_g\bar{r}k^3L)$.

HYDEN runs in total time of $O(kL((N_h/n)L+N_{a'}+N_g\bar{r}k^2))$. Notice that the input parameters d and e are missing from the formula — the reason is that the performance depends linearly on log d and e, both of which are accounted for in the $O(k)$ factor. HYDEN is sufficiently fast for designing a primer of length $k \leq 30$ for a set of hundreds of DNA sequences, each 1 Kbp long. Moreover, by modifying various parameters, one can control the tradeoff between the running time of the program and the quality of the solution it provides. We report concrete running times and parameters in **Subheading 5**.

```
H-Greedy (I; P):
P* ← P, improved ← "yes".
While improved = "yes" do:
    improved ← "no".
    Foreach degenerate character (b, i) in P do:
        P′ ← P without character b at position i.
        Foreach degeneracy (b′, i′) not in P do:
            P″ ← P′ with character b′ added at position i′.
            m (P″) ← Coverage of P″.
            If d (P″) ≤ d and m(P″) > m(P′) then P* ← P″
    If m(P*) > m(P) then P ← P*, improved ← "yes".
Output P.
```

Fig. 7. The greedy hill-climbing procedure used by HYDEN. *m(P)* denotes the coverage of primer *P*.

HYDEN is a generalization of the $(k - \delta)$ approximation of MC-DPD* that we presented in **Subheading 3.2**. If a set of binary strings of length k is supplied to the program, and $e = 0$, the alignment phase does nothing (the strings are already aligned), the second phase yields the approximation (H-CONTRACTION is identical to CONTRACTION when $|\Sigma| = 2$), and the final greedy phase may further improve the solution. We have no theoretical guarantee on the performance of HYDEN in the general case and, specifically, for genomic sequences of arbitrary length. Nevertheless, as we shall see, the results it produced in practice for the OR subgenome were highly satisfactory.

4.2. Designing Multiple Primer Pairs

The HYDEN program can design several primer pairs. The first pair is constructed by running the algorithm we have just described twice, for designing primers on the 5′ and on the 3′ side of the DNA sequences (the distance between the two regions can be set according to the specific requirements of the experiment). After the first primer pair is selected, all matching sequences are removed, and a second pair is designed using the remaining sequences. We repeat this process according to the number of primer pairs requested by the user. This iterative procedure, described independently in **ref. *10***, is a heuristic for solving MP-DPD. It is useful when more than one primer pair is required to reach satisfactory coverage. Another heuristic for solving MP-DPD is the program MIPS, which was reported to outperform HYDEN when applied to the task of designing multiplex PCR experiments for single-nucleotide polymorphism (SNP) genotyping ***(10)***. As noted by Souvenir et al. ***(10)***, the problems solved by the two algorithms are quite different—mainly, MIPS constructs a set of primers for one PCR experiment with multiple primers, whereas HYDEN designs primer pairs for separate experiments (one pair per experiment). If one wishes to use HYDEN for multiplex PCR, a better approach would be to design a set of 5′ primers and a set of 3′ primers separately, because each 5′ primer may pair with each 3′ primer, whereas HYDEN constructs one 3′ primer per 5′ primer. Each set could be constructed using an iterative procedure similar to the one described above (but on one side only), until sufficient coverage is reached. It would be interesting to compare the performance of this version of HYDEN to that of MIPS. Note that when using several different primers in the same PCR, one has to make sure the primers will not hybridize with one another. Both MIPS and HYDEN ignore this crucial issue; so, additional tools should be used to check whether the designed primers might cross-hybridize.

4.3. Running HYDEN

HYDEN receives an input file that contains a set of DNA sequences in Fasta format and a list of command-line parameters that specify the number of primers to design, their length and degeneracy, the regions within the sequences to use for designing the 5′ and 3′ primers, the maximal number of mismatches to allow between the primers and the sequences they match, and the parameters N_a, N_g, and N_h ($N_{a'}$ is automatically set to $5N_a$). For a more detailed description, see the “Readme.txt” file supplied with HYDEN.

HYDEN is distributed with a sample input file, called “HGP_50genes.fasta”, which contains 50 human OR genes of length roughly 1 Kbp. **Figure 8A** demonstrates an execution of HYDEN on these sequences. Given the parameters shown in the figure, HYDEN designs two pairs of primers. The first 300 bp (last 350 bp) in each sequence are used for designing the 5′ (3′) primers. The primers are of length 25 and maximum degeneracy 5,000 (5′) or 30,000 (3′). Each primer is allowed to have up to two mismatches with each sequence it covers, and a total of three mismatches are allowed between a pair of primers and each covered sequence. HYDEN reports the progress of the algorithm and summarizes its results in a table. **Figure 8B** shows the summary table from HYDEN’s output on the sample data. The first primer pair matches 34 (68%) of the 50 input sequences. Together, the two primer pairs HYDEN designed cover 40 (80%) genes.

A. Command-line parameters:

```
hyden.exe  –dna  HGP-50genes.fasta  -mprimers  2
           –from5   0  –to5   300  –len5  25  –deg5   5000  –mis5  2
           –from3  –1  –to3  –350  –len3  25  –deg3  30000  –mis3  2
           –mis  3  –nentropy  30  –nalgs  500  –nimprove  50
```

B. Output summary table:

```
Pair  | 5′ primer                    | 3′ primer (inverted)           | coverage  | % (acc.)
------+------------------------------+--------------------------------+-----------+---------
  0   | CTNSAYDCNCCYATGTAYTTYTTHC    | TCCTBADRSTRTARATNANNGGDTT      |    34     |  68 %
  1   | MCCCCATGTAYTTYTTYCYBDSMAN    | ATKRCWGHYTTBAMHWCHTYATTYC      |     6     |  80 %
------+------------------------------+--------------------------------+-----------+---------
```

Fig. 8. Example of an execution of HYDEN. (**A**) Command line for running HYDEN on the sample input file. (**B**) Output table of HYDEN, listing the two primer pairs it designed.

5. Applications

5.1. Deciphering the Human Olfactory Subgenome

HYDEN was originally developed and implemented as part of DEFOG, an experimental scheme for DEciphering Families Of Genes *(3)*. DEFOG provides a powerful means for analyzing the composition of a large family of genes with conserved regions and is thus especially useful in species for which little genomic data are available. DEFOG consists of several computational and experimental phases. First, given a subset of known gene sequences, HYDEN is used to design degenerate primer pairs. The primers are then used in PCR to amplify fragments of genes, known as well as unknown, of the same family. The fragments are cloned, and an oligofingerprinting (OFP) process *(14)* characterizes the clones by their patterns of hybridization with a series of very short (8-mer) oligonucleotides. Another algorithm, called CLICK *(15)*, clusters the clones into groups corresponding to the same gene according to their hybridization patterns. Finally, representatives from each cluster are sequenced and compared to the known gene sequences. The DEFOG methodology was developed jointly with the groups of H. Lehrach (MPI Berlin) and D. Lancet (Weizmann).

The DEFOG scheme was applied to the human OR subgenome. The human genome contains more than 1,000 OR genes, of which more than 60% are considered pseudogenes *(16,17)*. OR genes have a single coding exon of about 1 Kbp and code for seven transmembrane (TM) domain proteins *(18)*. They have several highly conserved regions, primarily in TM segments 2 and 7. In contrast, TM segments 4 and 5 show a high degree of variability—a crucial feature for recognizing a huge variety of odorants *(19)*.

Our experiment began with an initial collection of 127 OR genes, whose full DNA coding sequences of size 1 Kbp were known at the time ***(20)*** (the project began before the completion of the Human Genome draft sequence). This collection comprised our *training set*, on which HYDEN designed the primers. Altogether, we designed 13 primers—6 for the 5′ side and 7 for the 3′ side, of lengths $k = 26, 27$ and various degeneracies between 4,608 and 442,368. The primers on each side are quite similar to one another and differ mainly in their degeneracy, except for four special primers—one pair was designed at different positions, closer to the 5′ and 3′ ends of the genes, and another pair was designed on a subset of genes that were poorly matched by the other primers. These four primers were constructed to "fish out" genes that, for some reason, are not amplified by the other primers. A typical run of HYDEN on 300 bp segments of the 127 OR genes, with $k = 26$, $d = 20,000$, and $e = 2$ ($N_h = 50$, $N_a = 3000$, and $N_g = 100$), takes < 10 min, distributed evenly among the three phases of the program, on a 1.5 GHz Pentium4 PC.

We selected 20 different primer pairs from the 13 primers we designed and used them in (separate) PCRs. The degeneracy of a pair of primers is defined as the product of the degeneracies of both primers. The degeneracy of the pairs we selected ranged between 2.1×10^7 and 1.4×10^{10}. To the best of our knowledge, this is the highest degeneracy ever used successfully in PCRs—previous applications usually used degeneracies lower than 10^5. We also experimented with even higher degeneracies (up to 2.2×10^{11}), but their yield was usually very poor, perhaps because the concentration of each individual primer is too low to allow successful PCR amplification. Most primer pairs covered 70–80% of the training-set genes with up to three mismatched bases in both sides combined (we used a threshold of three mismatches, because early experiments have shown that it predicts successful PCR amplification reasonably well).

We obtained a total of 13,580 clones from all the PCR experiments. We then applied the OFP process and the CLICK clustering software, which partitioned the clones into 239 clusters and 121 singletons (single clone clusters). We selected representative clones from each cluster and successfully sequenced a total of 924 clones. The extremely degenerate primers we designed proved very effective: They achieved high sensitivity, amplifying a total of 300 unique OR genes, and extremely high specificity, yielding only 0.4% (4 of 924) non-OR products. The *sequencing efficacy* of the primers, defined as the percentage of distinct genes that were obtained from each primer pair out of the total number of clones sequenced for that pair, was very high—for 10 of 12 primer pairs with degeneracy over 10^9, sequencing efficacy was 79–93%, and for all

8 pairs with lower degeneracy, it was 57–79%. Another advantage of using highly degenerate primers is the large number of *new* genes they amplify—in our case, 231 of 300 genes were new. Altogether, the DEFOG experiment almost tripled the size of our initial OR repertoire, from 127 genes to 358. The full experimental details and an analysis of the performance of the primers are reported in **refs**. *2* and *3*.

After the publication of the first draft of the human genome, we analyzed the performance of various primer pairs on all full-length OR sequences that were computationally detected in the draft. This set consisted of 719 genes *(16)*. These genes served as a *test set*, with which we checked how well the coverage of our primers extends from the training set to a larger collection of genes. **Figure 9** shows the 3-mismatch coverage of several primer pairs, both for the training set and for the test set. As expected, there is a sharp and steady increase in the test-set coverage as the degeneracy increases—from 10% coverage for non-degenerate primers to 50–65% for the primers we used and 74% for a pair with degeneracy 4×10^{12}. In practice, one cannot use arbitrarily high degeneracies, for two reasons. First, highly degenerate primers have low specificity, and so, they might amplify many non-related sequences. Second, as

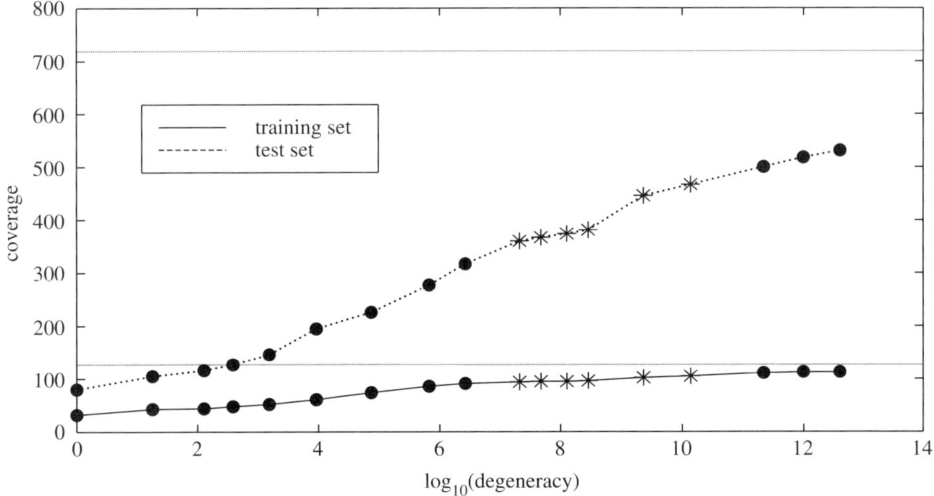

Fig. 9. Training-set and test-set 3-mismatch coverage of primer pairs with various degeneracies in the human olfactory receptors project. Primers that were actually used in the DEFOG experiment are marked by asterisks. The horizontal lines mark the size of the training and test sets.

mentioned earlier, PCR gives a poor yield when the degeneracy is very high. Additional analyses of the performance of the primers in the DEFOG project are given in **ref.** *2*.

5.2. The Canine Olfactory Subgenome

Encouraged by the results we obtained for the human OR subgenome, we launched a project with the group of D. Lancet (Weizmann) for analyzing the canine OR subgenome ***(21)***. We used a simplified version of DEFOG, in which we skipped the OFP and clustering phases (i.e., clones were selected for sequencing arbitrarily, rather than based on the fingerprints clustering). As very few canine OR genes were fully known at the time, we ran HYDEN on the set of 719 *human* ORs and designed several primer pairs with degeneracy between 1.2×10^6 and 2.2×10^{10}. Despite the significant differences between the human and the canine olfactory systems, the human-based primers amplified many ORs from the dog genome. The 1,200 clones we sequenced contained 246 distinct OR genes (the full dog OR repertoire is estimated to contain some 1,200 genes), again demonstrating the advantages of using highly degenerate primers for amplifying many related sequences. About 14% of the canine OR genes we obtained are pseudogenes, similar to the ratio in mouse (20%) ***(22,23)***, but far from the ratio in human (> 60%) ***(16,17)***. This reflects the fact that both dog and mouse are macrosmatic animals, that is, have a very acute sense of smell, whereas human is microsmatic.

5.3. Olfaction Versus Vision Among Primates

Another interesting project that utilized HYDEN is described in **ref.** *24*. In that study, degenerate primer pairs designed based on human ORs were used to sequence 100 OR genes in human and in 18 primate species, for which the genome sequence was not available, including apes, Old World monkeys (OWMs), and New World monkeys (NWMs). As expected, the proportion of OR pseudogenes in human was found to be very high (above 50%). In great apes and OWMs, roughly 30% of the sequenced ORs are pseudogenes, whereas in NWMs, this ratio is significantly lower—only 18% are pseudogenes. However, there is one exception: one NWM species, the howler monkey, was found to have a similar proportion of OR pseudogenes (31%) to that of OWMs and apes. Gilad et al. ***(24)*** noticed that another phenotype that is shared only by the howler monkey, OWMs, and apes is full trichromatic color vision. Thus, the deterioration of the olfactory subgenome repertoire and the acquisition of full trichromatic vision occurred independently in two separate evolutionary branches: in the common ancestor of OWMs and apes and in the

New World howler monkey. This suggests an association between two senses on an evolutionary genetic scale: as vision improved in some of the primate species, they became less dependent on their sense of smell, which led to its decline.

Acknowledgments

This work was partially supported by the German-Israeli Foundation for Scientific Research (GIF) under grant G-0506-183.0396 and by the Israel Science Foundation (grant 309/02).

References

1. Kwok, S., Chang, S., Sninsky, J., andWang, A. (1994) A guide to the design and use of mismatched and degenerate primers. *PCR Methods and Applications*. **3**, S39–47.
2. Linhart, C. and Shamir, R. (2005) The degenerate primer design problem: theory and applications. *Journal of Computational Biology* **12**, 431–456.
3. Fuchs, T., Malecova, B., Linhart, C., Sharan, R., Khen, M., Herwig, R., Shmulevich, D., Elkon, R., Steinfath, M., O'Brien, J., Radelof, U., Lehrach, H., Lancet, D., and Shamir, R. (2002) DEFOG: a practical scheme for deciphering families of genes. *Genomics* **80**, 295–302.
4. Pearson, W., Robins, G., Wredgs, D., and Zhang, T. (1996) On the primer selection problem in polymerase chain reaction experiments. *Discrete Applied Mathematics* **71**, 231–246.
5. Doi, K. and Imai, H. (1997) Greedy algorithms for finding a small set of primers satisfying cover length resolution conditions in PCR experiments. In *Proceedings of 8th Workshop on Genome Informatics*, pp. 43–52. Tokyo, Japan.
6. Bailey, T. and Elkan, C. (1995) Unsupervised learning of multiple motifs in biopolymers using expectation maximization. *Machine Learning* **21**, 51–80.
7. Lawrence, C., Altschul, S., Boguski, M., Liu, J., Neuwald, A., and Wootton, J. (1993) Detecting subtle sequence signals: a Gibbs sampling strategy for multiple alignment. *Science* **262**, 208–214.
8. Linhart, C. (2002) The degenerate primer design problem Masters thesis, School of Computer Science, Tel Aviv University, November 2002. Available at `http://www.cs.tau.ac.il/~chaiml/biology/dpd_thesis.ps.gz`.
9. Cormen, T. H., Leiserson, C. E., and Rivest, R. L. (1990) *Introduction to Algorithms*. MIT Press, Cambridge, Mass.
10. Souvenir, R., Buhler, J., Stormo, G., and Zhang, W. (2003) Selecting degenerate multiplex PCR primers. In *Proceedings of 3rd Workshop on Algorithms in Bioinformatics (WABI 2003)*, pp. 512–526.
11. Blum, M., Floyd, R., Pratt, V., Rivest, R., and Tarjan, R. (1973) Time bounds for selection. *Journal of Computer and System Sciences* **7**, 448–461.

12. Keich, U. and Pevzner, P. (2002) Finding motifs in the twilight zone. In *Proceedings of 6th Annual International Conference on Research in Computational Molecular Biology (RECOMB 2002)*, pp. 195–204.
13. Linhart, C. and Shamir, R. (2003). HYDEN – A software for designing degenerate primers. Available at `http://www.cs.tau.ac.il/~rshamir/hyden`.
14. Radelof, U., Hennig, S., Seranski, P., Steinfath, M., Ramser, J., Reinhardt, R., Poustka, A., Francis, F., and Lehrach, H. (1998) Preselection of shotgun clones by oligonucleotide fingerprinting: an efficient and high throughput strategy to reduce redundancy in large-scale sequencing projects. *Nucleic Acids Research* **26**, 5358–5364.
15. Sharan, R., Maron-Katz, A., and Shamir, R. (2003) CLICK and EXPANDER: a system for clustering and visualizing gene expression data. *Bioinformatics* **19**, 1787–1799.
16. Glusman, G., Yanai, I., Rubin, I., and Lancet, D. (2001) The complete human olfactory subgenome. *Genome Research* **11**, 685–702.
17. Zozulya, S., Echeverri, F., and Nguyen, T. (2001) The human olfactory receptor repertoire. *Genome Biology* **2**, RESEARCH0018.
18. Buck, L. and Axel, R. (1991) A novel multigene family may encode odorant receptors: a molecular basis for odor recognition. *Cell* **65**, 175–187.
19. Pilpel, Y. and Lancet, D. (1999) The variable and conserved interfaces of modeled olfactory receptor proteins. *Protein Science* **8**, 969–977.
20. Fuchs, T., Glusman, G., Horn-Saban, S., Lancet, D., and Pilpel, Y. (2000) The human olfactory subgenome: From sequence to structure and evolution. *Human Genetics* **108**, 1–13.
21. Olender, T., Fuchs, T., Linhart, C., Shamir, R., Adams, M., Kalush, F., Khen, M., and Lancet, D. (2004) The canine olfactory subgenome. *Genomics* **83**, 361–372.
22. Young, J., Friedman, C., Williams, E., Ross, J., Tonnes-Priddy, L., and Trask, B. (2002) Different evolutionary processes shaped the mouse and human olfactory receptor gene families. *Human Molecular Genetics* **11**, 535–546.
23. Zhang, X. and Firestein, S. (2002) The olfactory receptor gene superfamily of the mouse. *Nature Neuroscience* **5**, 124–133.
24. Gilad, Y., Wiebe, V., Przeworski, M., Lancet, D., and Pääbo, S. (2004) Loss of olfactory receptor genes coincides with the acquisition of full trichromatic vision in primates. *PLoS Biology* **2**, E5.

12

An Iterative Method for Selecting Degenerate Multiplex PCR Primers

Richard Souvenir, Jeremy Buhler, Gary Stormo, and Weixiong Zhang

Summary

Single-nucleotide polymorphism (SNP) genotyping is an important molecular genetics process, which can produce results that will be useful in the medical field. Because of inherent complexities in DNA manipulation and analysis, many different methods have been proposed for a standard assay. One of the proposed techniques for performing SNP genotyping requires amplifying regions of DNA surrounding a large number of SNP loci. To automate a portion of this particular method, it is necessary to select a set of primers for the experiment. Selecting these primers can be formulated as the *Multiple Degenerate Primer Design* (MDPD) problem. The *Multiple, Iterative Primer Selector* (MIPS) is an iterative beam-search algorithm for MDPD. Theoretical and experimental analyses show that this algorithm performs well compared with the limits of degenerate primer design. Furthermore, MIPS outperforms an existing algorithm that was designed for a related degenerate primer selection problem.

Key Words: Multiplex PCR; degenerate primers; SNP genotyping.

1. Introduction

Single-nucleotide polymorphisms (SNPs) are single-base differences among DNA sequences from the same species. **Figure 1** shows an example of an SNP between two sequences. It is estimated that there are roughly 3 million SNPs in the human genome *(1)*. Research investigating associations between SNPs and various diseases, along with studies of differences in how individuals respond to common therapies, promises to revolutionize medical science in the coming years *(2)*. It is, however, a daunting task to identify the specific genetic variations occurring in individuals, which is necessary to determine

From: *Methods in Molecular Biology, vol. 402: PCR Primer Design*
Edited by: A. Yuryev © Humana Press, Totowa, NJ

CGGTACT**T**GAGGCTA Person 1
CGGTACT**C**GAGGCTA Person 2

Fig. 1. A simple example of Single-nucleotide polymorphism (SNP).

the associations of SNPs with important phenotypes. Many techniques have been proposed for *SNP genotyping*, the task of determining the SNP content of a given genome. For these assaying techniques to be effective in large-scale genetic studies that genotype hundreds or thousands of SNPs, they must be scalable, automated, robust, and inexpensive *(3)*.

One technique for SNP genotyping involves the use of Multiplex PCR (MP-PCR) to amplify the genomic region around each SNP *(3)*. MP-PCR, like all PCR variants, uses oligonucleotide primers to define the boundaries of amplification. MP-PCR requires a pair of forward and reverse primers for each region to be replicated. The large-scale amplifications required by SNP genotyping may include hundreds or even thousands of such regions. Selecting such a large set of primers by current methods, especially by trial and error *(3)*, can be time-consuming and difficult.

1.1. Related Work

The *Primer Selection Problem* *(4)* seeks to minimize the number of primers needed to amplify a set of regions in a collection of DNA sequences. This problem has been shown to be NP-hard *(5)* through reductions from the set cover and graph coloring problems *(6)*. Although NP-hardness implies that an efficient exact algorithm for this problem is unlikely to exist, a number of heuristics have been proposed for it, including a branch-and-bound search algorithm *(7)*. Other proposed algorithms incorporate biological data about the primers into the search *(8,9)*.

To perform PCR, both forward and reverse primers are needed for each region to be amplified. Therefore, in a typical MP-PCR experiment, the number of primers needed is equal to twice the number of such regions. The algorithms mentioned above reduce the number of required primers to 25–50 of this value; however, even this reduced number remains high for the large-scale amplifications required by SNP genotyping.

The desire to amplify many regions with few primers leads to the use of degenerate primers *(10)*, which make use of degenerate nucleotides *(11)*. For example, consider the degenerate primer *ACMCM*; here, *M* is a degenerate nucleotide that represents the set of bases either of the bases *{A, C}*. This degenerate primer therefore describes the set of four ordinary primers *ACACA*,

ACACC, ACCCA, and *ACCCC}*. The *degeneracy* of a primer is the number of distinct DNA sequences described by it. The example shown above is a primer with degeneracy 4.

The use of degenerate primers in PCR introduces two new problems. First, the effective concentration of the desired primers is decreased by the presence of additional, undesired primers produced by combinatorial synthesis. Second, these undesired primers can lead to the amplification of unwanted regions of DNA. To minimize these effects while realizing the benefits of degenerate primer design, it is important to use primers of relatively low degeneracy.

The *Degenerate Primer Design (DPD) Problem* asks whether there exists a single degenerate primer of degeneracy at most some threshold d that can amplify designated regions from at least some number m of a set of input sequences. There are two variants of the DPD Problem. In maximum coverage DPD (MC-DPD), the goal is to find the maximum number of sequences that can be amplified by a primer of degeneracy at most d. In minimum degeneracy DPD (MD-DPD), the goal is to find a primer of minimum degeneracy that amplifies all of the input sequences. Both MC-DPD and MD-DPD have been shown to be NP-hard ***(12)***.

1.2. Formal Problem Description

We use notation from **ref.** ***12*** to describe the MDPD problems. In what follows, lower-case symbols (e.g., l, b, and i) represent numerical values, counting variables, or individual (possibly degenerate) characters in a sequence. Upper-case symbols (e.g., P and S) denote primers, sequences, or subsequences. Finally, calligraphic symbols (e.g., $\mathcal{S}$ and $\mathcal{C}$) represent sets of sequences or primers.

Let $\Sigma = \{A, C, G, T\}$ be the alphabet of DNA. A *degenerate primer* P is a string with a set of possible characters at each position, that is, $P = p_1 p_2 \ldots p_l$, where $p_i \subseteq \Sigma$, $p_i \neq \emptyset$, and l is the length of primer P. The *degeneracy* of P is $d(P) = \prod_{i=1}^{l} |p_i|$. For example, consider the degenerate primer $P' = \{A\}\{A, C\}\{A, C\}\{A, C\}$. The length of P' is $\ell = 4$, and its degeneracy $d(P')$ is 8. Using the IUPAC ambiguity codes ***(11)***, for degenerate nucleotides, P' may be written *AMMM*, where M is the degenerate symbol for the set $\{A, C\}$. Degenerate primers can be constructed by *primer addition*. For any two primers, P^1 and P^2, their sum P^3 equals $(p_1^1 \cup p_1^2)(p_2^1 \cup p_2^2) \ldots (p_l^1 \cup p_l^2)$.

For any sequence S_i in an input set $\mathcal{S}$, we say that a degenerate primer P covers S_i if there is a substring F of length l in S_i such that, for each character f_i in F, $f_i \in p_i$.

Problem 1 [Multiple Degenerate Primer Design (MDPD)]. *Given a set of n sequences over an alphabet* Σ *and integers l and k, is there a set of primers* $\mathcal{P}$*, each of length l, that covers all the input sequences, where* $|\mathcal{P}| \leq k$*?*

There are two optimization problems that are variants of the MDPD problem, which add additional constraints to the final solution $\mathcal{P}$.

Problem 2 [Primer-Threshold MDPD (PT-MDPD)]. *Given a set of n sequences over an alphabet* Σ *and integers l and* α*, find the smallest set of primers* $\mathcal{P}$*, each of length l, that cover all the input sequences, where* $\forall P_i \in \mathcal{P},\ d(P_i) \leq \alpha$.

Problem 3 [Total-Threshold MDPD (TT-MDPD)]. *Given a set of n sequences over an alphabet* Σ *and integers l and* α*, find the smallest set of primers* $\mathcal{P}$*, each of length l, that cover all of the input sequences, where* $\sum_{P_i \in P} d(P_i) \leq \alpha$.

The solution to PT-MDPD is a set of degenerate primers, each with degeneracy at most a specified threshold value α, whereas the solution to TT-MDPD is a set of primers whose sum of degeneracies is at most α. These problems have both been shown to be NP-complete ***(13)***.

2. Materials

The MIPS algorithm described in the **Subheading 3** was implemented using the C++ programing language. All experiments were performed on a 1.6 GHz AMD Athlon workstation with 2 GB RAM, running Red Hat Linux 7.3. The MIPS software is freely available on the Web at http://www.cse.wustl.edu/~zhang/projects/software.html.

3. Methods

The *Multiple, Iterative Primer Selector (MIPS)* algorithm makes a tradeoff between optimality and tractability to overcome the inherent difficulty of the MDPD problems. In this section, we describe the MIPS algorithm, analyze its complexity, present experimental results that validate its performance, and discuss practical considerations in applying the algorithm to real genomic sequences.

MIPS can operate in either of two modes, MIPS-PT or MIPS-TT, which solve the PT-MDPD and TT-MDPD problems, respectively. This section focuses on MIPS-TT, but we will highlight how MIPS-PT differs where appropriate.

3.1. Algorithm Overview

MIPS tries to find a minimal set of degenerate primers that together cover all the input sequences. To ensure minimality, it first tries to cover the input with one primer, then two, and so forth. For each number of primers p, MIPS constructs a set of p primers that covers as many of the input sequences as possible (hopefully, all of them). This set is constructed in a series of phases, each of which first produces a set of *candidate* primers, then chooses one of these candidates for the solution set.

Define a *k-primer* to be a degenerate primer that covers k input sequences. Each phase of MIPS first generates an initial set of candidate 2-primers, then iteratively builds up sets of candidate k-primers for $k = 3, 4, \ldots$. At each step, every k-primer is extended into one or more $(k+1)$-primers by generalizing it as needed to cover one additional input sequence, possibly increasing its degeneracy. Iteration stops when all input sequences have been covered, or when no primer can be extended without exceeding the global degeneracy threshold α. At this point, each candidate primer covers some number k_{last} of sequences. MIPS retains the k_{last}-primer of minimum degeneracy, removes the input sequences it covers from consideration, and begins the next phase on the remaining sequences. The algorithm proceeds in this fashion until all input sequences have been covered.

In a phase of MIPS that begins with c sequences to be covered, each k-primer P can generalize into c different $k+1$-primers, one for each additional sequence to be covered by P. Allowing this generalization to proceed unchecked would produce a number of candidate primers that grows exponentially with c, rendering the search intractable. To maintain tractability, MIPS retains only the b least degenerate $k+1$-primers obtained by generalizing a given set of candidate k-primers. This set of retained primers is known as a *beam*, and the search strategy employed is called *beam search* ***(14)***. Beam search differs from greedy or best-first search in that multiple nodes, in this case degenerate primers, are saved for extension instead of just one. MIPS' strategy of progressively extending a beam of degenerate primers similar to the consensus motif-finding model ***(15)***.

The number b of primers retained in the beam is a parameter that may be tuned by the user. Increasing b considers more primer candidates and may improve the quality of the solution, but it also lengthens the running time of the algorithm, as discussed in **Note 1**.

If a phase of MIPS terminates without covering all remaining input sequences, then no k-primer in the beam can be extended to a $(k+1)$-primer without exceeding the degeneracy threshold. As noted above, the algorithm selects the best degenerate k_{last}-primer, P_0, from the beam, then starts a new phase on the reduced set of input sequences not covered by P_0. In MIPS-PT, the degeneracy threshold for the reduced problem is the same as the original threshold, α, whereas in MIPS-TT, the threshold is reduced by the degeneracy of P_0.

For MIPS-PT, iteratively choosing the best candidate from each phase will eventually yield a set of primers that together cover all input sequences. However, this is not necessarily the case for MIPS-TT. After P_0 chosen, the new degeneracy threshold may be too low to cover the remaining sequences. In this case, MIPS-TT discards P_0, *backtracks* to the previous level $k_{last} - 1$ of the most recently completed phase, and adds the best $k_{\text{last}} - 1$ primer P'_0 to the solution. Note that P'_0 is in general less degenerate than P_0; so, the remaining degeneracy bound is less severely constrained in the next phase, which must now cover one additional sequence.

Figure 2 illustrates an execution of MIPS-TT. In the trees shown, the depth of a node represents the number of covered input sequences, whereas the number in each node represents the number of degenerate primers that will be used to cover those sequences. Each node can be expanded into two child nodes. The left child represents covering an additional sequence using an existing degenerate primer, whereas the right child represents covering an additional

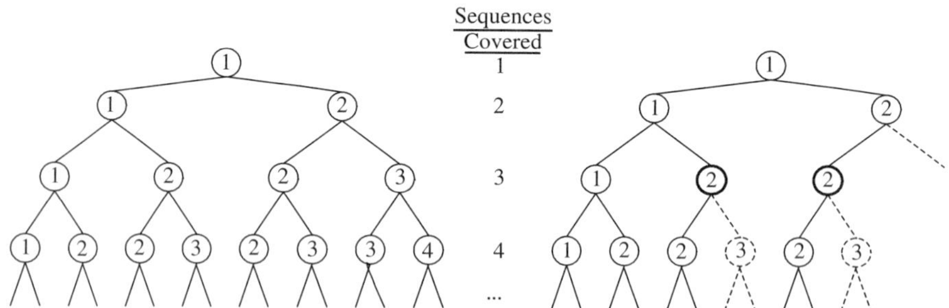

Fig. 2. Pruning of the search space by multiple, iterative primer selector total threshold (MIPS-TT). The node depth represents the number of covered input sequences, whereas the label represents the number of degenerate primers that will be used to cover those sequences. The left child represents covering an additional sequence using an existing degenerate primer, whereas the right child represents covering an additional sequence using a new degenerate primer. The left tree shows a full search. The right tree shows the pruning that takes place in MIPS-TT. Consider the two bold nodes. These states cover the same number of sequences with the same number of primers. MIPS-TT only expands the node whose total score is better.

sequence using a new degenerate primer. The left tree in **Fig. 2** shows a full search. The right tree shows the pruning that takes place in MIPS-TT during the backtracking phrase. Consider the two bold nodes. At these states, the same number of sequences are covered with the same number of primers. MIPS-TT will therefore only expand the node whose total score is better. While this greedy choice may not be optimal, it avoids the exponential expansion seen on the full tree by not exploring the nodes represented by dotted circles.

3.2. The MIPS Algorithm and Its Complexity

Algorithm 1 presents detailed pseudocode for MIPS. The top-level MIPS routine tries to find solutions with progressively larger numbers of primers p and implements backtracking within a single value of p. This main routine calls a subroutine, shown in Algorithm 2, that performs the actual beam search. For efficiency, the beam described in the previous section is represented as a priority queue, which efficiently maintains only the b least degenerate candidate primers seen so far. The beam search routine in turn calls a subroutine, shown in Algorithm 3, to populate the beam for k-primers for each successive k.

Algorithm 1 MIPS($\mathcal{S}, \alpha$)

Input: Sequence set $\mathcal{S}$ and degeneracy bound α
Initialize global variables (*see* **Note 2)**
$ALLOWABLE(0,0) = \alpha$.
for $p = 1$ to the number of degenerate primers that will be used **do**
  Let c = the maximum number of sequences that the (p-1) primers covered
  while $c > 0$ **do**
    $MIPS\_SEARCH(\mathcal{S} - COVERED(p-1,c), ALLOWABLE(p-1,c), p, c)$
    if this search covers S, print solution and exit
    else c=c-1

We now examine the theoretical bounds of MIPS and compare these values to the computing resources consumed in practice. See **Table 1** for a list of variables used in the section and what they represent.

3.2.1. Space

Define a fragment of the input to be an length l substrings of one of its sequences. The implementation of MIPS stores each fragment in the input

Table 1
Parameters of the Multiple, Iterative Primer Selector (MIPS) Algorithm

Variable	Represents
n	Number of input sequences
m	Average sequence length
b	Beam size
l	Primer length

as a separate string, which requires space $O(nml)$. The implementation also uses multiple $n \times n$ matrices to store degenerate primers that could eventually become part of the final solution, depending on the amount of backtracking required. These matrices use $O(n^2)$ storage. Therefore, the total amount of space used is $O(n^2 + nml)$.

Algorithm 2 MIPS_SEARCH($\mathcal{S}, \alpha, p, c$)

```
Input: Sequence set S, degeneracy bound α, primer number p, sequences covered c.
Output: total number of sequences covered
Initialize priority queue Q of size b;
Perform pair-wise comparisons and create lookup table T (see Note 3).
for all sequence S_i ∈ S do
  for all substring S_i[j, l] do
    Let c = {x|⟨f, x⟩ ∈ T and f is a k-length substring of S_i[j, l] }
    for all fragment C_k ∈ c do
      D = S_i[j, L] + C_k
      Insert D into queue Q
Let c′ = c
while queue Q is not empty do
  Let P = the best element of Q
  if degeneracy(P) < degeneracy(BEST(p,c)) then
    BEST(p, c′) = P
    ALLOW ABLE(p, c′) = α− degeneracy(P)
    COVERED(p, c′) = COVERED(p − 1, c)∪ covers(P, S)
    Q = ONE_PASS(Q, S, α)
  c′ = c′ + 1
return (c′ + 1)
```

Algorithm 3 ONE_PASS(Q, $\mathcal{S}$, α)

Input: Priority queue Q, set of sequences $\mathcal{S}$, degeneracy bound α.
Output: Priority queue Q'
for all primer $P \in Q$ **do**
  for all sequence $S_i \in \mathcal{S}$ **do**
    if $S_i \notin$ covers(P) **then**
      for all substring $S_i[j, l]$ **do**
        $D = S_i[j, l] + P$
        if degeneracy(D) $< \alpha$ **then**
          Insert D into queue Q'
return Q'

3.2.2. Time

We analyze time complexity in a bottom-up fashion. First, the ONE_PASS procedure, shown in Algorithm 3, compares each candidate k-primer in a beam with all fragments of the uncovered input sequences. ONE_PASS considers $O(bnm)$ primer extensions, because there are $O(nm)$ total input fragments and b candidate primers in the beam. Adding a fragment to a primer requires comparing each character of the fragment to the corresponding position in the primer; hence, the overall time complexity of ONE_PASS is $O(bnml)$.

We next consider the MIPS_SEARCH procedure, which implements a single phase of the algorithm. MIPS_SEARCH, shown in Algorithm 2, generates beams of k-primers for successive k and stores the best (least-degenerate) primer in each beam to facilitate later backtracking. MIPS_SEARCH uses ONE_PASS to build each new beam and could, in the worst case, build n beams. Its overall time complexity is therefore $O(bn^2ml)$.

Finally, the number of times that MIPS_SEARCH is executed by MIPS depends on the number of backtracking steps required. This number is directly related to the number of primers in the final solution. In the best case, if a solution is obtained using only one degenerate primer, there will be only one call to MIPS_SEARCH. In the worst case, if the solution requires n primers (one primer for each input sequence), there will be $n^2/2$ calls to MIPS_SEARCH. In general, if p is the number of primers in the final solution, then MIPS_SEARCH will be called $O(pn)$ times. This brings the overall time complexity of MIPS to $O(bn^3mlp)$.

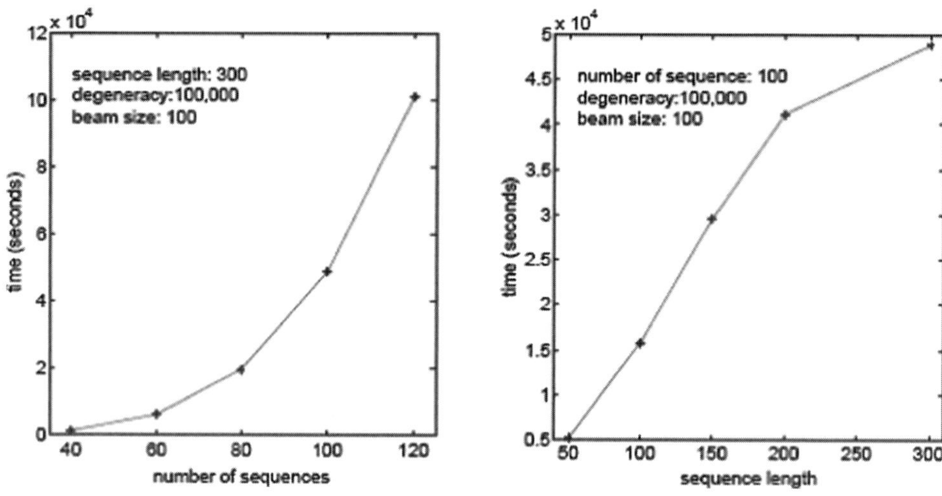

Fig. 3. Timing graphs for various input sizes.

The graphs in **Fig. 3** show how the running time of MIPS changes when various parameters of the input set are manipulated. These graphs correlate with the theoretical predictions of time dependencies. All experiments were run on a computer with an AMD 1.6 GHz CPU and 2 GB RAM running Red Hat Linux 7.3.

3.3. Experimental and Comparative Analysis

MIPS has been tested on both human and randomly generated DNA sequences. The human data set is a database of regions in the human genome surrounding 95 known SNPs. The sequences varied in length from a few hundreds to well over a thousand nucleotides. The location of the SNP on each sequence was marked to constrain the locations of the forward and reverse primers. To ensure effective PCR product analysis, primers could not be placed within 10 bases of the SNP, and the entire PCR product length could not exceed 400 bases.

3.3.1. Human Data

In an unpublished laboratory experiment, a set of degenerate primers of length 20 was manually constructed. Each primer was a mixture of 8 specific bases and 12 fully degenerate nucleotides (e.g., *AGTCG-GTANNNNNNNNNNNN*). For this experiment, the total degeneracy would be $\approx 4^{12}$. MIPS was originally designed to automate this procedure and reduce the total degeneracy and/or number of primers used. In practice, the desired accuracy in the experiment determines the actual parameter values used for MIPS. **Table 2** summarizes the results. For the 95 sequences, 190 primers would be needed in the general case. MIPS-PT decreased the total number of primers to 29 or 15% of the original unoptimized value when each primer was limited to a degeneracy of at most $4^9 = 262,\ 144$.

3.3.2. Artificial Data

The HYDEN algorithm ***(12)*** is a heuristic designed for finding approximate solutions to DPD problems. Recall that DPD is a set of problems, where the general goal is to find a *single* degenerate primer that either covers the most sequences while having a degeneracy value less than a specified threshold or covers all sequences with minimum degeneracy. The MDPD problem generalizes DPD to allow multiple primers.

HYDEN can solve the PT-MDPD problem indirectly by iteratively solving the MC-DPD problem on smaller and smaller sets. The algorithm first finds a single degenerate primer pair that covers as much of the input as possible subject to a fixed degeneracy bound, then recurs on the remaining, uncovered sequences. For the reasons described in **Note 4**, iteratively solving MC-DPD is not the most effective way to solve the PT-MDPD problem.

Figure 4 shows the number of primers that HYDEN and MIPS-PT needed to cover synthetic sets of DNA sequences of varying lengths, using varying degeneracy thresholds. The DNA sequences were i.i.d. random strings of equal length with equal base frequencies. Each program searched for degenerate primers without allowing any mismatches at any positions.

Table 2
Results on a Data Set of 95 Human Single Nucleotide Polymorphism (SNP) Regions

PT-MDPD		TT-MDPD	
Degeneracy	# Primers	Degeneracy	# Primers
$4^6 \approx 4K$	53	$4^9 \approx 262K$	44
$4^7 \approx 16K$	44	$4^{10} \approx 1M$	37
$4^8 \approx 64K$	36	$4^{11} \approx 4M$	30
$4^9 \approx 262K$	29	$4^{12} \approx 16M$	23

PT-MDPD, Primer Threshold Multiple Degenerate Primer Design; TT-MDPD, Total-Threshold Multiple Degenerate Primer Design.

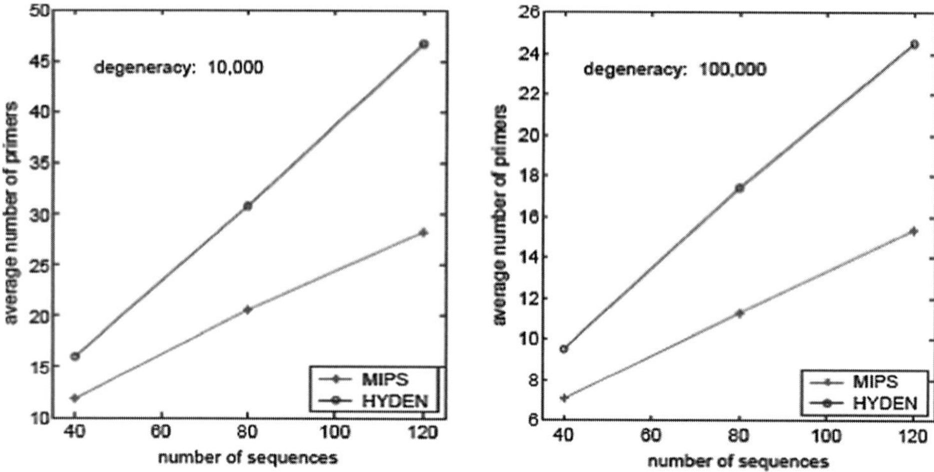

Fig. 4. The number of degenerate primers selected by HYDEN and MIPS for 20 randomly generated datasets in solving primer-threshold multiple degenerate primer design (PT-MDPD) for degeneracy thresholds of 10,000 (**A**) and 100,000 (**B**).

In general, HYDEN's solutions to PT-MDPD required more primers than those obtained by MIPS. For problems of more than 100 sequences with a degeneracy threshold of 100,000, the difference in solution sizes was as large as 60%.

3.4. Degenerate Primer Design

The use of degenerate primers introduces problems in biological assays. This section examines one important problem, mispriming. The following discussion demonstrates how mispriming is amplified when the background base composition is non-random (such as in the human genome), re-examines the quality of the MIPS solutions in light of mispriming, and suggests improvements to avoid the problem.

The key issue incurred by using degenerate primers is that such primers bind to more sequences than are necessary to do their job. For this discussion, we define *target primers* to be primers that are intended to be used in the PCR assay and *auxiliary primers* to be primers arising from degenerate primer design that may or may not bind to the input sequences but are not intended to be used in the PCR assay. The two main problems associated with degenerate primer usage are a decrease in the effective concentration of the target primers and an increase in the possibility of amplifying an unexpected region, or mispriming. To quantify these problems, the following questions are paramount:

- What is the expected efficacy of a given degenerate primer? In other words, for the set of primers that a given degenerate primer represents, what is the ratio of target primers to auxiliary primers?
- Given the presence of auxiliary primers, what is the expected number of undesired PCR products?

3.4.1. Degenerate Primer Efficacy

Multiplex primer design demands that many input sequences share sites complementary to some common (possibly degenerate) primer. In the general case, the sequences to be co-amplified are not related; so, their complementarity to a common primer is largely a matter of chance. Exploration of the chance-imposed limits of multiplexing requires calculating the number of unrelated DNA sequences likely to be covered by a single PCR primer of a given degeneracy.

Let $\mathcal{S}$ be a collection of n DNA sequences of common length m. Call a primer P an (l, α, k)-primer for $\mathcal{S}$ if it has length l and degeneracy at most α and covers at least k sequences of $\mathcal{S}$. A natural way to quantify the limits of multiplexing is to compute the probability that an (l, α, k)-primer exists for $\mathcal{S}$. However, this probability is difficult to compute, even assuming that $\mathcal{S}$ consists of i.i.d. random DNA with equal base frequencies. It is therefore more practical to compute the *expected* number of (l, α, k)-primers for $\mathcal{S}$. If this expectation is much less than one, Markov's inequality implies that $\mathcal{S}$ is unlikely to contain any such primer.

It is not necessary to count the total number of (l, α, k)-primers for $\mathcal{S}$, but only the number of *maximal* primers. A primer P of degeneracy at most α is said to be *maximal* if increasing P's degeneracy at any position would cause its total degeneracy to exceed α. The expected number of maximal (l, α, k)-primers for $\mathcal{S}$ is in general less than the total number of (l, α, k)-primers, but a primer of this type exists for $\mathcal{S}$ if and only if a maximal primer exists. Hence, the former expectation is more useful than the latter for bounding the probability that at least one (l, α, k)-primer exists.

3.4.1.1. Occurrence Probability for One Fixed Primer

Let P be a primer of length l, such that the jth position of P permits $|p_j|$ different bases. Let $\mathcal{S}$ be a collection of n i.i.d. random DNA sequences of common length m with equal base frequencies, and let T be a single l-mer at a fixed position in some sequence $S_i \in \mathcal{S}$. Say that P *matches* T if P would hybridize to T. Then

$$\Pr[P \text{ matches } T] = \prod_{j=1}^{l} \frac{|p_j|}{4} = \frac{d(P)}{4^l}.$$

The probability that P covers S_i, that is, that it matches *at least one* l-mer of S_i, depends in a complicated way on P's overlap structure, but if S_i is not too short and $d(P)/4^l \ll 1$ (both of which are typically true), then using Poisson approximation ***(16)***,

$$\Pr[P \text{ occurs in } S_i] \approx 1 - e^{-\frac{d(P)}{4^l}(m-l+1)}.$$

Let q be the probability that P matches somewhere in a single sequence of length m, and let $c(P)$ be P's coverage of $\mathcal{S}$, that is, the number of sequences of $\mathcal{S}$ in which P matches at some position. Because the sequences of $\mathcal{S}$ are independent, the probability that P matches in at least k sequences given by the binomial tail probability

$$\Pr[c(P) \geq k] = 1 - \Pr[B(n, q) < k],$$

where $B(n, q)$ is the sum of n independent Bernoulli random variables, each with probability q of success.

3.4.1.2. Computing the Expectation

Let $\Pi(l, \alpha)$ be the set of all maximal primers of length l and degeneracy at most α. To count the expected number $E_{l,\alpha,k}$ of l, α, k-primers for $\mathcal{S}$, we observe that

$$E_{l,\alpha,k} = \sum_{P \in \Pi(l,\alpha)} \Pr[c(P) \geq k].$$

Enumerating all $P \in \Pi(l, \alpha)$ to compute this expectation would be computationally expensive, but this enumeration is not needed for i.i.d. sequences with equal base frequencies. Given these assumptions about $\mathcal{S}$'s sequences, the probability that P matches a given l-mer does not change if the positions are rearranged (e.g., *AMC* versus *MCA*) or the precise nucleotides matched are changed (e.g., *RTG* versus *MCA*). Let W be a multiset of l values drawn from {1,2,3,4} that lists the degeneracies n_j (in any order) of a primer from $\Pi(l, \alpha)$. Then every primer described by the same W has the same probability of covering at least k sequences in $\mathcal{S}$. Hence, the desired expectation is given by

$$E_{l,\alpha,k} = \sum_{k} \#(W) \Pr[c(P) \geq k / P \text{ described by } W].$$

where the sum ranges over all feasible W for $\Pi(l, \alpha)$ and $\#(W)$ denotes the number of degenerate primers described by W. The probability is computed as described above.

To compute $\#(W)$, let W be a multiset with n_1 1s, n_2 2s, n_3 3s, and n_4 4s. If we fix *which* positions in P permit 1, 2, 3, and 4 nucleotides respectively, then there are $4^{n_1} \times 6^{n_2} \times 4^{n_3}$ ways of assigning nucleotide sets to these positions. Hence,

$$\#(W) = \binom{l}{n_1\ n_2\ n_3} 4^{n_1+n_3} 6^{n_2}.$$

Enumerating all feasible W for $\Pi(l, \alpha)$ is straightforward; so, the expectation can be computed.

3.4.1.3. Results

The theoretical estimates of the previous section can be used to evaluate whether a particular primer design algorithm performs well on the MC-DPD problem, that is, whether it finds degenerate primers with coverage close to the maximum predicted for a given set of input sequences. We evaluated the MIPS algorithm's performance on MC-DPD by comparing the primers it found in random DNA with those expected to exist in theory. For these experiments, we generated test sets of i.i.d. random DNA sequences with equal base frequencies with $n = 190$ and $m = 211$, so that the number and average length of the test sequences roughly matched those of the human DNA test sequences.

We used MIPS to find a single primer with maximum coverage in each test set, subject to varying degeneracy bounds α. **Table 3** compares the average coverage of primers found by MIPS in 20 trials to the largest coverage k such that $E_{l,\alpha,k}$ for test sets of the specified size is > 1. Primers with coverage exceeding this value of k are not expected to occur in the test sets, whereas primers with slightly smaller coverage may or may not occur frequently.

Table 3
Actual and Predicted Coverage of 20-mer Primers Found on Sets of 190 Random Sequences of Length 211

Degeneracy α	Avg coverage	Max predicted
1000	6.30	7
10000	10.55	12
100000	19.30	26

Avg coverage: average coverage of primer found over 20 random trials; Max predicted: largest coverage m such that $E_{20,\alpha,m} > 1$.

MIPS is adept at finding primers close to the maximum predicted coverage for relatively small degeneracies ($\alpha \leq 10000$). The gap between the best primers found by MIPS and those predicted to occur in theory grows with the degeneracy bound. It is unclear whether this fact represents a limitation of the algorithm or of the theoretical estimates, because primers with expectation greater than one may, with significant probability, still fail to occur. Moreover, the high degeneracies where MIPS might perform poorly are of less practical interest, because single primers with such high degeneracies are experimentally more difficult to work with.

Overall, MIPS appears to be operating close to the theoretical limit for MC-DPD problems of small degeneracy. Although our analysis does not directly address the MDPD problems, any large gap between the most efficient design and the designs produced by MIPS is unlikely to arise from failure to find single high-coverage primers when they exist.

3.4.2. Mispriming

Because of the presence of auxiliary primers, it is possible that a pair of primers binds to an undesired location and results in the amplification of an unwanted PCR product. *Mispriming* is the occurrence of this event where the unwanted product is indistinguishable, by size, from the targeted products.

Consider a set of degenerate primers with length l, such that the *total degeneracy* of the set is α. The following procedure is used to estimate the expected number of mispriming events when this primer set is applied to a genome of length g. The background model greatly influences the calculations; therefore, this calculation considers two different models: an i.i.d. random genome with equal base frequencies and the human genome.

A pair of l-mers cause a mispriming event if and only if they bind to the genome within δ bases of each other in the appropriate orientations to permit amplification of the sequence between them. Let i index the positions of the genome on its forward strand. Let the 0–1 random variable x_i indicate the event that an l-mer from our primer set is complementary to the forward strand at position i, and let $\bar{x}_i$ be the event that an l-mer is complementary to the reverse-complement strand at i. A mispriming event occurs at i if $\bar{x}_i \cap \bigcup_{j=i}^{i+\delta-1} x_j = 1$. Denote this event by the 0–1 indicator M_i. The total number of mispriming events M in a genome of size g is $\sum_{i=1}^{g} M_i$.

3.4.2.1. Mispriming in i.i.d. Random and Human Genome

In the first model, the genome consists of i.i.d. random sequence with equal base frequencies. For the expectation of a matching event to occur at a position i, we have that $E[x_i] = E[\bar{x}_i] = \frac{\alpha}{4^l}$.

Note that the matching events for each half of a primer pair are independent in an i.i.d. random DNA sequence when the two primers do not overlap. To simplify the calculations, the effects of overlapping primer boundaries are ignored. Using Poisson approximation to estimate the probability of the matching event on the forward strand:

$$E[M_i] = E[\bar{x}_i \cap \bigcup_{j=i}^{i+\delta-1} x_j] = E[\bar{x}_i]E\left[\bigcup_{j=i}^{i+\delta-1} x_j\right] \approx E[\bar{x}_i]\left(1 - e^{-\sum_{j=i}^{i+\delta-1} E[x_j]}\right).$$

Finally, setting $\rho = \alpha/4^l$, the expected mispriming rate is derived as

$$E[M] = \sum_{i=1}^{g} E[M_i] \approx g\rho\left(1 - e^{-\delta\rho}\right).$$

These calculations were tested using an artificially constructed "human-size" genome ($g \approx 3 \times 10^9$) of i.i.d. random sequence with equal base frequencies. The query set was obtained from results of MIPS-PT on the human data set used in **Subheading 3.3**. Finally, a PCR experiment was simulated using both the artificial genome and the real human genome (April 10, 2003) ***(17,18)***, assuming that the primers in the solution would bind only to their complementary sequences, thus ignoring inexact binding. In accordance with the calculations, a mispriming event was marked if (1) an instance of a matching event occurred in one strand and (2) another matching event occurred on the opposite strand within $\delta = 500$ bp. **Table 4** summarizes the total degeneracy of the solution, the predicted number of mispriming events, and finally the number of mispriming events seen in the simulation of the artificial and human genomes.

The model predicts the mispriming rate well for the artificial genome, but it fails to predict as well for the human genome. This result is not surprising, because the sequence of the human genome is strongly non-random. A key source of non-randomness is the presence of interspersed repeats and regions of low-complexity sequence ***(19,20)***. The presence of repetitive elements in the human genome can affect the mispriming rate of the MIPS solver by violating the implicit assumption that a degenerate primer's mispriming rate is solely determined by the degeneracy of the primer. The next section suggests a possible heuristic to select effective degenerate primers which do not misprime with such high frequency.

Table 4
Predicted and Actual Mispriming Rates in Simulated PCR Experiments with i.i.d. Random DNA and the Human Genome

Total degeneracy	Predicted	Random genome	Human genome
84720	0.009	0	82254
321456	0.133	1	112162
1262260	2.063	6	64938
4824870	30.12	81	201209

3.4.3. Reducing Mispriming Events

Consider an input sequence that contains fragments which are overrepresented in the human genome. If MIPS selects any of these fragments as the primer-binding site in the final solution, the likelihood of a mispriming event in the human genome increases. The solution to this problem, therefore, is to exclude these fragments from the final solution. To remove highly overrepresented fragments caused by interspersed repetitive elements in the input sequences, we preprocessed these sequences with RepeatMasker ***(21)*** using the primate filter. **Figure 5** shows the result of applying RepeatMasker to the human SNP data set.

A side effect of using the masked input set was that two of the sequences of the input set were rendered unusable. The masking process effectively reduces the size of the input sequences and therefore the possible binding sites. Two of the input sequences did not contain 20 consecutive unmasked bases, or any possible binding sites; so, they were omitted. **Table 5** summarizes the reduction in mispriming events when the input sequences are masked by comparing the mispriming rates of the results of MIPS-PT on the unmasked input set versus the masked input set.

Empirically, these results seem to indicate that simply removing overrepresented fragments from the input set renders the results of MIPS far more useful in practice by reducing the number of predicted PCR artifacts.

Another interesting result of masking the input sequences for this particular data set is the resulting solution from MIPS-PT. Intuitively, it is expected that reducing the size of the input set would likely increase both the size and total degeneracy of the final solution when compared with the original data set, because the likelihood of finding similar fragments is decreased. **Table 6** summarizes the number of primers selected and total degeneracy of the final solutions for both the original and masked input set. For each degeneracy

```
RepeatMasker summary:
==================================================
file name: RMemail6411.seq
sequences:            190
total length:       34874  bp  (34874 bp excl N-runs)
GC level:           40.59  %
bases masked:        2756 bp  (  7.90%)
==================================================
                number of       length      percentage
                elements*      occupied     of sequence
--------------------------------------------------
SINEs:                 14         1217 bp        3.49 %
      ALUs              8          601 bp        1.72 %
      MIRs              6          616 bp        1.77 %
LINEs:                  4          493 bp        1.41 %
      LINE1             3          392 bp        1.12 %
      LINE2             1          101 bp        0.29 %
      L3/CR1            0            0 bp        0.00 %
LTR elements:           5          551 bp        1.58 %
      MaLRs             2          289 bp        0.83 %
      ERVL              1           97 bp        0.28 %
      ERV_classI        2          165 bp        0.47 %
      ERV_classII       0            0 bp        0.00 %
DNA elements:           2          109 bp        0.31 %
      MER1_type         0            0 bp        0.00 %
      MER2_type         1           78 bp        0.22 %
Unclassified:           0            0 bp        0.00 %
Total interspersed repeats:       2370 bp        6.80 %
Small RNA:              2          218 bp        0.63 %
Satellites:             0            0 bp        0.00 %
simple repeats:         3           95 bp        0.27 %
Low complexity:         3           73 bp        0.21 %
==================================================
* most repeats fragmented by insertions or deletions
  have been counted as one element

The sequence(s) were assumed to be of primate origin.
RepeatMasker version 07/07/2001, default mode
run with cross_match version 0.990329
RepBase Update 6.3, vs 05152001
```

Fig. 5. Results of RepeatMasker on human SNP input dataset.

Table 5
Mispriming Rates in Simulated PCR Experiments with Original and Masked Input Sets

Degeneracy threshold	Original	Masked
$4^6 \approx 4K$	82254	164
$4^7 \approx 16K$	112162	1104
$4^8 \approx 64K$	64938	2043
$4^9 \approx 262K$	201209	17337

Table 6
Comparison of MIPS-PT Results on Original and Masked Input Sets

	Original		Masked	
Threshold	# Primers	Degeneracy	# Primers	Degeneracy
$4^6 \approx 4K$	53	84720	49	128144
$4^7 \approx 16K$	44	321456	42	319872
$4^8 \approx 64K$	36	1.262×10^6	34	1.299×10^6
$4^9 \approx 262K$	29	4.824×10^6	28	4.277×10^6

threshold tested, MIPS-PT actually selected *fewer* primers for the masked data set, and on two occasions, the total degeneracy of those primers was also *less* than that of the original set.

3.4.4. Alternate Strategies to Reduce Mispriming

Using RepeatMasker on the input set dramatically reduces the number of expected mispriming events by eliminating input sequence fragments that are overrepresented in the genome. However, it is still possible that one or more of the degenerate primers selected binds to an overrepresented fragment which does not occur at all in the input set. For example, suppose *ACACACAC* is a repetitive element in the genome that is not in the input. If MIPS chooses a primer *MMMMMMMM*, where *M* is the degenerate nucleotide which represents *{A,C}*, then this solution will incur a large number of mispriming events.

It is undesirable to select any degenerate primer that would bind to an overrepresented sequence fragment, whether that fragment is part of the input. We therefore want a method to determine whether a given degenerate primer is likely to cause a large number of mispriming events, so that we may exclude that primer from consideration *before* it is selected as part of a final solution. A simple approach would be to maintain a list of all possible degenerate primers along with their frequencies in the human genome. A scoring function could then be generated to calculate the likelihood of a degenerate primer being involved in a mispriming event. However, there are over 10^{24} degenerate primers of length 20; so, maintaining such a data structure is currently infeasible.

A more practical workaround would be to dynamically calculate a mispriming likelihood for each primer as it is encountered during beam search. It is feasible to estimate the probability of a degenerate sequence appearing in a complex background model such as the human genome using a high-order Markov model and a dynamic programming algorithm, similar to the Viterbi algorithm *(22)*.

4. Notes

1. For the average desktop computer, beam sizes larger than a few hundred result in impractical running times. For the input set we used, which contained 95 human DNA sequences, using beam sizes over 100 produced solutions that did not significantly improve as the beam size increased. Empirically, a beam size close in value to the number of sequences in the input set seems to produce a solution that is balanced in running time and quality.
2. The MIPS algorithm uses the following $n \times n$ matrices for backtracking and record-keeping.

   a. *BEST*(i,j) represents the ith candidate primer which covers j (of n) input sequences.
   b. *COVERED*(i,j) represents the set of sequences covered if the primer described by *BEST*(i,j) is part of the final solution.
   c. *ALLOWABLE*(i,j) represents the remaining degeneracy if the primer described by *BEST*(i,j) is part of the final solution.

3. The pairwise comparison of two sequence fragments is the dominant operation and a rate-limiting step of the MIPS algorithm. A majority of these comparisons are between two fragments that share few, if any, nucleotides. To avoid comparisons between dissimilar fragments, the exhaustive pairwise comparison is replaced with a FASTA-style *(23)* similarity lookup. For some fixed word size w, every length w substring of each fragment is added to a lookup table and associated with the fragment from which it originated. Two fragments are compared only if a table lookup shows that they share at least one word of size w. For DNA, a FASTA word length of $w = 6$ is recommended *(24)*.
4. The comparison results between MIPS and HYDEN can be partially explained by the differing design requirements of the DPD and MDPD problems. Even when

applied iteratively, the goal of the DPD problems is to have a result which could be divided into distinct PCR experiments. The goal of the MDPD problems is to have a set of primers for one large-scale PCR experiment. Specifically, to solve the DPD problem, the HYDEN algorithm must ensure that for any given degenerate forward primer that is discovered, exactly one degenerate reverse primer is used to cover the sequences covered by the forward primer. Therefore, a given degenerate forward primer is restricted as to which sequences it is reported to cover based on the presence of a suitable degenerate reverse primer, and vice versa. Moreover, the HYDEN algorithm has the additional restriction that any given degenerate primer is limited to covering a set of either forward or reverse primers, but not both.

Acknowledgments

The authors thank Pui Kwok for describing the problem and useful discussions. This research was supported by funding from NSF (grants IIS-0196057, ITR/EIA-0113618, DGE-0202737, and IIS-0535257) and NIH (grants GM08802 and HG00249). Preliminary results of this research and early versions of the paper appeared in **refs**. ***13*** and ***25***.

References

1. G. Marth, R. Yeh, M. Minton, R. Donaldson, Q. Li, S. Duan, R. Davenport, R. Miller, and P. Kwok. Single-nucleotide polymorphisms in the public domain: how useful are they? *Nature Genetics*, 27: 371–372, 2001.
2. F. S. Collins and V. A. McKusick. Implications of the human genome project for medical science. *JAMA*, 285:2447–2448, 2001.
3. P. Kwok. Methods for genotyping single nucleotide polymorphisms. *Annual Review of Genomics and Human Genetics*, 2:235–258, 2001.
4. W.R. Pearson, G. Robins, D.E. Wrege, and T. Zhang. On the primer selection problem in polymerase chain reaction experiments. *Discrete and Applied Mathematics*, 71:231–246, 1996.
5. M.R. Garey and D.S. Johnson. *Computers and Intractability: A Guide to the Theory of NP-Completeness*. Freeman, New York, NY, 1979.
6. K. Doi and H. Imai. A greedy algorithm for minimizing the number of primers in multiple PCR experiments. *Genome Informatics*, 10:73–82, 1999.
7. W. R. Pearson, G. Robins, D. E. Wrege, and T. Zhang. A new approach to primer selection problem in polymerase chain reaction experiments. In *Third International Conference on Intelligent Systems for Molecular Biology*, pp. 285–291. AAAI Press, Cambridge, United Kingdom, 1995.
8. P. Nicodeme and J. Steyaert. Selecting optimal oligonucleotide primers for multiplex PCR. In *Proceedings of Fifth Conference on Intelligent Systems for Molecular Biology ISMB97*, HAAI Press, Halkidiki, Greece pp. 210–213, 1997.

9. K. Doi and H. Imai. Complexity properties of the primer selection problem for PCR experiments. In *Proceedings of the 5th Japan-Korea Joint Workshop on Algorithms and Computation*, The University of Tokyo, Tokyo, Japan pp. 152–159, 2000.
10. S. Kwok, S.Y. Chang, J.J. Sninsky, and A. Wang. A guide to the design and use of mismatched and degenerate primers. *PCR Methods and Applications*, 3:S39–S47, 1994.
11. Cornish-Bowden. IUPAC-IUB symbols for nucleotide nomenclature. *Nucleic Acids Research*, 13:3021–3030, 1985.
12. C. Linhart and R. Shamir. The degenerate primer design problem. *Bioinformatics*, 18 (Suppl. 1):S172–S180, 2002.
13. R. Souvenir. An iterative beam search algorithm for degenerate primer selection. Master's thesis, Washington University, Saint Louis, MO, December 2003.
14. R. Bisiani. Search, beam. In S.C. Shapiro, editor, *Encyclopedia of Artificial Intelligence*, pp. 1467–1468. Wiley-Interscience, New York, NY, 2nd edition, 1992.
15. G.Z. Hertz and G.D. Stormo. Identifying DNA and protein patterns with statistically significant alignments of multiple sequences. *Bioinformatics*, 15: 563–577, 1999.
16. M.S. Waterman. *Introduction to Computational Biology*. Chapman & Hall, London, UK 1995.
17. The Genome Sequencing Consortium. Initial sequencing and analysis of the human genome. *Nature*, 409:860–921, 2001.
18. UCSC genome browser. Web site http://genome.ucse.edu.
19. A.F.A Smit. Origin of interspersed repeats in the human genome. *Current Opinion in Genetics and Development*, 6(6):743–749, 1996.
20. A.F.A Smit. *Structure and Evolution of Mammalian Interspersed Repeats*. PhD thesis, USC, 1996.
21. A.F.A. Smit and P. Green. RepeatMasker. Available at http://ftp.genome.washington.edu/RM/RepeatMasker.html.
22. G.D. Forney, Jr. The Viterbi algortihm. In *Proceedings of IEEE*, volume 61, pp. 268–278, 1973.
23. W.R. Pearson and D.J. Lipman. Improved tools for biological sequence analysis. In *PNAS*, volume 85, pages 2444–2448, 1988.
24. D. Gusfield. *Algorithms on Strings, Trees, and Sequences: Computer Science and Computational Biology*, chapter 15, p. 377. Press Syndicate of the University of Cambridge, 1997.
25. R. Souvenir, J. Buhler, G. Stormo, and W. Zhang. Selecting degenerate multiplex PCR primers. In *Proceedings of Workshop on Algorithms in Bioinformatics (WABI-03)*, 2003.

13

Primer Design for Multiplexed Genotyping

Lars Kaderali

Summary

Single-nucleotide polymorphism (SNP) genotyping can be carried out by annealing an oligonucleotide primer directly adjacent to the polymorphism and carrying out a single base extension using a polymerase reaction with labeled dideoxynucleotide triphosphates. This can be multiplexed by attaching a unique tag at the 5′-end of each oligonucleotide primer and binding the corresponding antitag to a DNA microarray or microbead. After the polymerase reaction, the tag–antitag system can be used to demultiplex the experiment. However, such an assay requires careful primer and tag design to avoid any crossreactivity among the primers, tags, antitags, and template sequence. A procedure for designing the primers is described in this chapter.

Key Words: Genotyping; minisequencing; multiplexing; primer design.

1. Introduction

Single-nucleotide polymorphisms (SNPs) comprise a substantial portion of all common human variation *(1,2)*. In the future, the routine analysis of SNPs will play a major role in medical diagnosis; it will enable studies of human genetic diversity, providing new insights into the history of human populations *(3)*; and it will be a major tool in the identification of genes that confer risk of common diseases *(4)*. Such applications require the simultaneous screening of hundreds to thousands of polymorphisms, constituting a pressing need for fast, robust, and cost-efficient SNP scoring methods.

In the minisequencing approach to genotyping, an oligonucleotide primer is annealed directly adjacent to the site of interest. A single base extension is then

From: *Methods in Molecular Biology, vol. 402: PCR Primer Design*
Edited by: A. Yuryev © Humana Press, Totowa, NJ

carried out using a polymerase reaction with labeled dideoxynucleotide triphosphates (ddNTPs), allowing a base call for the mutation of interest based on the label of the ddNTP inserted by the polymerase. This assay can be multiplexed by appending a unique tail to the 5′-end of each minisequencing primer and by covalently binding the reverse complementary tail to a microarray or microsphere surface. After the single-base extension reactions, demultiplexing of the experiment can be carried out by annealing the tailed primers to the complementary tags on the chip or bead. Bead color or array position then indicate the polymorphisms identity, and the nucleotide label reveals the genotype *(5)*. **Figure 1** illustrates the process.

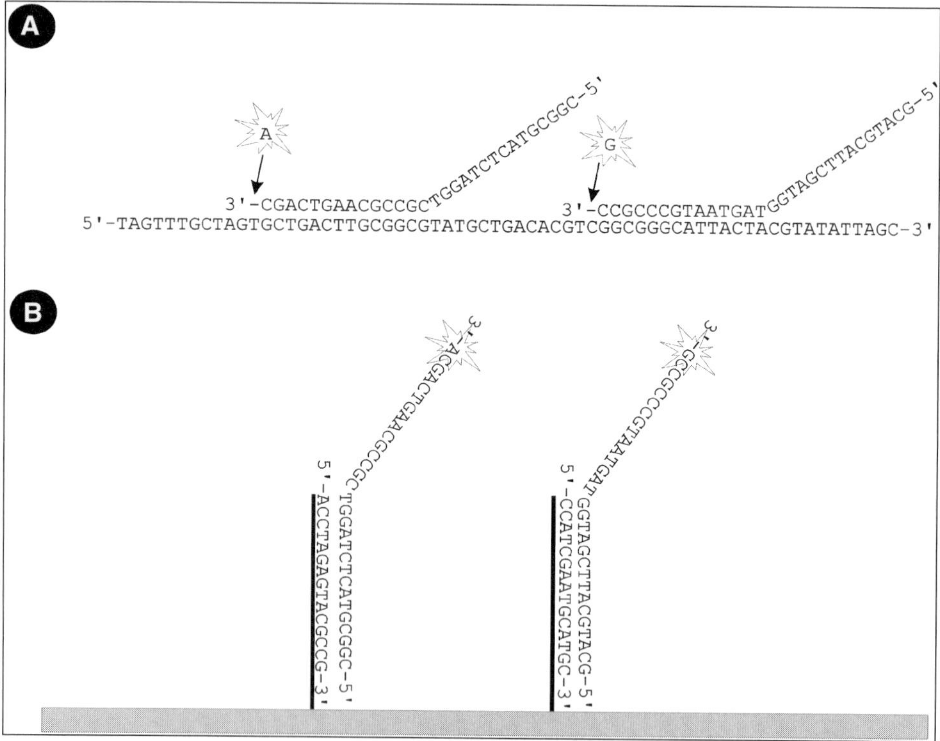

Fig. 1. Illustration of multiplexed minisequencing experiment. Tailed primers are annealed to template sequence and extended by one single base, bearing a colored label (**A**). Thereafter, the extended primer binds to a specific location on a microarray (**B**). Location on the array corresponds to SNP identity, and color of label reveals genotype.

Such a multiplexed assay poses complex requirements on primer design. First of all, a minisequencing primer is required for each polymorphism of interest. This primer must be specific to its target site, and all minisequencing primers must work under the same reaction conditions. In particular, the primer must not false prime, and different primers (from the same or distinct polymorphism sites) must not bind to one another, interfering with primer–template interactions.

In addition, a tag–antitag system is required for demultiplexing the assay. Here as well, each tag must be specific to its corresponding antitag, and all tag–antitag pairs must form under the same reaction conditions. The tags used in the assay must not bind to any of the minisequencing primers, and the formation of secondary structure of the primer–tag pairs (foldback of the pairs onto themselves) or heterodimer formation (binding of two distinct pairs with one another) must be avoided.

As the genotyping results depend strongly on proper primer design, precise predictions of interactions between nucleotides involved are a necessity. Mere string comparisons are unlikely to yield the required accuracy, which is why we will use the nearest-neighbor thermodynamic model to compute interactions in the following *(6)*. One problem in the application of the nearest-neighbor model is it requires advance knowledge of which nucleotide basepairs are going to form in the reaction. Unfortunately, in the case of false priming or primer dimer formation, this is not known. We have previously proposed a thermodynamic alignment algorithm to solve this problem, the algorithm will be described in **Subheading 2.1** and will then be used to carry out ensuing calculations *(7)*.

After the discussion of the thermodynamic alignment computations, we will give a step-by-step walkthrough of the separate tasks to be carried out for design of the minisequencing primers in **Subheading 2.2** and address the problems associated with multiplexing the primers. Then, in **Subheading 2.3**, the generation of tag–antitag systems for demultiplexing the assay is described, and pairing of primers and tags is discussed. Both of these sections should give sufficient detail for a programmer to implement the algorithms and will give the laboratory user a profound understanding of what is addressed by the approach and what is not. **Subheading 3** provides additional comments and notes on the algorithms described in **Subheading 2**.

The algorithms presented here have been implemented in the SBEprimer software package *(8)*, which is available for download from http://www.zaik.uni-koeln.de/AFS/Projects/Bioinformatics/sbeprimer.html or can be obtained directly from the author.

2. Methods

The methods described below outline (1) the computation of the change in free energy associated with binding of two arbitrary DNA strands, (2) the selection and multiplexing of primers for the genotyping reaction, and (3) the design and use of a tag–antitag system to demultiplex the assay.

2.1. Thermodynamic Alignment for Free-Energy Calculations

A main component of the following primer design algorithms is the computation of the change in free energy associated with the binding reaction of two arbitrary (more or less complementary) single-DNA strands. The nearest-neighbor thermodynamic model is a widely used model for the prediction of nucleic acid duplex stability. Its main use in primer design is to predict the stability of the desired primer–template pairs. For details on the model, *see* Chapter 1 or **ref.** ***6***.

For the prediction of false binding sites or primer dimer formation, the nearest-neighbor model is rarely used, because it requires knowledge about which basepairs do form in the duplex. Algorithms to simulate unimolecular folding of DNA or RNA strands based on the nearest-neighbor model do exist and could in principle be applied to false priming prediction as well ***(9)***; however, the running time of these methods is usually prohibitive when several thousand potential interactions need to be evaluated. This is why most primer design programs use algorithms such as BLAST to detect false priming.

In contrast, since, in the genotyping setting previously described, false priming will directly result in false basecalls, it would be highly desirable to have a quick approximate evaluation of thermodynamic properties of potential crosshybridization products.

The idea proposed here is to combine the popular Smith–Waterman alignment ... algorithm with the nearest-neighbor model. Using free energy as the alignment objective function, one maximizes the free energy of duplex formation by computing the alignment of two DNA sequences. Gaps in the alignment then correspond to unpaired bases in a duplex, as may occur in bulge loops or other secondary structure.

A good introduction to the Smith–Waterman alignment algorithm can be found in **ref.** ***10***. Roughly speaking, the idea is to compute an optimum alignment of length n, by extending shorter alignments of length $n - 1$ in a way maximizing the alignment score. In our case, the alignment score is the alignment free energy, because this will yield the most stable duplex two DNA single strands can form with one another. Only small modifications of the original Smith–Waterman algorithm are required to take into account the nearest neighbors from the thermodynamic model.

The aim of the thermodynamic alignment is to minimize the Gibb's free energy, for given temperature. Writing ΔG for the change in free energy, the target function for a given alignment z of sequences x and y can be calculated additively, as described in **ref. 6**.

$$\Delta G(z) = \sum_{i=1}^{k} \Delta G_{x_i x_{i+1}/y_i y_{i+1}},$$

where k is the length of the alignment, x_i and y_i are the ith base of sequence x and y, respectively (extended by gaps "-" if necessary), and $\Delta G_{x_i x_{i+1}/y_i y_{i+1}}$ is the change in free energy according to the nearest-neighbor model for pairing of neighboring bases x_i, x_{i+1} with bases y_i, y_{i+1}.

Finding the alignment with minimum free energy (and thus largest negative change in free energy, ΔG) is efficiently solvable using standard dynamic programing techniques, as in the Smith–Waterman alignment algorithm. When considering just the last position of an alignment of two sequences $x = x_1\ x_2 \ldots x_n$ and $y = y_1\ y_2 \ldots y_m$, there are three possibilities what such an alignment may look like: (1) the alignment may end with a pairing of x_n with y_m or (2) it may end with a pairing of x_n with a gap or (3) it may end with a pairing of y_m with a gap, compare **Fig. 2A**.

In the first case, omitting the last pair, one is thus looking at a prefix alignment of sequences $x_1 x_2 \ldots x_{n-1}$ with $y_1 y_2 \ldots y_{m-1}$, followed by the $x_n : y_m$ pair. In the second case, the prefix alignment aligns sequences $x_1 x_2 \ldots x_{n-1}$ with $y_1 y_2 \ldots y_m$, followed by the x_n:gap pair, whereas, in the third case, the prefix alignment is of subsequences $x_1 x_2 \ldots x_n$ with $y_1 y_2 \ldots y_{m-1}$, followed by the gap: y_m pair.

The Smith–Waterman alignment algorithm makes use of this observation by constructing a full alignment of two sequences recursively from alignments of prefixes. Writing the sequences x and y along the axes of a table S, the maximum alignment score of an alignment of prefixes $x_1 x_2 \ldots x_i$ with $y_1 y_2 \ldots y_j$ can be found at position (i, j) in the table and can be computed by adding the score of the last pairing with the score for the corresponding prefix alignment (*see* **Fig. 2B**). The optimum alignment score in position (i, j) in the table simply is the maximum of the three possibilities (1), (2), and (3). Mathematically, this can be expressed by the equation:

$$S(i, j) = \max \begin{cases} S(i-1, j-1) + w_{i,j} \\ S(i-1, j) + w_{i,\text{gap}}, \\ S(i, j-1) + w_{\text{gap},j} \end{cases}$$

where $w_{i,j}$ is the score for pairing x_i with y_j, and $w_{i,\text{gap}}$ and $w_{\text{gap},j}$ are the scores for pairing x_i or y_i with a gap, respectively.

(a)

X_1	X_2	...	X_{n-1}	X_n
Y_1	Y_2	...	Y_{m-1}	Y_m

(1)

X_1	X_2	...	X_{n-1}	X_n
Y_1	Y_2	...	Y_m	-

(2)

X_1	X_2	...	X_n	-
Y_1	Y_2	...	Y_{m-1}	Y_m

(3)

(b)

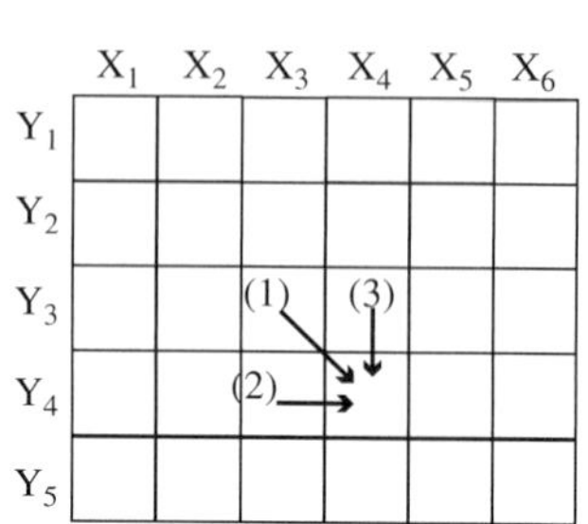

Fig. 2. **(A)** Three possible ways to construct an alignment of sequences $x = x_1x_2\dots x_n$ and $y = y_1y_2\dots y_m$ from prefixes. The alignment may end **(1)** with a pairing of x_1 with y_1, **(2)** with a pairing of x_1 with a gap, or **(3)** a pairing of y_1 with a gap. **(B)** Principle of Smith–Waterman alignment algorithm: alignments of prefixes are extended by considering the three possibilities how a longer alignment may end. The figure shows the dynamic programing matrix, in which the scores for all prefix alignments are stored. For example, for an alignment of $x_1x_2\dots x_4$ and $y_1y_2\dots y_4$, corresponding shorter alignments are extended by computing the score for the full alignment along the arrows indicated in the figure.

To compute a thermodynamic alignment, we use the Gibbs free Energy as alignment score. However, as the nearest-neighbor model looks not only at single basepairings, but makes use of the neighboring pairs as well, the Smith–Waterman algorithm needs to be adapted slightly. This is done by storing at each location in the table all the three possibilities how an alignment can end, compare the cases (1), (2), and (3) above, and adding a third index u to the table to distinguish the three cases. In the iteration, one can then choose the best score for the neighbors as well. The dynamic programing recursion then becomes

$$\Delta G_{i,j,0} = \min_{u=0,1,2} \left(\Delta G_{i-1,j-1,u} + [\Delta G(x_i, y_j)]_u\right),$$

$$\Delta G_{i,j,1} = \min_{u=0,1} \left(\Delta G_{i-1,j,u} + [\Delta G(x_i, -)]_u\right),$$

$$\Delta G_{i,j,2} = \min_{u=0,2} \left(\Delta G_{i,j-1,u} + [\Delta G(-, y_j)]_u\right),$$

where $\Delta G_{i,j,u}$ are the already calculated values of intermediary alignments of the first i bases of sequence x with the first j bases of sequence y; the third index u refers to the three possible types of alignment (0 for aligning x_i with y_j, 1 for x_i with a gap, and 2 for y_j with a gap), and $[\Delta G(x_i, y_j)]_u$, $[\Delta G(x_i, -)]_u$ and $[\Delta G(-, y_j)]_u$ are the respective values which have to be added according to the nearest-neighbor model.

Algorithm 1 shows the resulting program code. Note that this is a slight modification from usual alignment calculations because parameters stem from two neighboring base pairs and not from single base pairings.

Algorithm 1. Thermodynamic Alignment Algorithm

```
Input sequences x and y to align
Let a=length (x), b=length (y)
Allocate Memory for Table G[0..a, 0..b, 0..2]
Initialize Table: G[i, j, k]=0 for all i, j, k
for i from 1 to a do
  for j from 1 to b do
   G[i, j, 0]=min_u G[i-1,j-1,u]+[ΔG(x_i, y_j)]_u
   G[i, j, 1]=min_u G[i-1,j,u]+[ΔG(x_i, −)]_u
   G[i, j, 2]=min_u G[i,j-1,u]+[ΔG(−, y_j)]_u
  end for
end for
Return minimum of values in G[a,*,*] and G[*,b,*]
```

2.2. Step-by-Step Primer Design for Genotyping

The actual procedure to select primers for the genotyping reaction carries out a number of iterative steps, as follows:

1. For each SNP, determine whether the plus- or minus-strand primer adjacent to the polymorphism will false prime. If this is the case, eliminate the corresponding primer from further consideration.
2. Generate a list of primers of different length from the plus and minus strand (provided they have not been eliminated in **step 2**). Check that the primers are within length restrictions provided by the user and that the primer melting temperature is above the annealing temperature used in the genotyping assay.
3. Check each primer candidate for homodimer and hairpin formation. Eliminate primer candidates not satisfying requirements set.
4. Compute all possible heterodimer interactions between different primer candidates for all SNPs.

5. Choose maximal sets of mutually compatible primers, selecting one primer for each polymorphism and multiplexing as many SNPs as possible.

I will now detail each of the steps above in the following.

2.2.1. False Priming Check

There are two possible locations where a primer for any given SNP can be placed. As the minisequencing primer must bind directly adjacent to the 3′-end of the polymorphism, one can only chose which of the two strands of the DNA duplex the primer should be located on. The only further choice one has concerns the length of the primer toward its 5′-end, the location of the 3′-end being fixed.

Regarding potential false priming of a primer, one is mainly concerned with a primer forming stable interactions with the DNA template at an undesired location at the primers 3′-end. If such interactions occur and are sufficiently stable, the polymerase reaction will extend the primer at its 3′-end, potentially resulting in a false genotyping basecall. We will thus check whether bases complementary to the four 3′-end bases of any primer candidate occur anywhere in the template sequence. If this is the case, the thermodynamic alignment algorithm presented above is used to compute the free energy associated with the reaction of primer and template at the respective location, and the primer is discarded if this interaction is over a user-defined stability threshold.

The method used to quickly identify potential false priming sites uses a table, storing information about each primer's 3′-end bases. Basically, the idea is to create $4^4 = 256$ lists, corresponding to all possible DNA 4-mers. In each list, we store all primer candidates of maximum length as specified by the user containing the respective 4-mer at the 3′-end. Note that short primers at the same location need not be considered, because their sequence will be a subsequence of the maximum length primer. One then shifts a 4-base window over all template sequences and, by accessing the corresponding list, can quickly identify those primers that may potentially false prime at the respective template location. For such pairs of primers and template subsequence, the thermodynamic alignment algorithm is then used to compute a free energy change for the binding reaction. Note that, to account for stable prefixes of the two sequences as well, one should look for the minimum free energy in the entire table and not just the last row and last column. This will ensure that mismatches at the primer's 5′-end will be ignored and provides an additional margin of safety in the primer design process. If any stable interaction is found, the corresponding primer should not be used for the genotyping assay—independent of how long the primer is toward its 5′-end.

2.2.2. Primer Candidate Generation

After the exclusion of primer candidates prone to false priming, the next step is to consider primer length. Primers passing the first step described above are now used to generate a list of primer candidates, by changing the length of the primer at its 5′-end between user-specified minimum and maximum primer length parameters. Each primer candidate generated this way is inserted into a list of potential primers, storing along with the primer which polymorphism it is intended for, and whether it is from the plus or the minus strand of the template sequence.

In addition to restrictions on the length of a given primer, users may have preferences regarding the primer's melting temperature. At a minimum, for the assay to work, the melting temperature of the primer with its intended binding location adjacent to the SNP must be larger than the annealing temperature used in the experiment—otherwise, the primer will not bind to the template, the polymerase will not be able to extend the primer, and hence, no minisequencing reaction will occur. A user may wish to include an additional "margin of safety" on the minimum melting temperature and may also have preferences concerning the maximum temperature allowed.

The next step to be carried out is thus the computation of the melting temperature of a primer with its perfect complement in the Watson–Crick sense. This is easily done using the nearest-neighbor model, as described in **ref. 6**. Primer candidates with melting temperature not within the range required should then be removed from the list.

2.2.3. Evaluation of Primer Secondary Structure and Homodimer Formation

Primer candidates are then evaluated for secondary structure and homodimer formation. A primer folding back onto itself or binding to another primer for the same polymorphism will not be available for the minisequencing reaction, resulting in a weak signal. We would thus like to avoid such competing reactions.

The check for homodimer formation is easily carried out by using the thermodynamic alignment algorithm, by computing the alignment of a primer with itself. Special attention must be paid concerning the order in which the sequences are input to the algorithm—the first sequence given to the alignment algorithm must be the primer sequence in 5′ to 3′ order, the second sequence must be the same sequence in 3′ to 5′ order, but without Watson–Crick complementing the bases. Primer candidates showing homodimer formation with negative free energy change at the reaction annealing temperature (or a lower temperature if an additional margin of safety is desired) should be removed from the list of potential primers.

2.2.4. Heterodimerization and Primer Multiplexing

The next step to be carried out is to check which primers can be used together in the same experiment. The objective here is to find one minisequencing primer for each polymorphism, such that the chosen primers for different SNPs work together in one experiment. As we have already checked for false priming, homodimer formation, similar melting temperatures, primer length, and so on, the remaining step is to ensure that primers multiplexed together do not bind to one another, thus leading to false signal. If no single set of primers exists satisfying this condition, we would at least like to carry out as few experiments as necessary.

Mathematically, this problem can be described as a graph-coloring problem. A graph simply is a set of points, called nodes, connected by lines, called edges in graph theory. Graph coloring is the problem of assigning colors to the nodes, such that no two connected nodes have the same color, while minimizing the overall number of distinct colors used.

To model the primer-multiplexing problem as a graph-coloring problem, simply create one node for each primer candidate. In addition, the nodes should be labeled with the SNP location, and the sequence of the primer they correspond to. Two nodes should be connected with an edge, if the corresponding primers bind to one another and if in addition the primers belong to two different polymorphisms. **Figure 3** shows a sample graph created this way. Numbers in the nodes correspond to polymorphisms; thus, two nodes with the same number are two different primer candidates for the same SNP.

The generation of the graph requires the computation of all interactions between any pair of two primer candidates. Clearly, this is easily carried out using the thermodynamic alignment algorithm. As before, two primers are assumed to interact with one another, if the change in free energy associated with their nucleation at the annealing temperature is negative.

If we now try to color the nodes in the graph thus obtained with as few colors as possible, all nodes receiving the same color can be used together in one experiment—this is easily seen from the fact that nodes corresponding to primers that will interact are connected with an edge, but such nodes cannot be given the same color.

In fact, there is no need to assign a color to every node in the graph, because only one primer is required for each polymorphism. It is thus sufficient to assign a color to only one of the nodes corresponding to the same polymorphism, and

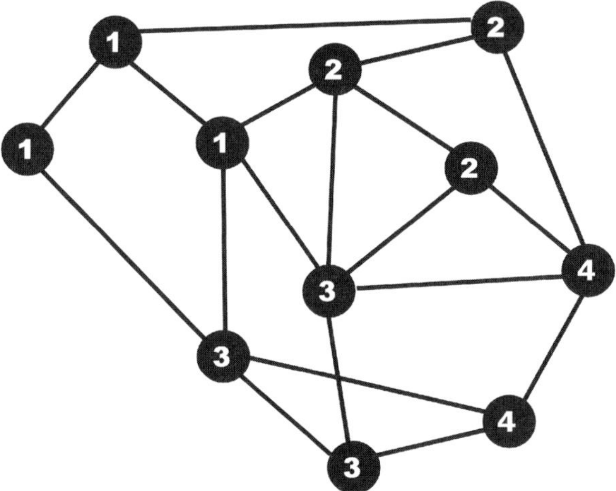

Fig. 3. Sample graph-coloring problem instance. Each node in the graph corresponds to a primer candidate, and numbers in the nodes correspond to polymorphism IDs. Two nodes with the same number correspond to two primer candidates for the same SNP. Edges between nodes indicate that the corresponding primers are predicted to bind to one another. The problem now is to choose one primer for each polymorphism, such that all primers chosen work together in one reaction; thus, choose one node from each group of nodes with the same number, such that the nodes chosen are not connected by edges.

which one can freely be chosen. The remaining nodes can be removed from the graph, simplifying the coloring problem (and probably reducing the number of colors required) considerably.

A simple argument will help us decide which nodes to leave in the graph. Consider all nodes corresponding to primer candidates for a given polymorphism (e.g., all nodes with number 1 in **Fig. 3**). The number of other nodes this node is connected to corresponds to the number of other primers the primer candidate considered will bind to. Most likely, the best primer candidate to keep for a given SNP is the one with fewest neighbors, that is, the one connected to the fewest other nodes. Therefore, we simply leave only this node in the graph, removing all other nodes (and corresponding edges) with the same SNP ID. If several nodes have minimal number of neighbors, just randomly choose one.

The resulting graph is then easily colored using a standard graph coloring algorithm, as can be found in most computer science textbooks. A very simple but still quite effective procedure simply starts at the node connected to most other nodes, assigns color 1 to it, and then iteratively assigns the smallest available color to the neighboring nodes, proceeding thus until the entire graph is colored. In this algorithm, the colors are assigned numbers from 1 to k, and by assigning the smallest permissible color to a given node, the algorithm tries to minimize the overall number of colors used.

Figure 4 shows the graph from **Fig. 3**, after removing nodes and with colors assigned to the remaining nodes. **Algorithm 2** summarizes the primer multiplexing step.

Given the colored graph, we can now conduct the experiment or experiments. The number of distinct colors in the graph corresponds to the number of separate experiments that need to be conducted. Each node remaining in the graph bears the label of one particular SNP and the primer sequence to use for that SNP, and all nodes with the same color can be multiplexed together in one experiment.

2.3. Tag-Generation for Multiplex Applications

The final ingredient required to demultiplex the assay is a tag–antitag system. The tags are attached to the 5′-end of the primers; each primer receives

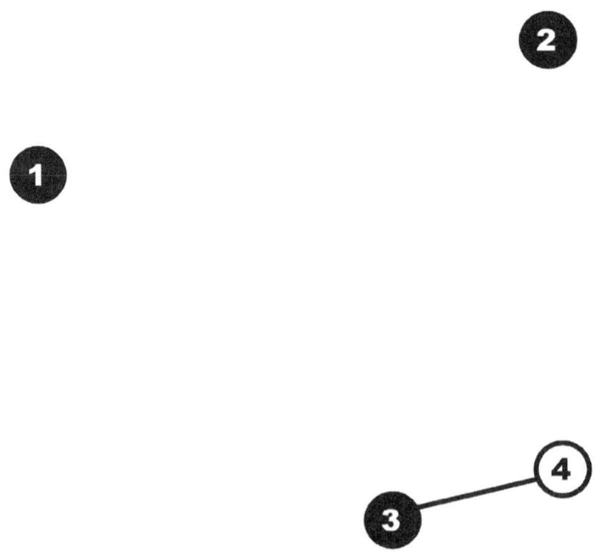

Fig. 4. The graph from **Fig. 3**, after removing all primer candidates (nodes) except for one with minimum number of interactions for each SNP. The primers for SNP 1, 2, and 3 work together in one experiment; a separate experiment is required for SNP 4.

Algorithm 2. Primer Multiplexing

```
Create graph with a node for every primer candidate
for all pairs of nodes do
   compute free energy change for most stable
   interaction of primers corresponding to the
   nodes, using algorithm 1.
   if free energy is negative
    Connect nodes with an edge
   end if
end for
Count number of neighbors for each node
For each SNP, remove all nodes except one node
with minimum number of neighbors. If several nodes
satisfy this condition, choose one randomly.
while not all nodes are colored
   choose remaining node n with largest number of
   neighbors (randomly if several nodes have
   same number of neighbors)
   color node n with smallest available color
end while
```

a different tag sequence. The corresponding antitags are then bound to a microarray or miscrosphere surface and are used to sort the primers to a specific location on the chip or to a specifically colored microsphere, thus demultiplexing the genotyping results.

2.3.1. Random Generation of Primer Tags

Requirements for the tags are as follows:

1. Each tag must bind to its corresponding antitag.
2. A tag must not bind to a foreign antitag.
3. No two tags should bind to one another.
4. No two antitags should bind to one another.

A straightforward method to generate a tag–antitag system is given in **Algorithm 3**. The approach taken simply is to randomly generate sequences and check whether they satisfy the above conditions, until enough tags have been generated.

Algorithm 3. Tag Generation

```
01 start with empty set of tags T
02 while more tags needed
03   repeat (generate sequences)
04    randomly generate DNA sequence S and its
      complement S^c
05    check whether S or S^c form a homodimer
06   until neither S nor S^c form a homodimer
07   if T is empty
08    add S and S^c to T
09   else
10    for all sequences t in T
11     check whether t and S or t and S^c bind
12    end if
13    if the check in lines 07–09 were negative for
      all t, add S and S^c to T.
14 end while
```

2.3.2. Pairing of Primers and Tags for Multiplexing

In principle, tags and antitags can be prefabricated and used for multiple different assays. However, one needs to check for possible interactions between primers and tags/antitags. The following problems need to be break addressed:

1. Binding of a primer–tag-pair to an undesired antitag, leading to wrong signal. This may happen, for example, if a primer binds to an antitag.
2. Binding of a primer–tag pair to another primer–tag pair, potentially leading to a wrong extension in the polymerase reaction.
3. Foldback of a primer–tag pair onto itself, causing wrong extension and lower signal because of the competing reaction.

Pairing of primers and tags under consideration of these conditions is relatively straightforward. Starting with a random pairing of primers and tags, one simply uses **Algorithm 1** as described in the previous sections to check for homodimer formation, heterodimer formation and crossreactivity between primer–tag pairs and antitags. If any problem is encountered, the involved tag/antitag pair is removed and replaced by a different tag/antitag pair. This is simply iterated until a feasible combination is found.

3. Notes

Note 1. Thermodynamic Alignment for Free Energy Calculations

- As DNA binds in reverse complementary order, sequence x should be input in 5′ to 3′ order, whereas sequence y must be given in 3′ to 5′ order.
- Steps 3 of **Algorithm 1** creates a table to store intermediary results of the computation. At position $G(i, j, s)$ in the table, the optimum change in free energy of the reaction of prefixes of the first i bases of sequence x with the first j bases of sequence y is stored. The third index k refers to the way the alignment ends—whether two bases are paired with one another ($k = 0$), a gap is paired with the last base in sequence $x(k = 1)$ or a gap is paired with the last base in sequence $y(k = 2)$. This distinction is necessary when the alignment is extended to prefixes of the first $i + 1$ and $j + 1$ bases, respectively, of sequences x and y.
- The loops in lines 5 and 6 go over the full length of sequences x and y and make sure that the alignment of the sequences is constructed step by step by elongating the prefixes of sequences that are aligned. Lines 7–9 carry out the respective computations as described in **Subheading 2**.
- Line 12 returns the largest change in free energy found, corresponding to the most stable alignment of the two input sequences.

A couple of additional remarks are in place concerning the entire computation at this point.

- First of all, initial gaps in the alignment should not be considered—this is achieved by initializing all fields in the table with zero in line 4.
- The same holds for terminal gaps. Terminal gaps are ignored by looking not just in cells $G(a, b, ^*)$ for the minimum score, but by considering all cells $G(0, b, ^*)$ to $G(a, b, ^*)$ and $G(a, 0, ^*)$ to $G(a, b, ^*)$.
- Dangling end thermodynamic parameters (compare **ref.** ***11***) can be made use of by appending an additional symbol at the beginning and end of the sequences x and y and by using the special parameters when this symbol occurs in the alignment.
- The algorithm as presented just returns the minimum free energy for an alignment of the two sequences x and y, without reporting the actual alignment. If needed, however, it is possible to recover the alignment from the dynamic programing table G, compare **ref.** ***10***.

Note 2. Step-by-Step Primer Design for Genotyping

A. Evaluation of Primer Secondary Structure and Homodimer Formation

A dedicated check for hairpin formation of the primer candidate could be included at this point. We have tested several simple heuristics for this purpose but have determined that, whenever one of these indicated a possibility of primer foldback, the homodimer check would also show homodimer formation. We thus consider a separate check for hairpin formation unnecessary.

B. Heterodimerization and Primer Multiplexing

- The graph-coloring algorithm described in this section is not guaranteed to find an optimum solution. In fact, finding an optimum solution to the problem is a very difficult problem for larger graphs and requires a large amount of running time. From the graph shown in **Fig. 3**, the reader should be able to manually find a solution such that all primers can be used together in one experiment. However, for real problems, this usually is not that easy, because the graph is much larger, and thousands of primer candidates need to be evaluated. In practice, the algorithm presented here has demonstrated good performance.
- Different and more complex graph-coloring algorithms do exist and can be adapted to this problem. Our results indicate that, in most cases, however, this does not yield significant improvements concerning the number of experiments that need to be conducted.

Note 3. Tag-Generation for Multiplex Applications

A. Random Generation of Primer Tags

- In line 4 of **Algorithm 3**, the random generation of sequences should be carried out base by base, drawing each base randomly from the set {A,C,G,T}. New bases should be added at the end of the sequence until the melting temperature (computed using the nearest-neighbor model) reaches a predefined bound, for example, 65°.
- The check in line 5 for homodimer formation should be carried out using **Algorithm 1**, by computing the change in free energy for a reaction of the sequence just generated with its complement.
- Similarly, the binding check in line 11 should be carried out using **Algorithm 1**. Note that one of the input sequences must be given in 5′ to 3′ order, the other in 3′ to 5′ order.
- A dedicated check for hairpin formation is not carried out, see the first remark in **Note 2**.
- As, when pairing primers and tags, a number of tags or antitags generated may turn out not to work together with the primers, one should generate more tags then one has polymorphisms in the assay, to be able to exchange them with different tags if necessary (*see* **Subheading 2.3.2.**).

References

1. Cooper, D., Smith, B., Cooke, H., Niemann, S., and Schmidtke, J. (1985) An estimate of unique DNA sequence heterozygosity in the human genome. *Hum. Genet.* 69, 201–205.

2. Venter, J., Adams, M., Myers, E., Li, P., et al. (2001) The sequence of the human genome. *Science* 291, 1304–1351.
3. Syvänen, A.-C. (1999) From gels to chips: 'minisequencing' primer extension for analysis of point mutations and single nucleotide polymorphisms. *Hum. Mutat.* 13, 1–10.
4. Schafer, A. and Hawkins, J. (1998) DNA variation and the future of human genetics. *Nat. Biotechnol.* 16, 33–39.
5. Cai, H., White, P., Torney, D., Deshpande, A., Wang, Z., Marrone, B., and Nolan, J. (2000) Flow cytometry-based minisequencing: a new platform for high-throughput single-nucleotide polymorphism scoring. *Genomics* 66, 135–143.
6. SantaLucia, J., Jr. (1996). A unified view of polymer, dumbbell, and oligonucleotide DNA nearest-neighbor thermodynamics. *Proc. Natl. Acad. Sci. U. S. A.* 95, 1460–1465.
7. Kaderali, L. and Schliep, A. (2002) Selecting signature oligonucleotides to identify organisms using DNA arrays. *Bioinformatics* 18, 1340–1349.
8. Kaderali, L., Deshpande, A., Nolan, J.P., and White, P.S. (2003) Primer-design for multiplexed genotyping. *Nucleic Acids Res.* 31, 1796–1802.
9. Zuker, M. (2003) Mfold web server for nucleic acid folding and hybridization prediction. *Nucleic Acids Res.* 31, 3406–3415.
10. Waterman, M.S. (1995) *Introduction to Computational Biology*. Chapman & Hall, London.
11. Bommarito, S., Peyret, N., and SantaLucia, J., Jr. (2000) Thermodynamic parameters for DNA sequences with dangling ends. *Nucleic Acids Res.* 28, 1929–1934.

14

MultiPLX

Automatic Grouping and Evaluation of PCR Primers

Lauris Kaplinski and Maido Remm

Summary

In this chapter, we describe MultiPLX—a tool for automatic grouping of PCR primers for multiplexed PCR. Both generic working principle and step-by-step practical procedures with examples are presented. MultiPLX performs grouping by calculating many important interaction levels between the different primer pairs and then distributes primer pairs to groups, so that the strength of unwanted interactions is kept below user-defined compatibility level. In addition, it can be used to select optimal primer pairs for multiplexing from list of candidates. MultiPLX can be downloaded from http://bioinfo.ut.ee/download/. Graphical Web-based interface to most functions of MultiPLX is available at http://bioinfo.ut.ee/multiplx/.

Key Words: PCR; multiplex; primer design.

1. Introduction

As the scope of genomic studies is growing exponentially each year, the cost effectiveness of each working step is of even bigger importance. Multiplexing PCR is one approach to save time and sample DNA if large number of genomic fragments have to be amplified by PCR. Multiplexing introduces new set of factors affecting the success of PCR, namely unwanted interactions between different primer pairs. The number of potential interactions is proportional to the square of the number of primer pairs in single tube (multiplexing level). For small number of PCR and low multiplexing levels, groups are often created by hand using trial and error, but this is not reasonable solution if thousands of PCRs have to be multiplexed with high levels.

From: *Methods in Molecular Biology, vol. 402: PCR Primer Design*
Edited by: A. Yuryev © Humana Press, Totowa, NJ

To automate large-scale multiplexing, we have created a program called MultiPLX *(1)*. It analyzes many interactions between primer pairs and generates groups that effectively minimize the number of unwanted interactions in every tube.

MultiPLX can also be integrated with primer design to choose primers from a list of candidates, based on their suitability for multiplex PCR. We have also successfully used it for multiplexing hybrid primers and nested PCRs.

MultiPLX is a command-line tool, available for both Linux (UNIX) environment and Windows command line.

2. Working Principle

The most important factors that affect the success of multiplexed PCR are unwanted bindings between primers and products from different PCR sets. Nearest-neighbor (NN) thermodynamics can be effectively used to predict the binding strength (Gibbs free energy, ΔG) between oligonucleotides and thus estimate the success rate of multiplexed PCR. Program MultiPLX evaluates all possible pairwise combinations of PCR primer pairs and calculates up to five different thermodynamic "score" values (worst case ΔG) for each combination. These scores can be used to determine which pairs are compatible with each other and can be put into single group. The decision is based on user-selected cutoff values.

In addition to the thermodynamic binding energies, the differences between primer melting temperatures, the differences between product lengths and one user-defined score can be taken into account. See **Note 1** for more detailed explanation of different factors that affect the success of multiplexed PCR.

The workflow of MultiPLX has the following steps:

1. Reading, calculating, and saving score data.
2. Reading, calculating, and saving grouping data.

Different tasks can be combined, for example, if scores are already calculated, there is no need to recalculate these, but they can be read from file instead.

MultiPLX can also be integrated into primer design workflow; so, the final primers for each PCR are chosen from among the candidates, based on the compatibility with multiplexing.

3. Step-by-Step Overview

The generic syntax of MultiPLX is:

```
cmultiplx -primers FILE [PRIMERARGS] [SCOREARGS]
[GROUPARGS]
```

```
pcr_A    TTAATCACCCGGCTCCCAGC    GGAGGGCTGACACAGGGAGG    TTAATCACCCGGCTCCC
AGCCGTGTTTCTGCAGAAGGAGCTCTTTCTAATTCAGCTGCTCCAGCCAGGAGGAGCCATAAGTAGAACAGGTG
GAAGTGCGTGCTGACTCTGCTATTCCCTTCTCCATCCTGTATTAGACCAGTTTCCCTTATCATAAATAAAAAC
ACGAAACAAAGACAAATGAGGGAAGGTGGTCTCTCATTTATCACCCAGACCCAGCCTCCCTGTGTCAGCCCTCC
pcr_B    GCCCCAATTTCCATGCTAAAGCGA    AGGGGAGCCCATTCTCGGATTTGAG    GCCCCAAT
TTCCATGCTAAAGCGAAGCATCATAAGTAGCAGTAGAATCAGCTCATCCTGCATTTGTTTTTCCCCAGCGAGGA
GCAATCAATTACCTGAGAGATGTCCGTGGCTGTGGCTAAACGGTGACCCAACAAGGTGCAATTGGCAAAGTGTC
ACTTTTATTCATGGCCTCAAATCCGAGAATGGGCTCCCCT
pcr_C    GCCTTTGAACAGAAAGGCAGGA    CAACTTGGCATGGATTGGACG    GCCTTTGAACAGAA
AGGCAGGAGGATACTGTACAAGTTCAGAGACAGAGAGAGGATGGTAGGTGGTCAGTAGCAAGTGATGAATTCAT
GAAAATGAGAATCTGCCTCTAGGAGGAGAGTTTAAAAGGAATACCAAATTCATGAAAATGAGACTCTATCTCCA
GGAGAGGAGTTTCAAAGTAATACCAAGTGGCCAAAGAGCCTCAGCAGAGATATGAACGTGACCGTGTGCCAGAA
GAAGGAAGGGCCCAGCTGGCCAGATGTTACGTCCAATCCATGCCAAGTTG
```

Fig. 1. Sample from a PCR definition file. The columns are PCR name, left primer sequence, right primer sequence, and (optional) product sequence(s).

The option "-primers" specifies PCR definition file. It specifies all primer pairs and products, one pair per line (*see* **Fig. 1**). Values have to be tab-separated, different products from single primer pair space-separated. Specifying products is not obligatory, if they are not known, or will not be used in analysis.

3.1. Step A: Calculating Compatibility Scores

In compatibility score tables, the worst-case thermodynamic binding energies between different PCR primer pairs are stored. As calculation of these is very time-consuming procedure, MultiPLX allows experimentator to save once calculated score table into files. This way the once calculated scores can be used many times, for example, during adjusting parameters for optimal grouping solution.

The interactions between primers and products are calculated by using NN thermodynamic model. The binding energies of NN pairs are read from file, and the default file can be easily replaced if better NN parameters for specific experiment conditions are available.

See **Note 2** for the file format of the thermodynamic file.

There are two types of score files—one for primer–primer interactions (3 scores for each PCR–primer pair combination) and the other for primer–product (2 scores for each PCR combination) interactions.

See **Note 3** for the explanation of different score types.

The scores will be calculated with the following option:

```
-calcscores SCOREIDS
```

The parameter SCOREIDS is a sequence of the numeric identifiers for scores (explained below).

An example of calculating score files is as follows:

```
test>cmultiplx-primers primers-10.txt-calcscores 12345
-saveprimscores primer-scores.txt -saveprodscores
product-scores.txt
```

"primers-10.txt" is the primer definition file. The option "-calcscores 12345" instructs MultiPLX to calculate all five thermodynamic compatibility scores (non-thermodynamic scores are always calculated on-the-fly, so they cannot be saved to intermediate score file). If all thermodynamic scores are not needed, some digits can be left out. For example, "-calcscores 123" calculates only scores for primer–primer interactions.

"primer-scores.txt" and "product-scores.txt" are the output files where scores will be saved. Although they are text files, the information in these is normally not meant to be read by users.

The possible SCOREIDS can be listed with the command "`cmultiplx –listscores`".

```
test>cmultiplx -listscores
Score codes:
  1 - Primer-primer both 3' ends
  2 - Primer-primer 3' end with any region
  3 - Primer-primer any regions
  4 - Primer-product 3' end with any region
  5 - Primer-product any regions
```

Calculating primer–product scores takes much more processing time than calculating only primer–primer scores. We have also found that primer–product interactions have smaller influence to PCR quality than primer–primer interactions. For large data sets, especially if small multiplexing level is needed, these can be left out.

An example of different score types is given in **Fig. 2**.

3.2. Step B: Calculating Groups

Primer pairs are distributed into groups based on cutoff levels of score values. Any two primers are allowed to be in the same group only if all relevant scores are below cutoff values. Although this is simplification and does not distinguish

```
Score type 1 - alignment of the 3' ends of two primers

5'-TTAATCACCCGGCTCCCAGC-3'
              |||  | |
           3'-GAGTTTAGGCTCTTACCCGAGGGGA-5'


Score type 2 -  alignment of the 3' ends one primer with the internal
region of other primer

5'-TTAATCACCCGGCTCCCAGC-3'
                   |||
         3'-AGCGAAATCGTACCTTTAACCCCG-5'

Score type 3 -  alignment of the internal regions of two  primers

      5'-TTAATCACCCGGCTCCCAGC-3'
                  |||||||
3'-GAGTTTAGGCTCTTACCCGAGGGGA-5'
```

Fig. 2. Different score types.

the level of compatibility of the group as a whole, it makes computation much more straightforward.

In general, it is not possible to find the optimal solution without trying all possible combinations (unrealistic even on supercomputers). MultiPLX uses a greedy algorithm for grouping, which gives reasonably good solution.

The most important parameters for grouping can be set with the following options:

```
-stringency LEVEL
-cutoff# VALUE
-calcgroups MAXGROUPS MAXITEMSINGROUP
```

Examples of calculating groups are as follows:

```
test>cmultiplx -primers primers-10.txt -loadprimscores
primer-scores.txt -loadprodscores product-scores.txt -stringency
normal -calcgroups 1000 10 -savegroups groups.txt

test>cmultiplx -primers primers-10.txt -loadprimscores
primer-scores.txt -loadprodscores product-scores.txt
-stringency low -cutoff1 -5 -cutoff8 4 -calcgroups 1000
10 -savegroups groups.txt
```

Both the examples read PCR primers and products from the "primers-10.txt" file and previously calculated scores from "primer-scores.txt" and "product-scores.txt."

The first example performs grouping, using a generic balanced set of cutoff values (*-stringency normal*). The maximum number of groups is 1000 and the maximum number of elements in a group is 10 (*-calcgroups 1000 10*). Final groups will be saved to the "groups.txt" file. The format of grouping file is shown in **Fig. 3**.

The second example uses looser set of cutoff values (*-stringency low*) that normally would make bigger groups. Then, it overrides two of these values. The minimum allowed dG for primer–primer binding from 3′-ends is set to −5 kcal/mol (*-cutoff1 -5*) and the maximum allowed difference between the melting temperatures of primers to 4 °C (*-cutoff8 4*). Other parameters are identical to the first example.

```
#Libbdm build 1.0.1-06-05-2003
#Total 19 rows 6 column in table
MXMultiplexTable
```

Name	1	2	3	4	5	6
Group 1	1	59	47	82	63	
Group 2	61	54	38	50	65	
Group 3	10	37	67	77	33	
Group 4	94	95	57	98	55	46
Group 5	97	8	22	18	30	
Group 6	17	91	39	80	68	
Group 7	31	41	79	5	69	
Group 8	11	56	9	60	48	
Group 9	75	78	70	2	83	
Group 10	88	44	72	3	32	21
Group 11	20	58	14	28	52	
Group 12	90	29	87	7	100	
Group 13	34	96	71	27	53	76
Group 14	64	49	40	73	81	
Group 15	42	62	35	93	23	43
Group 16	99	15	13	86	12	24
Group 17	6	19	45	4	85	
Group 18	84	66	16	36	74	
Group 19	25	92	26	89	51	

Fig. 3. The format of grouping file. Groups are listed in rows and PCR pair names in columns.

Using "-" as the grouping file name will print the grouping information to screen instead of file.

3.2.1. Stringencies and Cutoffs

Stringency specifies generic set of balanced values for grouping cutoffs. The allowed levels are "low," "normal," and "high." Low stringency results in the biggest groups but with bigger probability of PCR failure because of incompatibilities between primers.

The cutoff values of different stringencies are listed in **Note 4**.

For more detailed control, or if non-thermodynamic scores have to be taken into account, individual cutoffs can be overridden with "-cutoff#" option. The number sign "#" has to be replaced with an individual cutoff code (thus resulting in option names like *-cutoff1, cutoff2*...). Both stringency and cutoff options may be used together, in which case-specific cutoff values replace the ones determined by stringency.

The list of possible cutoff codes can be displayed with the following option "-listcutoffs".

```
test>cmultiplx -listcutoffs Cutoff codes:
   1 - Primer-primer both 3' ends
   2 - Primer-primer 3' end with any region
   3 - Primer-primer any regions
   4 - Primer-product 3' end with any region
   5 - Primer-product any regions
   6 - Product length max difference (range)
   7 - Product length min difference (ladder)
   8 - Maximum primer melting temperature difference
   9 - Custom score (maximum allowed)
```

There is no direct way to request certain predefined group size or certain number of groups. If such solution is needed, one has to set the maximum group size or the maximum number of groups to desired value and experiment with different cutoffs, until acceptable solution is found. As the grouping is usually very fast (unless very big number of optimization iterations is used), this is only a minor inconvenience.

See **Note 5** for more information about grouping.

3.3. Integrating MultiPLX with Primer Design

Calculating groups for existing list of primers usually gives quite good multiplexing levels. Still, as only a single primer pair is available for any amplified region, MultiPLX can only rearrange primer pairs between different groups. Difficulties in grouping may appear if some primer pairs have very strong interactions (and thus low compatibility) with many other pairs. This difficulty can be alleviated by combining primer design and primer grouping.

Primer design usually involves picking a single pair out of many possible candidates for each target region. MultiPLX can read more than one primer pair candidate for each target region and select the one that gives the smallest number of unwanted interactions.

MultiPLX primer input file can list arbitrary number of candidate pairs for each target region. The target region is identified by name. The primer pair candidates from the same target region can share common primers. An example of PCR primer file with multiple candidate primer pairs is shown in **Fig. 4**.

There are two options specific to primer selection:

```
-maxcandidates NUMBER
-savefinalset FILENAME
```

An example of selecting primers for multiplexing is as follows:

```
test>cmultiplx -primers primer-candidates.txt
-maxcandidates 10 -calcscores 123 -saveprimscores
primer-scores.txt

test>cmultiplx -primers primer-candidates.txt
-maxcandidates 10 -loadprimscores primer-scores.txt
-stringency normal -calcgroups 1000 1000
-savegroups groups.txt -savefinalset final-primers.txt
```

```
pcr_A   TTAATCACCCGGCTCCCAGC      GGAGGGCTGACACAGGGAGG
pcr_A   TTAATCACCCGGCTCCCAGC      CGTAGCTATGGCATCGATT
pcr_A   GCATGCCTATAAGCGATGGAC     CGTAGCTATGGCATCGATT
pcr_B   GCCCCAATTTCCATGCTAAAGCGA  AGGGGAGCCCATTCTCGGATTTGAG
pcr_B   GCCTTTGAACAGAAAGGCAGGA    CAACTTGGCATGGATTGGACG
```

Fig. 4. Sample from a PCR definition file with several primer pair candidates for each target region. The columns are PCR name, left primer sequence, and right primer sequence. The product sequences are missing in this example, but if present, they can be listed in the 4th column. Notice that some primer sequences can be shared between alternative primer pairs.

In the first step of this example, primer candidates are read from the file "primer-candidates.txt." Up to 10 candidate primer pairs are selected from each target region (*-maxcandidates 10*). Primer–primer interaction scores are then calculated and written to "primer-scores.txt" file.

In the second step, the same number of candidate primer pairs is read again from the primer file and the scores for all these candidate primer pairs (calculated in previous step) are read from "primer-scores.txt." Primer pairs are then grouped using normal stringency (*-stringency normal -calcgroups 1000 1000*). Using large value (*1000*) for both the number of groups and the number of group members ensures that the size of groups is actually limited by primer compatibility. If there is more than one primer pair candidate for PCR, MultiPLX automatically chooses the most compatible one (the one having the smallest number of unwanted interactions with all other primer pairs). Groups are then written to "groups.txt" and the chosen set of primer pairs to "final-primers.txt."

It is important to use the same number of candidates (*-maxcandidates*) for both score table generation and grouping, because the values in score table are identified by the positions of primers and pairs in the primer file.

4. Specific Applications

4.1. Testing Groups

MultiPLX can be used for evaluating an existing multiplexing solution by pointing out these primer pairs that have high probability of unwanted pairings with other group members. Primer compatibility for existing groups will be evaluated by using the same score table and cutoff values of scores as for calculating groups. Existing groups have to be presented in the same file format as the groups generated by MultiPLX.

Existing groups can be evaluated with the following option:

```
-saveoffenders FILENAME
```

An example of evaluating existing multiplexing groups is as follows:

```
test>./cmultiplx -primers primers-100.txt -loadprimscores
primscores-100.txt -stringency low -cutoff1 -5.5
-groups testgroups.txt -saveoffenders -

1
95  72  PrimPrimEnd2  -5.8
95  32  PrimPrimEnd2  -5.5
2
```

```
58   82   PrimPrimEnd2   -5.6
3
8     7   PrimPrimEnd2   -5.9
17   75   PrimPrimEnd2   -5.8
4
39   44   PrimPrimEnd2   -5.7
5
6
38   64   PrimPrimEnd2   -5.7
86   88   PrimPrimEnd2   -5.9
7
96   61   PrimPrimEnd2   -5.8
61   18   PrimPrimEnd2   -5.5
```

In this example, the existing groups are read from the "testgroups.txt" file and evaluated using low stringency and user-specified cutoff1. The output lists the groups and all of their members, which have some scores above cutoff values.

4.2. Universal and Complex Primers

Universal PCR primers that are linked to both ends of a studied DNA fragment are frequently used in large-scale genomic applications. The compatibility of universal primers is usually tested in design phase and does not need to be re-evaluated by the MultiPLX program. However, if hybrid primers with universal 5′-end and target-specific 3′-end are used, the MultiPLX might be useful to test their compatibility and/or to group them.

As complex primers are longer than simple ones, the score calculation with MultiPLX is slower than that for simple primers. As the effect of universal (5′) end will be the same for all primers, it is possible to limit calculating scores to only target-specific part of primer. To do that, new primer definition file, where all universal primer fragments are removed, has to be constructed using some external tool.

This approach does not take into account the possibility of alignment between the middle part of one primer (partly specific, partly universal) with the end or the middle of another primer or product.

4.3. Custom Scores

MultiPLX can use one additional score file to introduce custom factors into multiplexing. For example, the number of predicted PCR products amplified by any given pairwise combination of PCR primers from template genome can be calculated using the GenomeTester package *(2)* and used as input for MultiPLX. The custom version of GenomeTester for this is accessible from the Webpage of the Department of Bioinformatics, University of Tartu (http://bioinfo.ut.ee/gt4multiplx/).

MultiPLX allows to use only a single custom score value. If more than one custom score are needed, these have to be combined into single value beforehand. Also, the value must have negative correlation with PCR success (i.e., big values are bad, small values are good).

5. Web Interface

MultiPLX can be accessed from the Webpage of the Department of Bioinformatics, University of Tartu (http://bioinfo.ut.ee/multiplx/).

6. Notes

Note 1: The Parameters Affecting the Success of PCR Multiplexing

The unwanted bindings can be broadly divided into different classes, based on whether they take place between two primers or one primer and one product and whether the 3′-end of primer is bound or free.

In addition to bindings between primers and products, it is often useful to consider also the melting temperatures of primers and the differences in product lengths.

A. Primer–Primer Interactions

For multiplexed PCR, there are four different interactions between any pair of PCR primers. Thus, all possible primer pairs have to be tested for unwanted interactions with all other primers in the same multiplexing group, and if their alignment is too strong, they are moved to another group.

All primer–primer interactions lower the concentration of free primers and thus the probability of binding the primer to the target site. Additionally, the interactions involving the 3′-end of one of both primers create a new possible elongation site.

B. Primer–Product Interactions

Primer–product interactions have similar effect as primer–primer interactions. But as product concentrations are much lower (at least in the beginning of PCR experiment), the effect is lower as well.

In the case of products, we do not have to differentiate the bindings between the 3′-end of the primer with any region of the product and the bindings between the 3′-end of a primer with the 3′-end of the product, as the overall effect is similar in both cases.

C. Additional Properties (Melting Temperature and Product Length)

In addition to unwanted bindings, it is often useful to limit the maximum difference between the melting temperatures of primers in a single group. The melting temperature of primers is correlated with the speed of PCR, and it is better to keep all individual PCR rates in multiplex group as similar as possible. The difference in product lengths has a similar effect; thus, it may be useful to limit the maximum difference between product lengths as well.

If final PCR products will be detected by gel electrophoresis, it may be necessary to have different product lengths in a single group; so, they will separate on gel and thus can be individually detected.

The melting temperatures of primers depend on PCR mixture parameters (the concentration of monovalent and divalent cations and the concentration of DNA). These parameters can be specified by the following command-line options:

```
-csalt SALT
-cmg MAGNESIUM
-cdna DNA
```

SALT is the concentration of monovalent salts in mM (default 50).
MAGNESIUM is the concentration of magnesium in mM (default 1.5).
DNA is the concentration of primers in nM (default 50).

Although the salt and DNA concentrations affect absolute melting temperatures of primers, they are usually irrelevant for multiplexing, as only temperature differences between melting temperatures are used. As the effect is identical for all primers, differences remain the same. So, if the exact experiment conditions are not known beforehand, defaults (or other reasonable values) are safe to use for multiplexing.

MultiPLX uses the following formula to calculate melting temperature:

$$Tm = dH/(dS+1.987*\ln(cDNA))+16.6*\log(Na^{+}[NORM])$$

Tm-melting temperature in kelvins.
DH-binding enthalpy in J/mol.
DS-binding entropy in J/(mol*K).
Na^+[NORM]-normalized salt concentration.

Na^+[NORM] is calculated from the following formula *(**3**)*:

$$Na^{+}[NORM]=(Na^{+}[mM]+120.0*\mathrm{sqrt}(Mg^{2+}[mM]-dNTP[mM]))/1000;$$

Na^+[mM], Mg^{2+}[mM], the concentrations of monovalent and divalent cations in mM/l.
dNTP[mM], the total concentration of nucleotide triphosphates in mM/l.

Note 2: Thermodynamic Data

Thermodynamic data table lists all dinucleotide combinations and corresponding thermodynamic NN parameters. As MultiPLX has to calculate pairings with mismatches, both Watson–Crick pairs and pairs with single mismatch have to be listed in the table. In addition to these, missing nucleotides can be specified to take the contribution of dangling ends into account.

All missing values are treated as 0. The default parameters are from published sources ***(4–10)***.

Thermodynamic data are specified for MultiPLX with the following option:

```
-thermodynamics FILENAME
```

A sample from thermodynamic data file is shown in **Fig. 5**.

Note 3. Score Types Used by MultiPLX

MultiPLX uses the following score types:

1. Maximum binding energy (deltaG) of two primers including 3′-ends of both primers (PRIMPRIMEND2).
2. Maximum binding energy of 3′-end of one primer with any region of another primer (PRIMPRIMEND1).
3. Maximum binding energy of any region of different primers (PRIMPRIMANY).
4. Maximum binding energy of 3′-end of one primer with any region of PCR product (PRIMPRODEND1).
5. Maximum binding energy of any region of a primer with any region of PCR product (PRIMPRODANY).
6. Maximum product length difference between compared PCR primer sets.
7. Minimum product length difference between compared PCR primer sets.
8. Maximum difference in primer melting temperatures between compared PCR primer sets.

Note 4. Cutoff Values for Different Stringencies

Stringencies represent generic balanced set (according to our knowledge) of cutoffs for grouping. Depending on the number of PCR primer pairs to be grouped, high stringency may give groups of about 1–4, normal stringency 3–8, and low stringency 7–15 PCR primer pairs.

```
# libdna thermodynamic table
#
# Format: PQ/RS dH dS
# P,Q,R,S - nucleotides
# dH Enthalpy (cal/mol)
# dS Enthropy (kcal/mol*K)
#
Energy+Enthropy
#
# Watson-Crick-Paris
# SantaLucia J. and Hicks D. 2004. The Thermodynamics of DNA
# Structural Motifs.
# Annu. Rev. Biophys. Biomol. Struct. 33:415-40
#
AA/TT -7600 -21.3
TT/AA
AC/TG -8400 -22.4
AG/TC -7800 -21.0
AT/TA -7200 -20.4
CA/GT -8500 -22.7
GT/CA -8400 -22.4
TG/AC
CC/GG -8000 -19.9
GG/CC -8000 -19.9
CG/GC -9800 -24.4
GC/CG -9800 -24.4
CT/GA -7800 -21.0
GA/CT -8200 -22.2
TC/AG
TA/AT -7200 -21.3
#
# A-C Mismatches
# Allawi H.T. and SantaLucia J. 1998. Nearest-neighbor
# thermodynamics of internal A-C mismatches in DNA: sequence
# dependence and pH effects.
# Biochemistry 37:9435-44
#
AA/CT 7600 20.2
TC/AA
AA/TC 2300 4.6
CT/AA
AC/CG -700 -3.8
GC/CA
AC/TA 5300 14.6
AT/CA
AG/CC 600 -0.6
CC/GA
CA/AT 3400 8
TA/AC
CA/GC 1900 3.7
CG/AC
CC/AG 5200 14.2
GA/CC
...
```

Fig. 5. Sample from a default thermodynamic file.

The default cutoff values of stringencies take neither product length differences nor melting temperature differences into account; so, if these should be used, specifying explicit cutoff values is needed.

The cutoff values of stringencies are given in **Fig. 6**:

Note 5. Grouping Algorithm

Primer pairs are selected one by one and tested against each member of each existing group. If a primer pair is compatible with an existing group, and it has fewer items than maximum number allowed, then the item is placed in that group. Otherwise, it is tested against the members of next group and so on. If it is incompatible with all existing groups and the maximum number of groups

```
Low strigency (-stringency low)

PrimPrimeEnd2:            -6.0
PrimPrimeEnd1:           -10.0
PrimPrimeAny:            -10.0
PrimProdEnd1:            -14.0
PrimProdAny:             -14.0
Max prod len diff:           -
Min prod len diff:           -
Max Melting temp diff:       -
Max custom score:            -


High stringency (-stringency high)

PrimPrimeEnd2:            -4.0
PrimPrimeEnd1:            -8.0
PrimPrimeAny:             -8.0
PrimProdEnd1:            -12.0
PrimProdAny:             -12.0
Max prod len diff:           -
Min prod len diff:           -
Max Melting temp diff:       -
Max custom score:            -

PrimPrimeEnd2:            -2.0
PrimPrimeEnd1:            -6.0
PrimPrimeAny:             -6.0
PrimProdEnd1:            -10.0
PrimProdAny:             -10.0
Max prod len diff:           -
Min prod len diff:           -
Max Melting temp diff:       -
Max custom score:            -
```

Fig. 6. Cutoff values of stringencies.

is not exceeded, a new group is created; otherwise, the grouping fails with an error message "ERROR: Cannot fit primers into groups."

The number of final groups is sensitive to the order of primers to be grouped. The order can be modified by the following option:

```
-initialorder VALUE
```

The allowed values are "file," "friends," and "random."

File keeps primer pairs in the same order that they are in the input file.

Friends orders primer pairs by the number of compatible pairs, starting from the smallest (fewest compatible pairs) value. Thus, the primers that have smaller probability to be compatible with others are distributed before the primers that have higher probability of success. Usually, this option gives the optimal solution (smallest number of groups).

Random distributes primers randomly. It can be useful, if non-deterministic grouping is desired, for example, for testing the compatibility of different primer selection methods with multiplexing.

For random order, grouping more than one iteration can be performed with the following option:

```
-groupiter NUMITERATIONS
```

The solution with the smallest number of groups among all iterations will be chosen as the final result.

Greedy grouping results in groups of very different sizes. Usually, the first groups are the biggest, because all primers are placed into the first compatible group.

The size of the groups can be made more uniform with the following option:

```
-optimize NUMITERATIONS
```

The optimization is on by default, and the default number of iterations is 10,000. To turn it off, the number of iterations has to be set to 0.

The optimization works by moving primer pairs from bigger groups into smaller groups and swapping the members between groups if no pair can be moved into smaller group. It does not guarantee a uniform group size, but under normal circumstances, the sizes of groups do not differ by more than one.

Acknowledgments

This work was supported by the grant EU19730 from Enterprise Estonia. The authors thank Katre Palm for a valuable help with English grammar.

References

1. Kaplinski, L., Andreson, R., Puurand, T. and Remm, M. (2005) MultiPLX: automatic grouping and evaluation of PCR primers. *Bioinformatics*, 21, 1701–2.
2. Andreson, R., Reppo, E., Kaplinski, L. and Remm, M. (2006) GENOMEMASKER package for designing unique genomic PCR primers. *BMC Bioinformatics*, 7, 172.
3. von Ahsen, N., Wittwer, C.T. and Schutz, E. (2001) Oligonucleotide melting temperatures under PCR conditions: nearest-neighbor corrections for Mg2+, deoxynucleotide triphosphate, and dimetyl sulfoxide concentrations with comparison to alternative empirical formulas. *Clinical Chemistry* 47, 1956–61.
4. Allawi, H.T. and SantaLucia, J. (1997) Thermodynamics and NMR of internal G-T mismatches in DNA. *Biochemistry* 36, 10581–94.
5. Allawi, H.T. and SantaLucia, J. (1998) Nearest-neighbor thermodynamics of internal A-C mismatches in DNA: sequence dependence and pH effects. *Biochemistry* 37, 9435–44.
6. Allawi, H.T. and SantaLucia, J. (1998) Nearest-neighbor thermodynamic parameters for internal G-A mismatches in DNA. *Biochemistry* 37, 2170–9.
7. Allawi, H.T. and SantaLucia J. (1998) Thermodynamics of internal C-T mismatches in DNA. *Nucleic Acids Res* 26, 2694–701.
8. Kaderali, L. (2001) *Selecting Target Specific Probes for DNA Arrays*. Universität zu Köln, Köln.
9. Peyret, N., Senevirtane, P.A., Allawi, H.T. and SantaLucia, J. (1999) Nearest-neighbor thermodynamics and NMR of DNA sequences with internal A-A, C-C, G-G and T-T mismatches. *Biochemistry* 38, 3468–77.
10. SantaLucia, J. and Hicks, D. 2004. The thermodynamics of DNA structural motifs. *Annu Rev Biophys Biomol Struct* 33, 415–40.

15

MultiPrimer

A System for Microarray PCR Primer Design

Rohan Fernandes and Steven Skiena

Summary

To construct full-genome spotted microarrays, a large number of PCR primers that amplify the required DNA need to be synthesized. We describe an algorithmic technique that allows one to use fewer primers to achieve this goal. This can reduce the expense of constructing full-genome spotted microarrays considerably. PCR primers are usually designed, so that each primer occurs uniquely in the genome. This condition is unnecessarily strong for selective amplification, because only the *primer pair* associated with each amplification needs be unique. We also describe the interface to our software, MultiPrimer, that computes a small set of primers for amplification of a given gene set.

Key Words: Polymerase chain reaction; PCR primer design; heuristic optimization; minimum primer set.

1. Introduction

Microarray technology *(1,2)* has revolutionized our understanding of gene expression. DNA microarrays are glass or nylon substrates on which DNA probes of known sequence are affixed. Through complementary binding with sample DNA, these probes enable large-scale gene expression and gene discovery studies. A single experiment on one microarray can provide a researcher information on thousands of genes simultaneously. Microarrays can be used either when one wants to assay expression of a large number of genes or when the sample to be studied is small. They are useful for studying gene expression from a single sample or for comparative gene expression studies. Applications

From: *Methods in Molecular Biology, vol. 402: PCR Primer Design*
Edited by: A. Yuryev © Humana Press, Totowa, NJ

of microarrays include cell cycle analysis *(3,4)*, studying the response of cells to environmental stress *(5)*, and the impact of gene knockouts in yeast *(6)*.

To date, whole genome sequences for 823 organisms are known *(7)*. Even though many of these are small viruses, more than 100 free-living species have already been sequenced. For most of these organisms, including many pathogenic bacteria and agricultural pathogens, we know relatively little of their biology except the sequence. To make effective use of the vast amount of genomic data available, we require to develop custom microarray design/fabrication technologies which are inexpensive enough for typical individual investigators to pursue. In this chapter, we present a polymerase chain reaction (PCR) primer design technique that can potentially reduce fabrication costs significantly.

A substantial percentage of the expense in constructing full-genome spotted microarrays comes from the cost of synthesizing PCR primers to amplify the desired DNA. For example, in the NIH-funded Microarray equipment grant of Futcher and Leatherwood to build spotted microarrays for the yeasts *Saccharomyce cerevisiae* and *Saccharomyce pombe*, about \$110,000 of a total \$220,000 budget was allocated to PCR primer synthesis.

Historically, PCR primers are designed, so that each primer occurs uniquely in the genome. This condition is unnecessarily strong for selective amplification, because only the *primer pair* associated with each amplification needs to be unique. Thus, by careful design in a genome-level amplification project, we can reuse literally thousands of primers in multiple roles, for a substantial reduction in cost.

This chapter is divided into two principal sections: In **Subheading 2**, we briefly review results first presented in *(8)*. We repeat the descriptions of the algorithms used for multiple-use PCR primer design along with some comments. Additionally, we describe added functionality and results because the publication of the previous work. In the final section, we describe the software for multiple-use PCR primer design, MultiPrimer, and how to use it.

2. Methods and Algorithms for Multiple-Use PCR Primer Design

PCR has revolutionized the practice of molecular biology, making it routine to create millions of copies of a single gene or any other portion of a genome. PCR requires the presence of two single-stranded DNA sequences called *primers*, which complement specific parts of either the forward or the reverse strand of the double-stranded DNA and enable duplication of the region in-between.

Primer design is the problem of constructing these delimiting elements of the reproduced region. Important criteria in primer design include melting temperature, PCR product size, secondary structure, and the uniqueness of each designed primer.

We briefly describe a more efficient approach for genome-level primer design. Amplifying a gene requires left and right primers that hybridize to nearby regions on the genome. Because the efficiency of PCR amplification falls off exponentially as the length of the product increases, PCR becomes ineffective for product sizes beyond 1200 bases or so. Although it is nice to have primers that hybridize to a unique region on the genome, this is not strictly necessary for successful PCR as hybridization outside the target region will not result in significant amplification unless both primers hybridize sufficiently closely to each other.

Now there is a potential to re-use primers to amplify multiple genes provided that we maintain *primer pair* uniqueness. The possible primer savings from such an approach is enormous. Let n be the number of genes to be amplified and m be the minimum number of primers required to amplify all of them. As m primers can result in $m(m+1)/2$ unique primer pairs, potentially $m = \sqrt{2n}$ primers suffice for amplification instead of n with the conventional design. For a yeast-sized genome of 6000 genes, this potentially reduces the number of primers needed from 12,000 to 78. Although this lower bound is exceptionally optimistic, the potential is very compelling and worthy of further investigation.

Multiple-use primer design requires the solution of a difficult combinatorial optimization problem. Given a set of k potentially amplifying primer pairs for n genes, $k \geq n$, find a minimal set of primers from these pairs such that we can amplify each gene using only combinations of primers from this set.

It is convenient to model this problem, which we call *minimum primer set*, as an edge-coloring problem on graphs. From here onwards, we will refer to a specific primer sequence simply as primer. Represent each candidate primer as a vertex of the graph. For every primer pair that uniquely amplifies a given gene, connect the two vertices with an edge and label (color) the edge with the name of the gene. Our problem is to find the smallest subset of vertices that induces a subgraph that contains edges with all possible gene colors.

An alternate formulation of the problem, called *budgeted primer set*, seeks the k primers (vertices) that cover the maximum number of different genes (colors) possible. We want to maximize our investment by selecting a set of primers that enables us to amplify as large a set of genes as possible. Thus, k is a budget for how many primers we are willing to synthesize, and we seek to amplify as many genes as possible under this constraint.

Unfortunately, we have shown that both of these primer design problems are NP-complete. Indeed, we have proved that the minimum primer design problem is hard even to approximate within a logarithmic factor.

Theorem 1 *The minimum primer set problem is inapproximable to less than* $a[1 - o(1)] \ln n - o(1)$ *factor.*

Theorem 2 *The budgeted primer set problem is NP-complete.*

The proofs of theorems 1 and 2 can be found in ***(8)***.

2.1. Multiple-Use Primer Design Heuristics

Our initial heuristic for minimum primer set was inspired by Charikar's heuristic ***(9)*** to approximate the densest subgraph problem ***(10)***, where the density of a subgraph is the ratio of number of edges induced by vertices in the subgraph to the number of vertices in the subgraph. The greedy algorithm successively strips the vertex with the minimum degree in the subgraph from it. In the end, we select the intermediate subgraph with the maximum average degree, which can be shown to be at least half as dense as the densest possible subgraph. Our graph is edge-colored, however, and we seek the color-densest graph, which complicates the problem considerably.

We note that the budgeted version of the problem is related in some sense to the densest k-subgraph problem, which has been well studied ***(11,12)***, with approximation bounds on greedy approaches that are not very encouraging.

In our first heuristic (*see* **Algorithm 1**) for minimum primer set, all the primers/vertices are initially weighted according to the number of genes/edge-colors they are incident to. At each iteration, we discard the lowest weight vertex, which does not result in any color/gene being lost in the subgraph. Such a vertex is referred to in **Algorithm 1** as the minimum-weight non-critical vertex. The algorithm and its performance on yeast data sets (*S. cerevisiae* and *S. pombe*) are described in detail in ***(8)***, and it can be invoked in MultiPrimer by the command `SOLVEGRAPH1`. This algorithm runs in time $O[|V| \cdot (|V| + |E|)]$ and requires space $O(|V| + |E| + |C|)$.

The Densest Subgraph heuristic works well in practice, but it is computationally expensive. Because its running time is quadratic in $|V|$, it performs especially slowly on the larger data sets. Hence, we explored the possibility of using a simpler, linear-time heuristic. The heuristic we designed is much faster than the previous one especially on larger data sets. The new heuristic (*see* **Algorithm 2**) uses the same candidate primer input graphs as the previous heuristic.

Algorithm 1 Densest Subgraph Algorithm

```
input : An edge-colored graph
output: A minimal subgraph with edge-color set same as the input graph
repeat
    for each color c do
        Compute the number of remaining edges colored c, n_c, in the
        graph;
    end
    Set the edge-weight of each color c edge as w_c = 1/n_c;
    For each vertex, set vertex-weight equal to the sum of the
    edge-weights of edges incident on that vertex;
    Remove the minimum-weight non-critical vertex, along with all
    incident edges, from the graph;
until no more edges are removable;
```

Algorithm 2 Linear-Time Greedy Heuristic

```
input : An edge-colored graph
output: A minimal subgraph with edge-color set same as the input graph
repeat
    for each vertex in the graph do
        Compute the colored degree of the vertex, that is, the number
        of different colored edges that it is adjacent to;
    end
    for each color c in the graph do
        Find an edge colored c, termed seed-edge, that is induced by
        vertices with maximum sum of colored degrees;
    end
    repeat
        for each edge e (with color c_e) in the graph do
            Check if replacing the edge of color c_e in the seed-subgraph
            with edge e will reduce the number of vertices;
            If the potential replacement reduces the number of vertices
            in the seed-subgraph let it proceed;
            If the potential replacement does not reduce the number of
            vertices but leaves it the same size, then let it proceed
            with probability 1/2;
        end
    until no improvement in number of vertices in an iteration;
    for each edge e (with color c_e) in the seed-subgraph do
        If the edge e shares edges with any other edge in the
        seed-subgraph, then remove it from the seed-subgraph;
        If edge e satisfies the previous condition, then remove all edges
        with color c_e from the original graph;
    end
until no improvement in number of vertices in an iteration;
```

For an *S. cerevisiae*, degenerate data set with melting temperature in the range 42–52 °C (an input graph with over 35 million edges), this heuristic produced a solution within 25 min on a Sun Ultra Sparc server with 3 GB of RAM running on SunOS 5.6. On the contrary, our implementation of the Densest Subgraph heuristic produced a solution only after two full days of computation. The new heuristic uses the same candidate primer input graphs as the previous heuristic. It can be accessed in MultiPrimer by using the `SOLVEGRAPH2` command.

A few notes are in order regarding this heuristic. The algorithm runs in time $O(|V|+|E|+|C|)$ if properly implemented. Steps 3 and 6 enable us to make an intelligent guess about which regions of the graph are dense in colors, and picking initial edges out of them promises a better solution. The result of these steps is a seed-subgraph that *spans* all the colors.

Steps 10–12 are the main optimizing steps of the heuristic. With the right data structures, step 10 takes $O(1)$ time for each edge of the graph. The introduction of randomization in step 12 contributed some improvement to the final solution.

Finally, in steps 16 and 17, we eliminate all colors costing us two vertices in our solution seed-subgraph (as they are no better than any other primer pair for the gene) and look for a better solution outside of it.

As our new heuristic is considerably faster than the previous one, we could now use it on larger data sets. Thus, we added methods to our software to allow the user to customize the maximum degeneracy of the primers to be used. These methods can be used by means of the `ADVANCED-DEGENERACY` command. We briefly describe the technique used to derive the *degenerate primers*. We first construct a hamming distance graph with vertices corresponding to the simple primers. Two primers are adjacent if they have a hamming distance of 1, i.e., they differ exactly in one base. Using breadth first search on this hamming graph, we merge nearby primers until we obtain the degenerate primer with the largest degeneracy within our defined maximum. Then, all edges that were adjacent to a simple primer vertex were added to the corresponding degenerate primer vertex.

In **Table 1**, we present primer design results for data sets using degenerate primers of increasing degeneracy. The data sets consisted of primer pairs for all genes of *S. cerevisiae* and *S. pombe*, with each primer between 8 and 12 bases long and melting temperatures in the range 47–57 °C. We observe that, for the *S. cerevisiae* data set, there is a dramatic reduction in the primer set size required as degeneracy increases from 1 (i.e., no degeneracy) to 128. In fact, the primer savings increase from 27 to 62%. The results for the *S. pombe* data set are similar.

Table 1
The Number of Primers Required by Our Linear-Time Design Heuristic as a Function of the Degree of Degeneracy

		Maximum degree of degeneracy								
Yeast	T_m	1	2	4	6	8	16	32	64	128
Cerevisiae	47–57	5511	4083	3604	3599	3420	3209	3059	2937	2812
Pombe	47–57	5058	3705	3367	3299	3142	2971	2841	–	–

Missing entries indicate running out of memory during computation.

3. Description of MultiPrimer

We provide an implementation of the algorithms described in the **Subheading 2** as a program called MultiPrimer available at http://algorithm.cs.stonybrook.edu/compbio/MultiPrimer ***(13)***. The program is available as C source code only, which has been tested on SunOS 5.6 using the gcc 3.4.2 compiler. A call to the SunOS 5.6 sort program (found usually in `/usr/bin/sort`) is made in the code, and this is the only external software dependency.

3.1. Input/Output File Types and Formats

MultiPrimer has various input and output file formats. Examples of all of these are available in the `demo` directory structure of the distribution. We will describe the most important ones briefly here.

1. **Primer Pair Input/Output File:** This file contains a list of genes along with associated primer pairs that amplify them. It contains records of the form:

   `<gene-name> <left-primer> <right-primer>`.

2. **Primer Index File:** This file contains the mapping of each primer code to a unique index number. The maximum index number appears on the first line. The second line specifies if the file contains any *degenerate* primers or is solely made up of *Non-degenerate* primers. The remaining lines have records of the form:

   `<primer-index> <primer-code>`.

3. **Gene Index File:** This file contains the mapping of each gene name to a unique index number. The maximum index number appears on the first line. The remaining lines have records of the form:

   `<gene-index> <gene-name>`.

4. **Primer–Gene Graph File:** This file contains the adjacency list of the graph that has primers as its vertices and genes as its edges. The first line contains the maximum index number of all the genes. The second line contains the maximum index number of the primers. The remaining lines of the file have records of the form:

   `<ps-index> <global-index>(<ps-index> <gene-index> <L/R>)*.`

   By `ps-index`, wemean `primer-index` mentioned previously. The term `global-index` differs from `ps-index` only when the primer–gene graph contains degenerate primers. Degenerate primers are a mix of *simple* primers that differ only in a few bases and thus have a bounded *order of degeneracy*. This means that several original *simple* primers are no longer part of edges in the primer–gene graph, because these edges now are attached to degenerate primer vertices. Thus, some primer vertices are deleted, whereas new vertices have to be added for degenerate primers. To handle this efficiently, new index numbers are generated for the primers that are part of edges in the degenerate primer–gene graph, and a link to the actual index is maintained by means of the `global-index` field.

Other types of output files are described in the documentation available in the `doc` directory of the MultiPrimer distribution.

3.2. Command File Format and Running MultiPrimer

MultiPrimer functions can be invoked using a simple 9-line command file as:

`MultiPrimer -x <command-file-name> .`

In this section, we will describe the format of the command file with some examples. Examples of command files are available in the `demo/param` directory of the MultiPrimer distribution.

This first line (L1) of the command file can have the following values:

`BUILDGRAPH:` This command instructs MultiPrimer to read data provided in the primer pair I/O file and output the primer–gene graph file.

`STATS`: This operation produces useful statistics derived from the primer-pair data files and the primer–gene graph file. This command can be run after the `BUILDGRAPH` operation.

`SIMPLE-DEGENERACY`: This command instructs MultiPrimer to read in the primer–gene graph file and creates a new primer–gene graph file with degenerate primers included. The degenerate primers can be a combination of simple primers that differ in one single base only.

`ADVANCED-DEGENERACY`: This command instructs MultiPrimer to read in the primer–gene graph file and creates a new primer–gene graph file with degenerate primers included. The degenerate primers can be a combination of simple primers with a specified maximum order of degeneracy.

`SOLVEGRAPH1`: This command instructs MultiPrimer to read the primer–gene graph file and run the densest subgraph heuristic on it to produce a solution subgraph as output with edges representing all genes.

`SOLVEGRAPH2`: This command instructs MultiPrimer to read the primer–gene graph file and run the linear-time greedy heuristic on it to produce a solution subgraph as output with edges representing all genes.

`PRINTSOLUTION`: This command instructs MultiPrimer to read the primer–gene graph file and the primer index and gene index files to produce a solution file with records consisting of primer pairs and associated gene.

`PRINTDEGSOLUTION`: This command is the same as the preceding command except that it is to be used to output a primer–gene solution sub-graph with possible degenerate primers.

The remaining lines of the command file are described, and default values are given in parentheses.

L2—The name of the primer-pair input file (Primerout.txt).
L3—The name of the primer–gene graph file (Graph.txt).
L4—The name of the primer-index file (primers.txt).
L5—The name of the gene-index file (colors.txt).
L6—The name of the output degenerate primer–gene graph file (degenerate.out).
L7—The name of the degenerate-primer-index file (degprimers.txt).
L8—The name of the primer-pair output file (outprimers.txt).
L9—The maximum order of degeneracy if `ADVANCED-DEGENERACY` is the command (no default value).

References

1. E. Maier, S. Maier-Ewert, D. Bancroft, and H. Lehrach. Automated array technologies for gene expression profiling. *DDT*, 8:315–324, 1997.
2. M. Schena, D. Shalon, R. Davis, and P. Brown. Quantitative monitoring of gene expression patterns with a complementary DNA microarray. *Science*, 270: 467–470, 1995.

3. R. Cho, M. Campbell, E. Winzeler, L. Steinmetz, A. Conway, L. Wodicka, T. Wolfsberg, A. Gabrielian, D. Landsman, D. Lockhart, and R. Davis. A genome-wide transcriptional analysis of the mitotic cell cycle. *Molecular Cell*, 2: 65–73, 1998.
4. P.T. Spellman, G. Sherlock, M.Q. Zhang, V.R. Iyer, K. Anders, M.B. Eisen, P.O. Brown, D. Botstein, and B. Futcher. Comprehensive identification of cell cycle-regulated genes of the yeast saccharomyces cerevisiae by microarray hybridization. *Molecular Biology of the Cell*, 9:3273–3297, 1998.
5. A.P. Gasch, P.T. Spellman, C.M. Kao, O. Carmen-Harel, M.B. Eisen, G. Storz, D. Botstein, and P.O. Brown. Genomic expression programs in the response of yeast cells to environment changes. *Molecular Biology of the Cell*, 11: 4241–4257, 2000.
6. T.R. Hughes, M.J. Marton, A.R. Jones, C.J. Roberts, R. Stoughton, C.D. Armour, H.A. Bennett, E. Coffey, H. Dai, Y.D. He, M.J. Kidd, A.M. King, M.R. Meyer, D. Slade, P. Y. Lum, S.B. Stepaniants, D.D. Shoemaker, D. Gachotte, K. Chakraburtty, J. Simon, M. Bard, and S.H. Friend. Functional discovery via a compendium of expression profiles. *Cell*, 102:109–126, 2000.
7. NCBI. Entrez-genome. `http://www.ncbi.nlm.nih.gov/entrez/query.fcgi?db=genome`, 2002.
8. R. J. Fernandes and S. S. Skiena. Microarray synthesis through multiple-use PCR primer design. *Bioinformatics*, 18:S128–S135, 2002.
9. M. Charikar. Greedy approximation algorithms for finding dense components in a graph. In *Proc. APPROX 2000*, pages 84–95. Lecture Notes in Computer Science 1913, Springer Verlag, Berlin, Germany 2000.
10. G. Gallo, M. Grigoriadis, and R. Tarjan. A fast parametric maximum flow algorithm and applications. *SIAM Journal of Computing*, 18:30–55, 1989.
11. Y. Asahiro, K. Iwama, H. Tamaki, and Tokuyama T. Greedily finding a dense subgraph. *Journal of Algorithms*, 34:203–221, 2000.
12. U. Feige and M. Seltser. On the densest k-subgraph problems. Technical Report CS97-16, The Weizmann Institute, Jan., 1997.
13. R.J. Fernandes and S.S. Skiena. Available at `http://algorithm.cs.stonybrook.edu/compbio/MultiPrimer`, 2006.

C

Allele-specific PCR

16

Modified Oligonucleotides as Tools for Allele-Specific Amplification

Michael Strerath, Ilka Detmer, Jens Gaster, and Andreas Marx

Summary

Allele-specific polymerase chain reaction (PCR), a method that reports nucleotide variations through either the presence or the absence of a DNA product obtained through PCR amplification, holds the promise to combine target amplification and analysis in one single step. Recently, it has been reported that the selectivity of allele-specific PCR can be significantly increased through the employment of chemically modified primer probes. Here, we report on significant developments of primer probe design and synthesis along this line.

Key Words: Allele-specific amplification; modified oligonucleotides; polymerase; PCR; SNP; DNA synthesis.

1. Introduction

A conceptually simple approach to detect single-nucleotide variations, for example, single-nucleotide polymorphisms (SNPs), within genes is allele-specific polymerase chain reaction (PCR) amplification ***(1–4)***. PCR primers are designed in a way that the 3′ end of one of them is located opposite the position of interest. In case of one allelic variant, a perfectly matched hybrid is formed (*see* **Fig. 1**). If another allele is present, the hybrid is mismatched at the 3′ end, and the PCR should be discriminated by the used DNA polymerase and the exponential product formation would be prevented.

To analyze PCR product formation, agarose gel electrophoresis or monitoring in real-time using a DNA double-strand specific dye (e.g., SybrGreen I) and an

From: *Methods in Molecular Biology, vol. 402: PCR Primer Design*
Edited by: A. Yuryev © Humana Press, Totowa, NJ

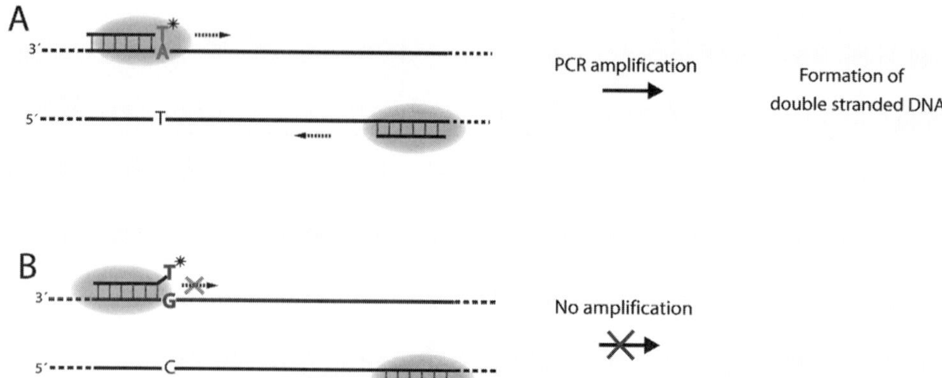

Fig. 1. Allele-specific amplification. **(A)** In case of a matched allele-specific primer probe, PCR amplification takes place. **(B)** In case of a single nucleotide variation, the primer probe forms a mispaired complex, obviating PCR product formation.

appropriate real-time PCR system is feasible. Owing to the binding of double-stranded DNA, a fluorescence signal occurs, which is directly bound to product formation in a linear fashion ***(5)***.

The reliability and specificity of the allele-specific PCR approach is dependent on the ability of the DNA polymerase to discriminate between the extension of a mispaired paired primer end and the canonically paired one. Tedious time-intensive optimization steps remain necessary to find buffer conditions, temperature profiles, and sequence compositions of primer strands for each allele-specific system ***(1–4)***.

We and others discovered that allele-specific PCR becomes more robust when certain covalent modification of the 2′-deoxyribose was implemented in the allele-specific primers ***(6–12)***. Using these modified primers, allele-specific amplification becomes more robust. It has been shown that allele-specific PCR directly from genomic DNA without the need of a pre-amplification step is feasible.

Significantly higher amplification selectivity is observed by the application of primer probes that bear small 4′-C modifications like the methoxymethylene group at the 3′ end (*see* **Fig. 2**, compound **1**) ***(12)***. A commercially available 3′–5′-exonuclease-deficient variant of a DNA polymerase from *Thermococcus litoralis* [Vent (exo-) DNA polymerase] was used for amplification.

Increased specificity in allele-specific amplification assays was also achieved by the use of allele-specific oligonucleotide primers bearing a 3′-nucleotide that is bridged with a 2′-O-4′-C methylene group (*see* **Fig. 2**, compound **2**)

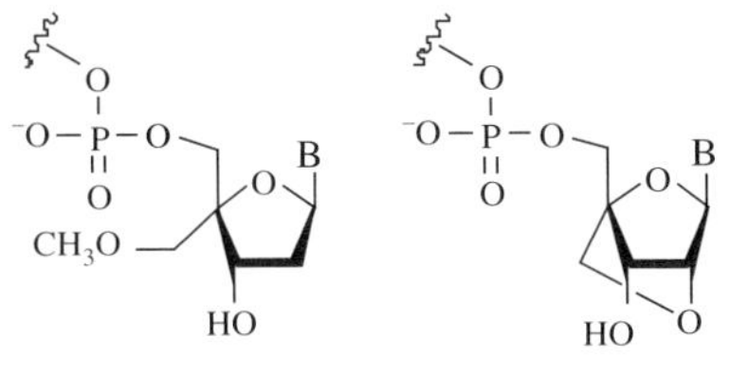

Fig. 2. Structures of modified primer strands employed in allele-specific PCR. B, nucleobase.

(11). These modified primer strands are commercially available as TrueSNP™ primers by proligo (http://www.proligo.com).

2. Materials

1. ABI 392 DNA/RNA synthesizer (Applied Biosystems).
2. Standard phosphoramidates and solid supports.
3. Biometra Seq 2 Sequencing Apparatus 40 × 30 cm.
4. TBE buffer: 89 mM Tris [tris(hydroxymethyl) aminomethane], 89 mM boric acid, and 2 mM ethylene-diamine-tetraacetic acid (EDTA).
5. Controlled pore glass (CPG) solid support (500 Å pore size, ChemGenes Corporation).
6. Solvents and chemicals of the highest purity grade (Fluka, Aldrich).
7. Flash chromatography: silica gel G60, 230–400 mesh (Merck).
8. Thin-layer chromatography: pre-coated plates (silica gel 60 F_{254}, Merck).
9. Thermomixer (Eppendorf).
10. Silanized glass wool (Serva).
11. UV spectrometer (e.g., nanodrop).
12. Vacuum concentrator (Eppendorf).

3. Methods

The methods describe the synthesis of DNA oligonucleotides that contain a 4′-C methoxymethylene-modified thymidine at the 3′-terminus following published protocols ***(12–15)*** and are depicted in Figure 3.

3.1. Preparation of Oligonucleotides Containing 4′-C Methoxymethylene-Modified Thymidines

3.1.1. 3′-O-tert-Butyldimethylsilylthymidine

Thymidine 2 (25 g, 103 mmol) and 4, 4′-dimethoxytriphenyl chloromethane (38.6 g, 114 mmol) were dissolved under argon in anhydrous pyridine. After

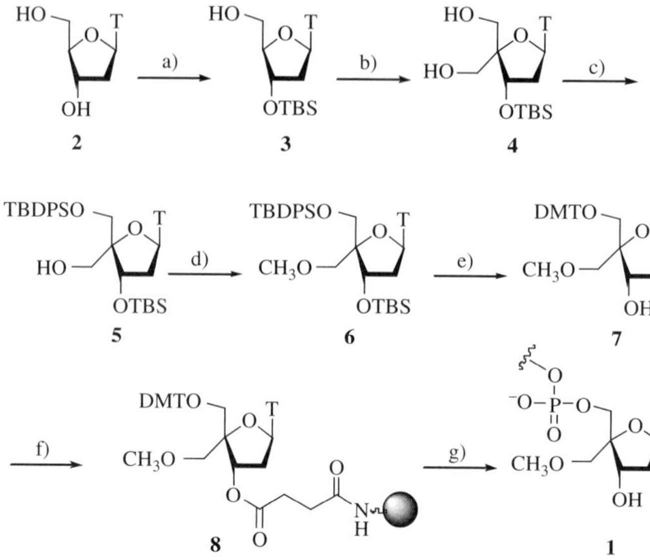

Fig. 3. d) NaH, MeI, THF; e) (i) 1 M TBAF, THF, (ii) DMTCl, DMAP, pyridine; f) EDC, DMAP, succinylated LCAA-CPG, pyridine; then 4-nitrophenol; then piperidine, then acetic anhydride/pyridine/THF (Cap A) and 1-methylimidazole/THF (Cap B); g) (i) oligonucleotide synthesis, (ii) 33% NH_4OH. TBS = *tert*-butyldimethylsilyl; TBDPS=*tert*-butyldiphenylsilyl, DMT = 4, 4′-dimethoxytrityl; DMAP = 4-(*N*, *N*-dimethylamino)pyridine; LCAA-CPG=long chain alkyl amine modified controlled pore glass; EDC=1-(3-dimethylaminopropyl)-3-ethylcarbodiimide hydrochloride; TBAF=tetra-*n*-butylammonium fluoride.

stirring for 8 h, the mixture was quenched with methanol (20 ml) and concentrated in vacuum. The residue was coevaporated with toluene (2 × 50 ml). The obtained oil was dissolved in CH_2Cl_2 (400 ml) and washed with saturated $NaHCO_3$ solution (2 × 400 ml), and the aqueous phase was extracted with CH_2Cl_2 (50 ml). The organic phases were combined, dried ($MgSO_4$), and evaporated to an oil. The residue was coevaporated with toluene (2 × 50 ml), dissolved in diethylether (30 ml), and 2 h refluxed. The colorless precipitate was filtered off (55.3 g, 101.5 mmol, 98.5%) and used directly in the next step without further purification.

5′-*O*-(4, 4′Dimethoxytrityl) thymidine was coevaporated twice with anhydrous pyridine (40 ml), dissolved in anhydrous dimethylformamide (DMF) (60 ml) under argon, and imidazole (20.4 g, 303.9 mmol) and *tert*-butyldimethylchlorsilane (16.8 g, 111.4 mmol) were added. The mixture was stirred for 18 h at room temperature, then poured into saturated aqueous $NaHCO_3$ solution (1 l), and extracted with CH_2Cl_2 (500 ml). The aqueous phase was extracted with CH_2Cl_2 (2 × 300 ml), and the organic phases were washed with water (2 × 1 l). The combined organic phases were dried ($MgSO_4$) and concentrated in vacuum. The residue was used without further purification.

To a solution of 5′-*O*-(4, 4′-dimethoxytrityl)-3′-*O*-*tert*-butyldimethylsilyl thymidine in tetrahydrofuran (THF) (60 ml) was added 80% acetic acid (330 ml). After stirring for 18 h at room temperature, the mixture was cooled to 0 °C and neutralized with 25% ammonium hydroxide solution (350 ml). Then the mixture was poured into water (500 ml) and extracted with CH_2Cl_2 (3 × 350 ml). The combined organic phases were washed with aqueous saturated $NaHCO_3$ solution (2 × 1 l), dried ($MgSO_4$), and concentrated in vacuum. The residue obtained was purified by flash chromatography (SiO_2, CH_2Cl_2/Et_2O, 10:1), which yielded compound **3** as a yellow foam (26 g, 73 mmol, 75%); $R_f = 0.33$ (ethyl acetate/cyclohexane, 3:1).

3.1.2. 3′-O-tert-Butyldimethylsilyl-4′-C-Hydroxymethylthymidine (4)

Compound **3** (26 g, 73 mmol) was coevaporated with toluene (50 ml) and dissolved in dimethylsulfoxide (DMSO) (40 ml). To the solution were added N, N'-dicyclohexylcarbodimide (DCC) (60.3 g, 292 mmol) and pyridinium trifluoroacetate (7.05 g, 36.5 mmol) under argon. After stirring for 18 h at room temperature, oxalic acid (3.65 g, 40.5 mmol) in methanol (20 ml) was added slowly. The colorless precipitate was filtered off and washed with CH_2Cl_2 (300 ml). After washing the organic phase with water (1.5 l), the aqueous phase was extracted with CH_2Cl_2 (2 × 300 ml). The organic phase was washed with water (2 × 1 l), and the aqueous phases were extracted with CH_2Cl_2 (100 ml). The combined organic phases were dried ($MgSO_4$) and evaporated. The residue was dissolved in dioxane (160 ml) without further purification. Then 33% aqueous formaldehyde solution (80 ml) and 2 N NaOH were added. After stirring for 2 h, the mixture was cooled to 0 °C, neutralized with tartaric acid (32 g), and evaporated. After the addition of water (500 ml), the aqueous phase was extracted with CH_2Cl_2 (4 × 300 ml) and the combined organic phases were dried ($MgSO_4$) and concentrated in vacuum. The obtained oil was dissolved in ethanol (100 ml) and cooled to 0 °C. Subsequently, $NaBH_4$ (3.0 g, 84 mmol) was added in small portions. After 15 min, tartaric acid (32 g) and acetic acid (10 ml) were added, and the mixture was concentrated in vacuum. The obtained residue was dissolved in water (500 ml) and extracted with CH_2Cl_2 (4 × 500 ml). The combined organic phases were dried ($MgSO_4$) and concentrated in vacuum. The residue obtained was purified by flash chromatography (SiO_2, CH_2Cl_2/EtOH;

3:1, 5:1, 10:1, ∞ : 1), which yielded compound **4** as a yellow foam (16 g, 41.3 mmol, 57%); $R_f = 0.19$ (ethyl acetate/cyclohexane, 5:1).

3.1.3. *3′-O-tert-Butyldimethylsilyl-5′-O-tert-Butyldiphenylsilyl-4′-C-Hydroxymethylthymidine* (5)

To a solution of compound **4** (16 g, 41.3 mmol) in anhydrous pyridine (400 ml) was added a mixture of 4, 4′-dimethoxytriphenyl chloromethane (15.38 g, 4.5 mmol) in pyridine (100 ml) slowly under argon at 0 °C. After stirring for 18 h at room temperature, methanol (20 ml) was added, concentrated in vacuum, and the residue coevaporated with toluene (2 × 50 ml). The obtained residue was used directly in the next reaction.

To a solution of the residue and imidazole (8.37 g, 123 mmol) in DMF (70 ml) was added *tert*-butyldiphenylchlorosilane (13.86 ml, 53.3 mmol) under argon. After stirring for 4 h at room temperature, the mixture was poured into water (500 ml) and extracted with CH_2Cl_2(3 × 500 ml). The organic phase was washed with water (2 × 500 ml), dried ($MgSO_4$), and concentrated in vacuum. The obtained yellow foam was dissolved in THF (50 ml) and 80% acetic acid (205 ml) and stirred for 8 h at 50 °C. Then the mixture was cooled to 0 °C and neutralized with 25% aqueous ammonium hydroxide solution (216 ml), poured into water (500 ml), and extracted with CH_2Cl_2(3 × 350 ml). The combined organic phases were dried ($MgSO_4$) and concentrated in vacuum. The residue obtained was purified by flash chromatography (SiO_2, cyclohexane/ethyl acetate, 3:1–1:1–1:∞), which yielded compound **5** as a colorless foam (13 g, 20.8 mmol, 51%); $R_f = 0.43$ (ethyl acetate/cyclohexane, 1:1).

3.1.4. *3′-O-tert-Butyldimethylsilyl-5′-O-tert-Butyldiphenylsilyl-4′-C-Methoxymethylene Thymidine* (6)

After coevaporation and drying under vacuum overnight, nucleoside **5** (325 mg, 0.52 mmol) was dissolved in THF (6 ml). At 0 °C, sodium hydride (44.8 mg, 1.12 mmol) was added, and the reaction mixture was stirred for 30 min. Then iodomethane (162 μl, 2.6 mmol) was added and stirring continued for 10 h at 0 °C. The reaction was quenched by the addition of methanol (2 ml) and allowed to warm up to room temperature. Hereafter, saturated aqueous sodium bicarbonate solution was added, and the aqueous phase extracted with dichloromethane. The combined organic phases were dried ($MgSO_4$) and concentrated in vacuum. Purification of the obtained residue by flash column chromatography (SiO_2, ethyl acetate/cyclohexane, 1:4 – 1:1) furnished compound **6** as a white foam (250 mg, 0.39 mmol, 75%); $R_f = 0.64$ (ethyl acetate/cyclohexane, 1:1).

3.1.5. 5′-O-(4,4′-Dimethoxytrityl)-4′-C-Methoxymethylene Thymidine (7)

To compound **6** (123 mg, 0.19 mmol) dissolved in THF (4 ml) was added 1 M tetra-*n*-butylammonium fluoride (TBAF) (0.43 ml, 0.43 mmol) at 0 °C and stirred for 30 min. Then the reaction mixture was allowed to warm up to room temperature, and stirring was continued for 3.5 h. A small amount of silica was added, and the mixture was evaporated to dryness. The impregnated silica was coevaporated with toluene and subjected to column chromatography (SiO_2, cyclohexane/ethyl acetate 1:9 – ethyl acetate/methanol 9:1) to yield the desired alcohol as a white foam (54.7 mg, 0.19 mmol, 99%); $R_f = 0.48$ (ethyl acetate/methanol 9:1). Subsequently, the alcohol (53.7 mg, 0.19) was dissolved in pyridine (1 ml), and 4, 4′-dimethoxytrityl chloride (DMTCl) and a catalytic amount of 4-(dimethylamino) pyridine (DMAP) were added at 0 °C, and the cooled solution was stirred for 30 min. Then the reaction mixture was allowed to warm up to room temperature while stirring was continued for 4 h. Hereafter, the reaction was quenched by the addition of methanol (1.5 ml) while stirring was continued for 30 min. The solvent was removed in vacuum, and the obtained residue purified by flash column chromatography (SiO_2, cyclohexane/ethyl acetate 3:7 + 1% Et_3N). Compound **7** was isolated as a white foam (60 mg, 0.10 mmol, 54%); $R_f = 0.28$ (cyclohexane/ethyl acetate 3:7).

3.1.6. Preparation of the CPG Solid Support (8)

Typically used solid support consists of amino-derivatized borsilca glass beads with a controlled pore size of 500 Å.

Activation of the glass beads is performed as followed. 2 g LCAA-CPG, 400 mg (4 mmol) succinicacidanhydride, and 48 mg (0.4 mmol) DMAP were combined in a round bottom flask and dried in vacuum. Subsequently, 20 ml anhydrous pyridine was added. The mixture was shaken for 12 h. The solid support was washed with successively 20 ml pyridine followed by 40 ml methanol and 60 ml dichlormethane. The succinylated CPG was dried under vacuum.

For qualitative control of the succinylation, an aqueous solution of ninhydrine to both the activated and the crude solid support was added. The crude support should become deeply blue, whereas the successful activated support should remain white or light blue. Compound **7** was coupled to 1 g succinylated LCAA-CPG using the following procedure. 1 g succinylated LCAA-CPG, compound **7** (0.1 mmol), 24 mg DMAP (each 0.1 mmol/1.0 g CPG), and 192 mg 1-(3-dimethylaminopropyl)-3-ethylcarbodiimide hydrochloride (EDC) (1.0 mmol/1.0 g CPG) were combined. Afterward, the flask was evaporated and flushed with argon. 50 ml Pyridine (10 ml/1.0 g CPG) and

400 μl NEt_3 (80 μl/1.0 g CPG) were added and shaken for 12 h. Afterward, 120 mg 4-nitrophenol (0.5 mmol/1.0 g CPG) was added, and shaking was continued for additional 24 h. Afterward, 5 ml piperidine (5 ml/1.0 g CPG) was added, and shaking was continued for 5 min. Then, the beads were immediately filtered off using a glass filter with pore size D3 (*see* **Note 1**). The beads were washed with 25 ml pyridine, followed by 50 ml methanol, and finally with 100 ml neutralized CH_2Cl_2 (using saturated $NaHCO_3$) (*see* **Note 2**). After drying, the beads were suspended in respectively 7.5 ml acetic anhydride/pyridine/THF and 1-methylimidazole/THF alternatively the commercially available (Cap A and Cap B) capping reagents supplied as chemicals for automated DNA-synthesis can be used. After shaking for 2 h, the beads were filtered off and washed as described above. Typical loading ranges of compound **8** are 5–40 μmol/g.

3.1.7. Quantification of the Nucleoside Coupled to the Solid Support

Prepare a mixture of 10.2 ml 60% $HClO_4$ and 9.2 ml methanol. Suspend 4 mg of the solid support in 5 ml of the mixture. The amount of coupled nucleoside can be calculated measuring the absorption value at 498 nm using the following equation:

$$x[\mu\text{mol/g}] = (A_{498}{}^{*}V_{\text{acid}}{}^{*}14.3)/\text{mCPG}$$

whereby V_{acid} is the volume of 60% $HCIO_4$ in ml and mCPG the mass of the employed CPG in g.

The measurement should be done at least in double experiments.

3.1.8. Synthesis of 4′-C-Modified Oligonucleotides 1

The synthesis of oligonucleotide **1** was carried out on a 0.2 – μmol scale on an Applied Biosystems 392 DNA synthesizer, using compound **8** (*see* **Note 3**) and commercially available 2-(cyanoethyl) phosphoramidites. A standard method for 2-(cyanoethyl) phosphoramidites was used, with the exception that the coupling times from the modified nucleotides were extended to 10 min. Yields for modified oligonucleotides are similar to those obtained for unmodified oligonucleotides. Depending on the method of purification the decoupling of the last trityl group should be performed or not. For high performance liquid chromatography (RP-HPLC), cleaning the trityl group is needed; for purification by polyacrylamide gel electrophoresis (PAGE), the trityl group should be cleaved off. After synthesis, the deprotection and cleavage of the oligonucleotides from the solid support can be performed by treatment with concentrated NH_4OH at 55 °C for 12 h followed by the removal of the NH_4OH by evaporation at 65 °C.

3.2. Purification

3.2.1. Purification of the Oligonucleotides

After deprotection, cleavage, and evaporation, the remaining was suspended in water (75 μl/100 nmol). Twice the amount of a loading buffer composed of 80% formamide, and 20 mM EDTA (pH 8.0) was added. The suspension was then loaded to a 12% polyacrylamid gel containing 8 M urea. Dimensions of 40-cm length and a thickness of 1–2 mm are recommended. The gel should be cooled during the run. A typically used current is 90 mA and 1 × TBE buffer in the upper and lower vessel. After PAGE, the oligonucleotides were localized using UV shadowing (*see* **Note 4**). Therefore, the gel was transferred onto a polyethylene film covering a standard TLC plate (e.g., Merck Silica Gel 60 F_{254}) containing a fluorescent dye for DNA detection. By UV radiation from above, spots of DNA become visible as shadows. After localization, the spot was cut out of the gel, and the gel fragment was crushed by pressing through a 5-ml syringe. The DNA was eluted by adding 1 ml water to an amount of 500 μl crushed gel and shacked overnight at 55 °C.

The suspension was filtered using a 5-ml syringe darned with silanized glass wool. The DNA oligonucleotides were recovered from the solution by evaporation followed by standard precipitation using 300 μl sodium acetate (0.3 M) and 900 μl ethanol for each 500 μl crushed gel. The oligonucleotides were quantified by absorption measurements at 260 nm using the following formulae:

$$\varepsilon[\mathrm{mMP}^{-1P}] = (15.4^{*}\mathrm{A} + 11.7^{*}\mathrm{G} + 8.8^{*}\mathrm{T} + 7.3^{*}\mathrm{C})^{*}0.9$$

$$c\left[\mathrm{mMP}^{-1P}\right] = \frac{\mathrm{ODB}_{260B}}{\varepsilon\left[\mathrm{mMP}^{-1P}\right]}$$

Total over all yields of purified oligonucleotides are in the 15–30% range. The integrity of all modified oligonucleotides should be confirmed by MALDI-To F MS.

3.2.2. Preparation of a PAGE Gel

The following stock solutions were prepared. A (25% Bis/AA in 8.3 M urea or Rotiphorese sequencing gel concentrate (e.g., Roth), B (8.3 M urea), C (8.3 M urea in 10 × TBE) and 10 × TBE (tris base 108 g, boric acid 55 g, and EDTA 0.5 M; pH 8.0 40 ml). For a typical 12% PAGE gel, a 250 ml mixture containing 120 ml A, 105 ml B, and 25 ml was prepared. For polymerization 1.8 ml 10% ammoniumperoxo disulfate and afterward 90 µl N, N, N′, N′-tetramethylethylenediamine (TEMED) were added immediately before casting the gel.

3.3. Design of Primer Probes and Reverse Primers

Owing to the application of the modified primers as allele-specific primer probes, the positions and therefore the sequence contexts of the 3′-primers are given. The only variable is the primer length at its 5′-end. To guarantee its uniqueness, the minimal length should exceed 18 bases. The calculated melting temperature should be in between 45 and 65 °C and should be equal for both primers (*see* **Note 5**). To reduce the possibility of primer–dimer formation, regions of accumulated G or C bases should be prevented. The design of the 5′-primer is not regulated in position or sequence context. For the purpose of SNP detection, amplicon lengths of 100–500 base pairs should be used. In most cases, it should be sufficient to follow the few abovementioned rules. Additionally, it could be helpful to use an oligonucleotide calculator to receive further information concerning self-complementary parts between the 5′- and 3′-primer.

4. Notes

1. The filter unit used for washing steps should not be smaller than pore size 3 to reduce the time needed for filtering and therefore to hold the incubation times.
2. For coupling to the solid support, neutralized dichloromethane should be used.
3. For automated synthesis of the primer probes using a DNA synthesizer, commercially available cartridges for unmodified solid support can often be reused but should be sealed using Teflon tape.
4. The time of UV light irradiation while the UV shadowing process should be reduced to a minimum.
5. Easy to use oligonucleotide calculators are available free of charge in the internet (e.g., http://www.basic.northwestern.edu/biotools/oligocalc.html).

Acknowledgments

We acknowledge financial support of this work by the Deutsche Forschungsgemeinschaft.

References

1. Gibbs, R. A., Nguyen, P.-N., and Caskey, C. T. (1989) Detection of single DNA base differences by competitive oligonucleotide priming. *Nucleic Acids Res.* 17, 2437–2448.

2. Newton, C. R., Graham, A., Heptinstall, L. E., Powell, S. J., Summers, C., Kalsheker, N., Smith, J. C., and Markham, A. F. (1989) Analysis of any point mutation in DNA. The amplification refractory mutation system (ARMS). *Nucleic Acids Res.* 17, 2503–2516.
3. Green, E. K. (2002) Allele-specific oligonucleotide PCR, in *PCR Mutation Detection Protocols* (Theophilus, B. D. M., and Rapley, R., eds.), Humana, Totowa, NJ, pp. 47–50.
4. Strerath, M., and Marx, A. (2005) Genotyping – from genomic DNA to genotype in one tube. *Angew. Chem. Int. Ed. Engl.* 44, 7842–7849.
5. Wilhelm, J., and Pingoud, A. (2003) Real-time polymerase chain reaction. *Chembiochem* 4, 1120–1128.
6. Strerath, M., and Marx, A. (2002) Tuning PCR specificity by chemically modified primer probes. *Angew. Chem. Int. Ed. Engl.* 41, 4766–4769.
7. Tews, B., Wilhelm, J., Summerer, D., Strerath, M., Marx, A., Friedhoff, P., Pingoud, A., and Hahn, M. (2003) Application of the C4′ alkylated deoxyribose primer system (CAPS) in allele specific real-time PCR for increased selectivity in discrimination of single nucleotide sequence variants. *Biol. Chem.* 384, 1533–1541.
8. Latorra, D., Arar, K., and Hurley, J. M. (2003) Design considerations and effects of LNA in PCR primers. *Mol. Cell Probes* 17, 253–259.
9. Strerath, M., Gaster, J., Summerer, D., and Marx, A. (2004) Increased single-nucleotide discrimination of PCR by primer probes bearing hydrophobic 4′C-modifications. *Chembiochem* 5, 333–339.
10. Latorra, D., Campbell, K., Wolter, A., and Hurley, J. M. (2003) Enhanced allele-specific PCR discrimination in SNP genotyping using 3′ locked nucleic acid (LNA) primers. *Hum. Mutat.* 22, 79–85.
11. Ugozzoli, L. A., Latorra, D., Puckett, R., Arar, K., and Hamby, K. (2004) Real-time genotyping with oligonucleotide probes containing locked nucleic acids. *Anal. Biochem.* 324, 143–152.
12. Gaster, J., and Marx, A. (2005) Tuning single nucleotide discrimination in PCR: syntheses of primer probes bearing polar 4′C-modifications and their application in allele-specific PCR. *Chem. Eur. J.* 11, 1861–1870.
13. Yang, C. O., Wu, H. Y., Fraser-Smith, E. B., and Walker, K. A. M. (1992) Synthesis of 4′-cyanothymidine and analogs as potent inhibitors of HIV. *Tetrahedron Lett.* 33, 37–40.
14. Yang, C. O., Kurz, W., Eugui, E. M. McRoberts, M. J., Verheyden, J. P. H., Kurz, L. J., and Walker, K. A. M. (1992) 4′-Substituted nucleosides as inhibitors of HIV: an unusual oxetane derivative. *Tetrahedron Lett.* 33, 41–44.
15. Marx, A., Erdmann, P., Senn, M., Körner, S., Jungo, T., Petretta, M., Imwinkelried, P., Dussy, A., Kulicke, K. J., Macko, L., Zehnder, M., and Giese, B. (1996) Synthesis of 4′-C-acylated thymidines. *Helv. Chim. Acta* 79, 1980–1994.

17

AlleleID

A Pathogen Detection and Identification System

Arun Apte and Siddharth Singh

Summary

Efficient clinical diagnosis of pathogens is important for the management of infectious diseases. Conventional methods have longer turnaround time and, in most cases, lower sensitivity. Nucleic acid-based methods for the detection of microorganisms are rapid, sensitive and are generally successful even when the culturing of microorganisms fails. Sequence-based molecular methods such as real-time PCR, microarrays, and band biosensors provide high sensitivity, rapid diagnostics, and higher specificity allowing differentiation between related strains. Although numerous chemistries are available for the molecular level identification of pathogens, the most common are qPCR and DNA microarrays. Both of these techniques have a high accuracy when used with specific primers and probes. Manual design of these primer and probes is both tedious and results in lower quality of results because of the inability to simultaneously handle multiple criteria for design. Here, we describe a program AlleleID that designs qPCR and microarray assays to identify and detect pathogens.

Key Words: Multiple alignment; pathogen identification; pathogen detection; species-specific design; taxa specific; cross-species microarray; qPCR assays; microarrays and diagnostic assay.

1. Introduction

Detection and accurate identification of pathogens have always been a challenge for biologists working in pathology. Historically, microbiological and serological methods were used to detect and identify pathogenic agents from samples. But because of the diversity and low genomic abundance of these pathogens, these traditional methods of detection and identification are giving way to sequence-based molecular methods.

From: *Methods in Molecular Biology, vol. 402: PCR Primer Design*
Edited by: A. Yuryev © Humana Press, Totowa, NJ

Sequence-based pathogen identification protocols are popular because they are fast and accurate *(1)*. To be useful in pathogen detection, a genomic sequence of the test organism should have characteristics including being conserved to a greater extent among the related organisms, having a low mutation rate, and having appropriate molecular properties such as secondary structures and melting temperature for the molecular biological assays being used.

AlleleID is a program that helps to address the complex task of pathogen identification with great accuracy. It designs probes and primers for qPCR and microarray assays for pathogen identification. In addition to pathogen detection, AlleleID can also be used for allelic discriminations assays, biodiversity analysis, and specific cDNA amplification. Researcher can design group-specific probes for detecting a particular pathogen and can design strain-specific probes to identify the genotype of these pathogens. AlleleID support various chemistries including molecular beacons, TaqMan®, FRET probes, and SYBR Green®.

2. AlleleID Designs Primers and Probes for Related Gene Assay

The pathogen-specific assay designs in AlleleID are collectively referred to as related gene assays (*see* **Fig. 1**). AlleleID detects species-specific or taxa-specific/cross-species-specific *(2)* regions of a given set of sequences by aligning them. It detects mismatches (which are used to discriminate a given sequence as belonging to a unique species) and conserved regions (which are common to all the sequences in the set hence can be used to discriminate a taxa or a group). It then uses these regions to design primer and probes.

2.1. Alignment of Sequences

AlleleID uses the ClustalW *(3)* algorithm for the alignment of nucleotide sequences. These alignments are then used to design related gene assays.

2.2. Algorithm for Related Gene Assay Design

AlleleID uses a proprietary algorithm for probe design. The basic steps are as follows:

1. AlleleID aligns the nucleotide sequences.
2. After aligning AlleleID, search engine looks for all conserved regions and mismatches in an alignment.
3. The program designs probes according to the assay-specific parameters (i.e., unique for species specific and conserved for taxa specific) as specified by a researcher.

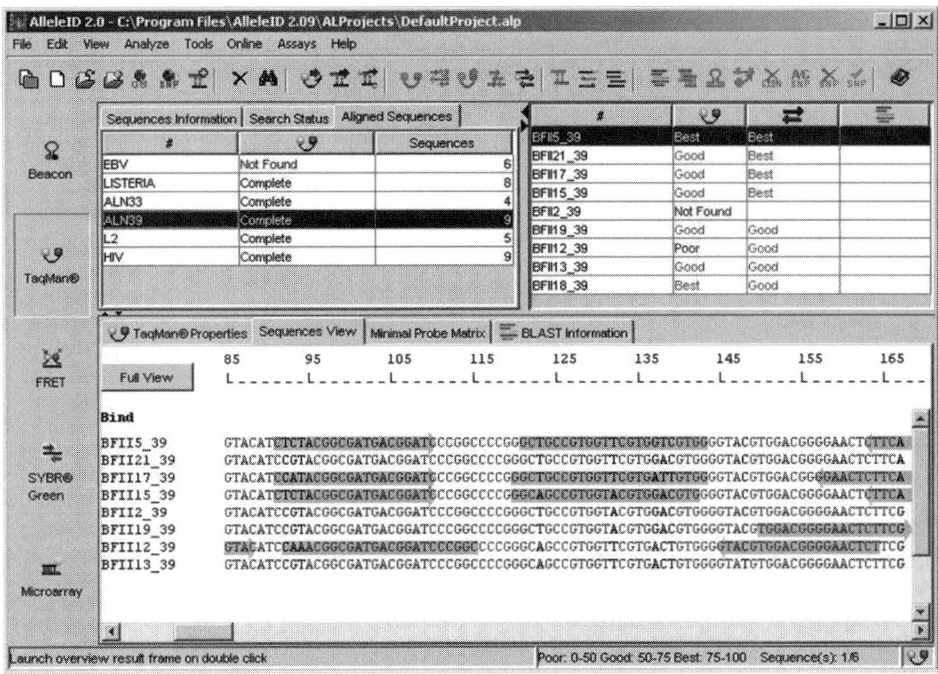

Fig. 1. AlleleID main screen; shown are the project management pane, search status pane and result property pane.

4. Once the probes meeting the selected criteria are picked, the program designs appropriate flanking primer pair to amplify the target templates bound by the probes.
5. There are two special cases when (1) a minimal probe set is designed for species-specific design and (2) mismatch tolerant probes are designed for taxa/cross-species-specific design. When the minimal probe design option is selected, the program picks the minimum number of probes that can uniquely identify all the sequences in the alignment. Then, the program will make unique combination of these primers and probes. The given combination is unique for a single sequence, thus ensuring specific identification. When mismatch-tolerant probe design option is selected for taxa/cross-species-specific design, the program will pick up the probes that can tolerate mismatches that appear in an alignment at its 5′-end. These probes are designed considering the difference in the T_m of individual probe sequences.
6. A robust rating system evaluates all the compatible probe and primer pairs to rank them according to their efficiency. The higher the rating of a design, the higher the probability of assay success.

2.3. TaqMan® Probe Design for Related Gene Assays

TaqMan® *(4)* probes rely on the 5′–3′ exonuclease activity of Taq polymerase, which cleaves a labeled probe when hybridized to a complementary target. It is the most widely used qPCR chemistry.

The probe consists of a fluorophore attached to the 5′-end of the probe and a quencher to the 3′-end. Having the fluorophore at the 5′-end and the quencher at the 3′-end, simplifies the synthesis of the probe. The probe binds the target during each annealing step of the PCR. The enzyme's double-strand-specific 5′–3′ nuclease activity displaces the 5′-end of the probe and then degrades it. Cleavage continues until the remaining probe melts off the amplicon. This process releases the fluorophore and the quencher into solution separately leading to an irreversible increase in the fluorescence from the reporter. As the polymerase will cleave the probe only while it remains hybridized to its complementary strand, the temperature conditions of the polymerization phase of the PCR must be adjusted to ensure probe binding. It is usually carried out at 8–10 °C below the T_m of the probe to ensure maximum 5′–3′ exonuclease activity of the polymerase, although this reduces the efficiency of the enzyme. Because of its simple chemistry and sensitivity, TaqMan® probes are widely used for qPCR assays.

2.3.1. Species-Specific TaqMan® Assay Design

AlleleID designs TaqMan® probes for species/strain level identification of pathogens. The program detects mismatch positions in an alignment that are unique for a species; it then designs a probe with the mismatch position at the center of the probe. This ensures the specific binding of the probe to the target. To design the species-specific assay, the researcher can choose between unique probes or minimal probes.

2.3.1.1. Unique TaqMan® Probes for Species-Specific Design

When the researcher selects to design unique probes for the assay, the program will try to design the probes that bind at positions that have mismatch bases unique to a species in an alignment. This type of design will help identify the sequences that differ even by a single base in the mixture of sequences (*see* **Fig. 2**).

2.3.1.2. Minimal TaqMan® Probe Set for Species-Specific Design

AlleleID has the ability to design a pool of probes to identify the sequences. A minimal probe set is the fewest number of probes required for the unique identification of a given set of sequences. Individual sequences in a sample are identified using a unique combination of probes, of this minimal set.

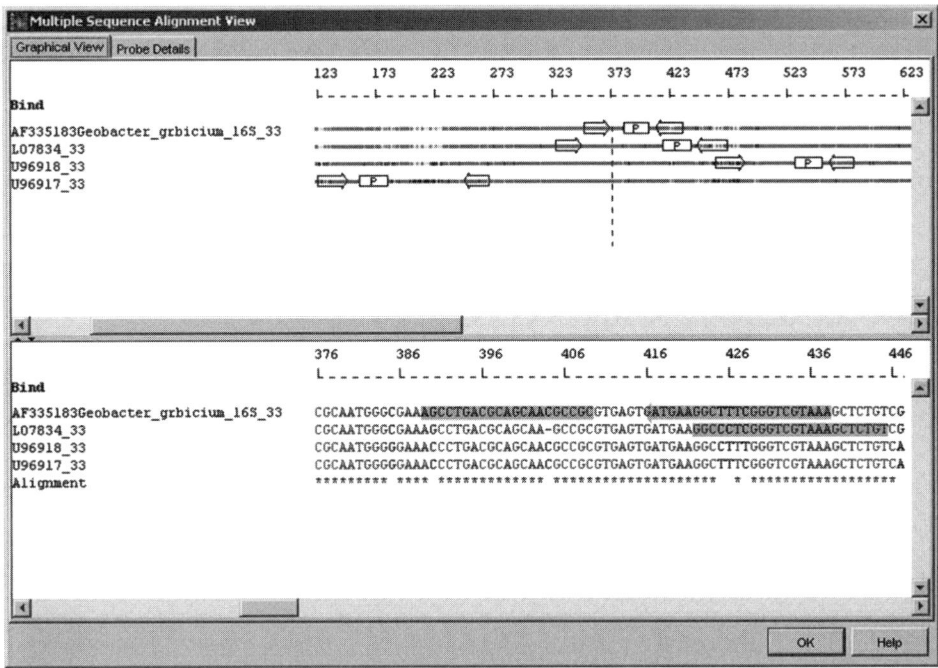

Fig. 2. Design of unique TaqMan® probes for Genobacter 16s rRNA sequences.

2.3.1.3. Unique Primers for Species-Specific Design

When the researcher selects unique primer pair, the program will try to design a primer pair that will specifically amplify the region, flanking the probe will avoiding cross hybridization with any other sequences in the set. This ensures maximum specificity because both primers and probes are unique to the target sequence.

2.3.1.4. Optimal Primers for Species-Specific Design

The optimal primer pair option designs the minimal number of primers that can amplify all the selected sequences in an alignment. It designs primer pairs that bind at the conserved regions across the alignment and amplify all the sequences. Selecting this option is highly cost effective as it drastically reduces the number of primers needed to amplify all the sequences across an alignment. For example, a researcher working with a sample of four different strains of *Mycobacterium tuberculosis* would need to design four primer pairs to amplify and detect each strain. Using the optimal primer option, the researcher would require only three primers to perform the same assay. A single sense primer binds to all the four sequences, whereas the two anti-sense primers could bind to two sequences each. These optimal primers can also tolerate mismatches at the 5′-end and hence can even bind to two or more sequences that are conserved yet have a few mismatches.

2.3.2. Taxa-Specific/Cross-Species-Specific TaqMan® Assay Design

AlleleID designs TaqMan® probes to detect and identify a group of pathogens *(5)*. The program makes use of the conserved regions in the alignment to design a common probe that can identify all individuals belonging to that group. To design the taxa/cross-species-specific assay, the researcher can choose either conserved or mismatch probe options and between conserved and optimal primer pair option.

2.3.2.1. Conserved TaqMan® Probe for Taxa/Cross-Species-Specific Design

When the researcher selects the conserved probe option, the program designs a common probe in the conserved region of the alignment. The program considers all the possible conserved regions of the alignment and the region where the designed probe has the best rating is identified (*see* **Fig. 3**).

2.3.2.2. Mismatch-Tolerant TaqMan® Probe Design

AlleleID has the ability to design TaqMan® probes with a limited number of mismatches tolerated at the 5′-end of the probe. When the researcher chooses to design mismatch-tolerant probes, the program considers the majority consensus for design (*see* **Fig. 4**).

2.3.2.3. Primer Design Options for Taxa/Cross-Species-Specific Assay

The program offers two options for designing primer pairs, conserved and optimal. A conserved primer pair can amplify all the sequences in an alignment; this option is useful for homologous sequences. As discussed in species-specific designs, the program also design optimal primer pairs.

2.4. Other Chemistries Supported in AlleleID for Related Gene Assay

In addition to TaqMan® probes, the researcher can also design molecular beacons *(6,7)* and microarray probes *(8)* for pathogen detection and identification. The same design options available for TaqMan® design are available for designing beacons and microarray probes.

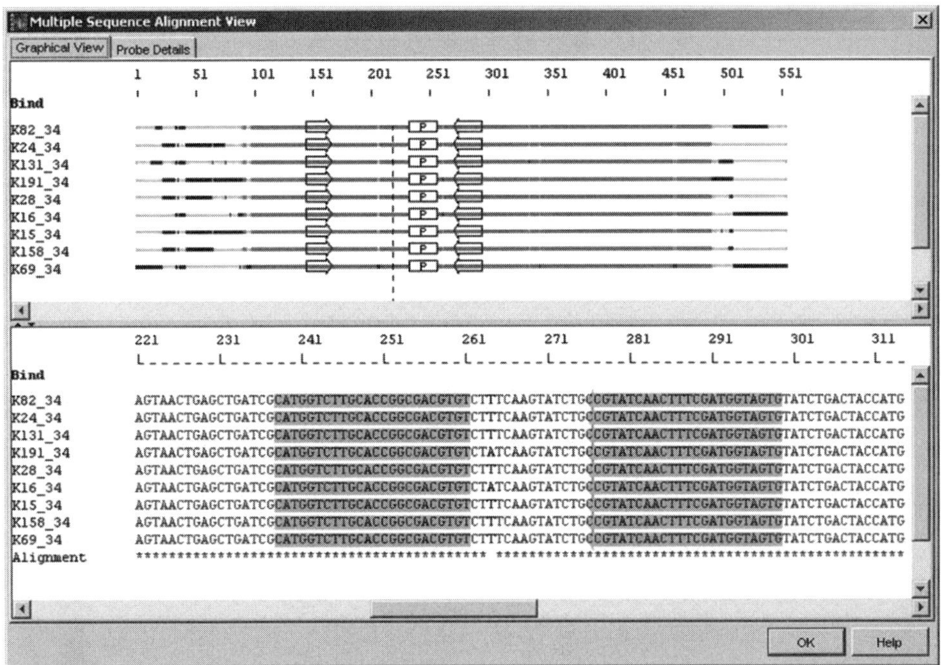

Fig. 3. Taxa-specific TaqMan® probe designed on 16s rRNA from nematode sequences.

2.5. Rating System

In AlleleID, the sophisticated search algorithm calculates all properties of every possible primer and probe within the permitted parameter range. The rating determines how well a potential oligo meets the search parameters relative to the tolerance limits specified for each parameter. The rating is based on the mean squared normalized error, a statistical approach to quantify and normalize the merits, and demerits of a group of parameters. The error is normalized to give a rating between 0 to100. The rating determines how well the designed oligo meets the search parameters relative to the tolerance limits specified for each parameter. The parameters used in calculating the rating are as follows:

1. T_m.
2. Runs and repeats.
3. Hairpin internal ΔG.
4. Hairpin 3′-end ΔG.

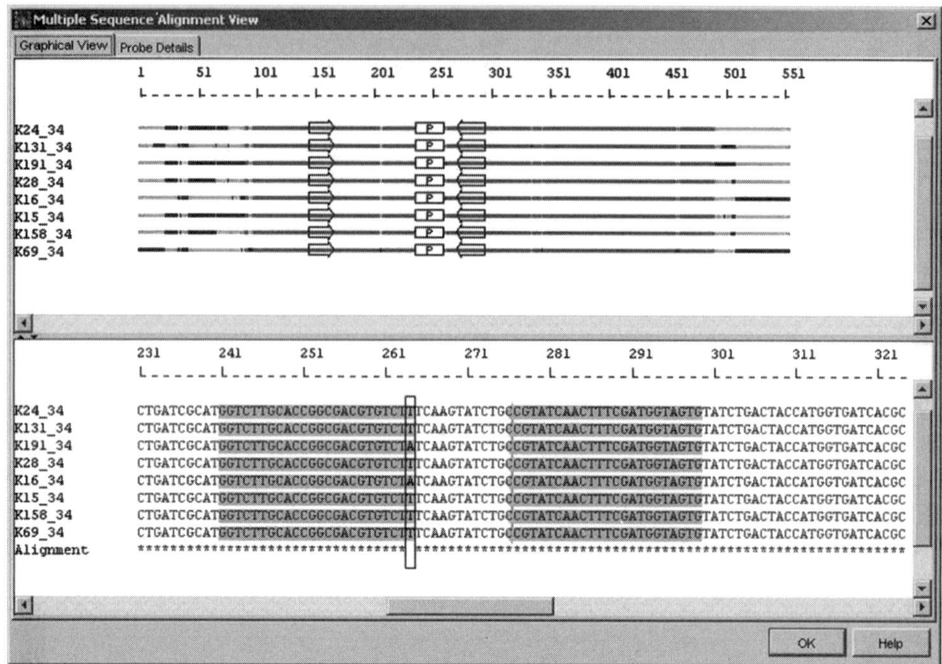

Fig. 4. Design of a universal TaqMan® probe designed for cross-species identification. Nematode 16s rRNA sequences were detected by a taxa-specific assay. The probe is in antisense direction.

5. Self-dimer internal ΔG.
6. Self-dimer 3′-end ΔG.
7. GC Clamp (for primers only).
8. T_aOpt (for primers only).

T_m is calculated based using nearest-neighbor thermodynamic calculations. AlleleID then calculates the self-dimer and hairpin stability of each primer. GC Clamp is the number of consecutive G/C bases at the 3′-end of primers.

$$\text{Rating} = 100\left[1 - \frac{1}{N}\sum_{n=1}^{N}\left(\frac{\text{Parameter}_n - \text{Target}_n}{\text{Tolerance}_n}\right)^2\right]$$

A demerit is determined for each of the abovementioned parameters. The demerit weight is calculated using the parameters set by the user. For example, if the self-dimer 3′-end ΔG value is set to 5.0, then the weight for this will be (0.999/5.0). If there is a self-dimer 3′-end ΔG value of 2.0, then the demerit will be Demerit = $[(0.999/5.0)^*2.0]^2$. If the value of any parameter is greater than the parameter tolerance, then the oligo automatically fails.

From the rating equation, it is evident that if all parameters were exactly on target or summation (S) of all the variables is 0, the rating would be 100. On the contrary, if summation of all the variables (parameters) is greater than 0, then the rating can vary from 100 to 0. It would be 0 when the summation of all the variables is equal to 1 or when all parameters were out at their tolerance limits. Tolerance is a permissible value of variation acceptable by the program for designing oligos from a pre-set standard (default) value. For instance, variation of ±5 °C is the tolerance value specified for Target T_m parameter of 55 °C. The tolerance plays two related but distinct roles. First, it is used to reject primers with parameters outside of the specified range. Then, for acceptable primers, the inverse tolerance is the rating weight given to each parameter. There are no negative rating values because all primers with one or more parameters out of limits are rejected.

The rating of a given probe depends on two factors:

1. How close the oligo is to the target value of each parameter.
2. How tightly the tolerances are specified.

The default tolerances provided in the program are chosen to represent the requirements of typical experiments. Highly rated probes have most or all parameters near their ideal values and are very likely to work well, but it is important to note that all designed primers and probes meet all the specifications, no matter what their rating. Target is a specified value for a particular given parameter. For instance, if we specify a value for Target T_m parameter as 55 °C, then this value is the Target value. For instance, one can specify the T_m tolerance, which corresponds to the acceptable range of temperature over and below the optimal T_m for which primers are acceptable. Default for the temperature tolerance is ±5 °C.

Example: Status—No primer pairs found—294 rejected (ProdLen: 188 T_m Match: 106) indicates the total number of primer rejected and the parameters causing the failure. To find a suitable oligo, one may change the search parameters based on the search status message. For the above example, one can increase both the product length and T_m to check if it is possible to design primers. The rating provides feedback on how closely a primer or probe meets all the target values. When setting the tolerance limits, the sole criteria should be acceptability. Although counter intuitive, specifying tolerances tightly does not assure better results. A low-rated probe would not necessarily fail. It may be worthwhile to consider its properties relative to your experimental needs and decide whether the probe is acceptable or whether the low rating is an artifact of narrow tolerances.

3. Applications of AlleleID in Pathogen Identification, Genotyping, and Biodiversity Studies

3.1. Species-Specific Assay Design for Major Histocompatibility Type Recognition

Major histocompatibility (MHC) proteins are the cell surface antigens present in most eukaryotes, which identify a particular tissue type. In humans, MHC typing plays an important role in tissue grafting. If MHCs are incompatible, then the graft is rejected. As the MHC genes are highly conserved, there are only very minor differences in the MHC sequences among the different species and individuals. AlleleID utilizes such differences in the sequences for genotyping probe design. AlleleID can design either unique probes or minimum number of probes to identify a sequence. When the researcher selects the unique probe design option, the program tries to design the probe for a mismatched base in an alignment that is unique to a particular MHC (*see* **Fig. 5**). In this case, each sequence can be identified with a unique probe without cross hybridizing with any other sequences in the alignment. If there is no such unique probe, the probe search fails. In this case, the minimal probe set design option can be used. Although a single probe can identify more than one sequence, the combination of such probes uniquely identifies a given sequence. As seen in **Table 1**, although probes P1 and P3 identify BFII 1 and BFII 3, BFII1 is uniquely identified by P1 and P3, and BFII 3 is uniquely identified by the combination of P1, P3, and P7. The minimal probe set is highly cost effective as it designs minimum number of probes are required for unique identification.

3.2. Application of Species-Specific Assay to Identify Pathogens

Processed foods are a common vector for pathogens. Processed food products can become contaminated at various stages in the manufacturing process. The predominant contaminating organisms such as *Clostridium botulinum, Campylobacter* sp., *Salmonella* sp., *Escherichia coli* O157:H7, and *Listeria* sp. can form spores. Detection of spores using traditional microbiological or serological methods may not be possible. In such cases, DNA-based detections enable early detection.

3.3. Detection of Listeria monocytogenes Strains using AlleleID

The *Listeria* genus is composed of numerous species and strains. Of these, *L. monocytogenes* is a very common human pathogen. *L. monocytogenes* can

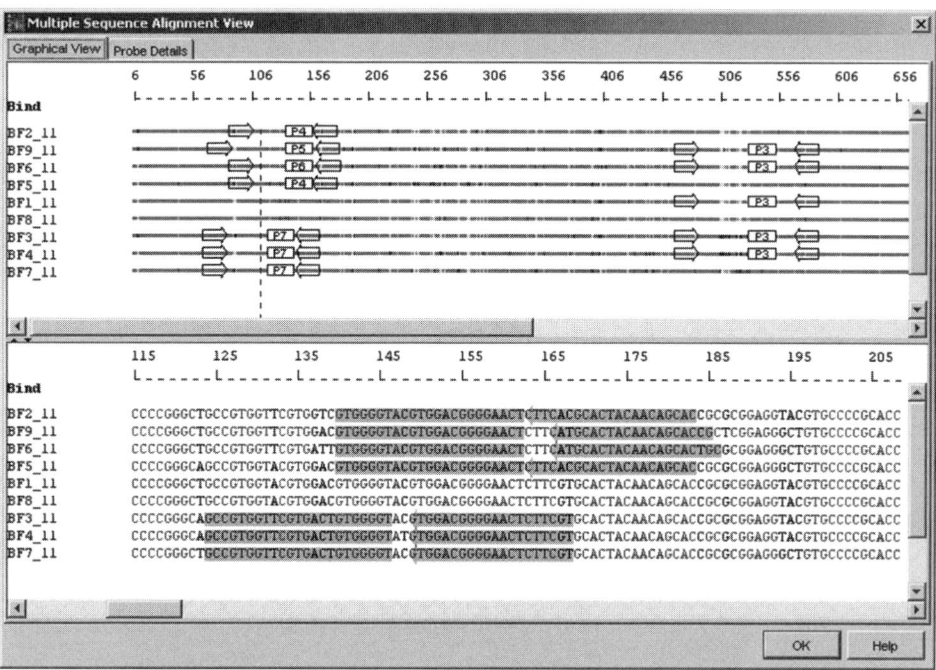

Fig. 5. Mismatch tolerant TaqMan® probe designed to identify 16s rRNA from nematode sequences. The rectangle indicates the presence of mismatched bases at the 5′-end of the antisense probe.

Table 1
Probe Set Combinations Which Uniquely Identify Any Sequence in the Set

Sequence	P1	P2	P3	P4	P5	P6	P7	Status
BFII 2	–	–	–	+	–	–	–	Y
BFII 9	–	–	+	–	+	–	–	Y
BFII 6	+	–	+	–	–	+	–	Y
BFII 5	+	–	–	+	–	–	–	Y
BFII 1	+	–	+	–	–	–	–	Y
BFII 8		+	–	–	–	–	–	Y
BFII 3	+	–	+	–	–	–	+	Y
BFII 4	–	–	+	–	–	–	+	Y
BFII 7	–	–	–	–	–	–	+	Y

grow under a wide variety of temperatures and pH making control through refrigeration difficult. The temperature range for the growth of *L. monocytogenes* is 1–45 °C. It is most commonly found in raw milk, meat, fruits, and vegetables as well as in frozen meat, poultry, and seafood. Infected individuals have a morbidity rate of almost 25%. In addition, it results in losses of over $100 million/year to the food industry from the recall and destruction of contaminated batches. The US FDA advices a zero tolerance for *L. monocytogenes* because of its morbidity rate. Detection and identification of *L. monocytogenes* in food by normal microbiological methods is time consuming and requires at least 3 days for culturing. This escalates storage costs for the products.

QPCR probes have been developed to detect the different gene sequences specific for *L. monocytogenes*. Invasion-associated protein (IAP) is one of the gene products, which is found only in *L. monocytogenes* and can be used to identify different strains. Strain level identification of the pathogen is essential because there is a difference in virulence among the different strains of the pathogens that belong to the same species. These strains vary only in small number of single-nucleotide polymorphisms (SNPs). Effective identification of these strains requires designing probes that uniquely identify such SNPs. AlleleID has a set selection option, allowing the researcher to design an assay that will test positive for a given set and negative for another specified set. The researcher can group sequences into a "bind to" set or an "avoid" set. Probes will be designed for all sequences in the bind to set while ensuring that no designed probes bind to the sequences in the avoid set. This makes the design of strain identification assays easy. *L. monocytogenes* strain CECT 4031 can be detected from other seven strains using the bind to and avoid sets. As shown in **Fig. 6**, a unique probe can be used to identify the CECT 4031 strain in a pool of eight strains. The designed probes does not bind to any of the other seven strain sequences in the avoid set.

3.4. Taxa-Specific/Cross-Species Assays for Biodiversity Studies

Identification of organisms in an ecosystem is a common need among evolutionary biologists. The taxa/cross-species design option of AlleleID facilitates the design of probes to categorize different organisms present in a sample. This can also be used to design the probes for phylogenetic variations.

Many organisms belonging to different taxa share the same habitat. To survive in a given habitat, these organisms have to undergo various modifications and adaptations. These modifications and adaptations may be at the phenotypic level or at the genotypic level or both. For example, numerous different bacteria live in anaerobic conditions. They all synthesize an enzyme

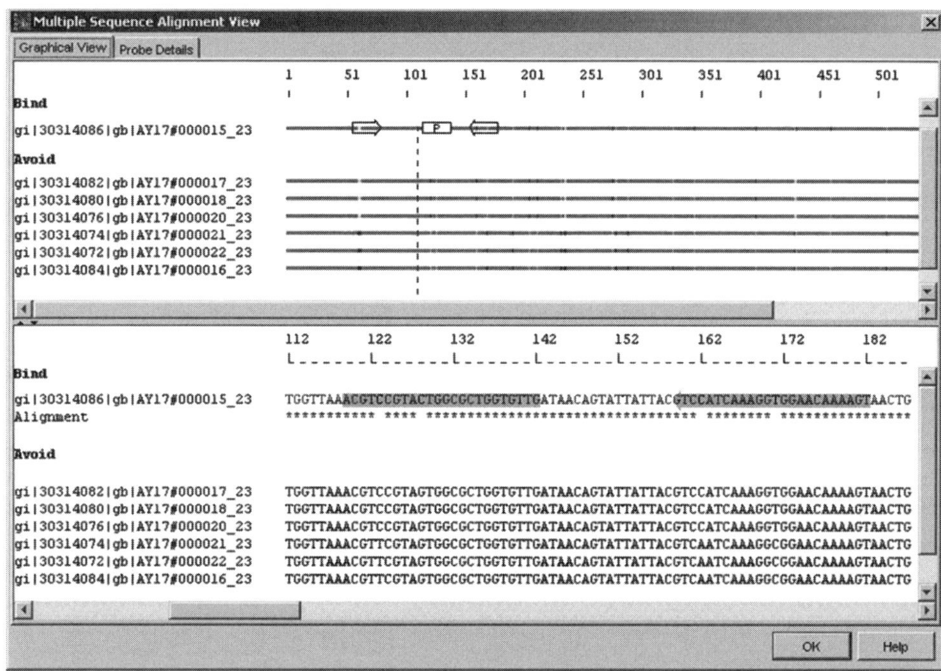

Fig. 6. A unique probe to uniquely identify the CECT 4031 strain from a pool of eight strains in a sample.

that makes use of NADP+ for electron transfer. A common probe can be used to identify such genes that are common to all the organisms in a given habitat. Phylogenetic analysis can be performed using the common probes designed by AlleleID. The researcher can design a set of common probes that specifically identify a given group. This can be extended to identify other organisms sharing the gene being targeted.

3.5. Design Microarray Probes for Genotyping

All the individuals belonging to the single species have nearly identical genetic composition. The differences within a species can be attributed to SNPs that give the special characteristics to individuals of a species. These SNP loci have been used to develop molecular biology assays to identify and characterize single individuals. Such SNPs can also be used to differentiate a wild-type allele from a mutant type. Microarray probes *(9)* can also be used for the identification of microbial pathogens by designing assays that target SNPs.

4. What If No Assay Designs Are Possible?

The default parameters for the related gene assay design are optimized for the majority of assay designs. However, should the default parameters not yield any search results AlleleID provides easy relaxation of the default parameters. Most of the parameters have a tolerance value (*see* **Fig. 7**) associated with them. These tolerance limits can be relaxed by the researcher to accommodate problematic sequences. Expert users can fine-tune a design by changing the advanced parameters (*see* **Fig. 8**).

5. Specificity Check for Assays Designed with AlleleID

False positives because of cross homology are a persistent problem for the design of sequence-based pathogen identification assays. AlleleID ensures that all primers and probes are checked for their specificity. The researcher

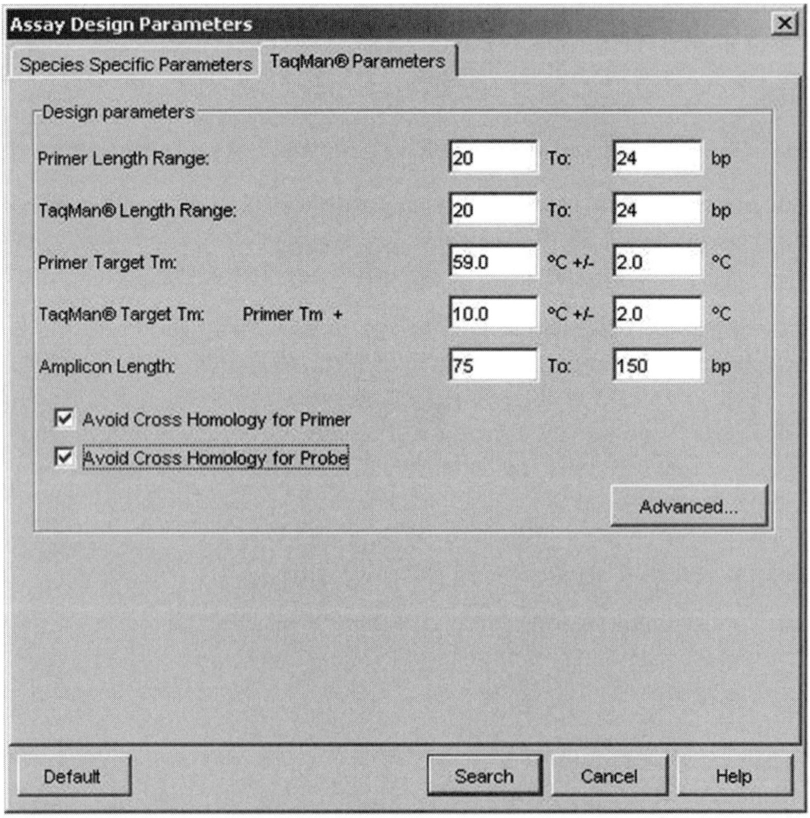

Fig. 7. Dialog showing TaqMan® search parameters and their respective tolerances.

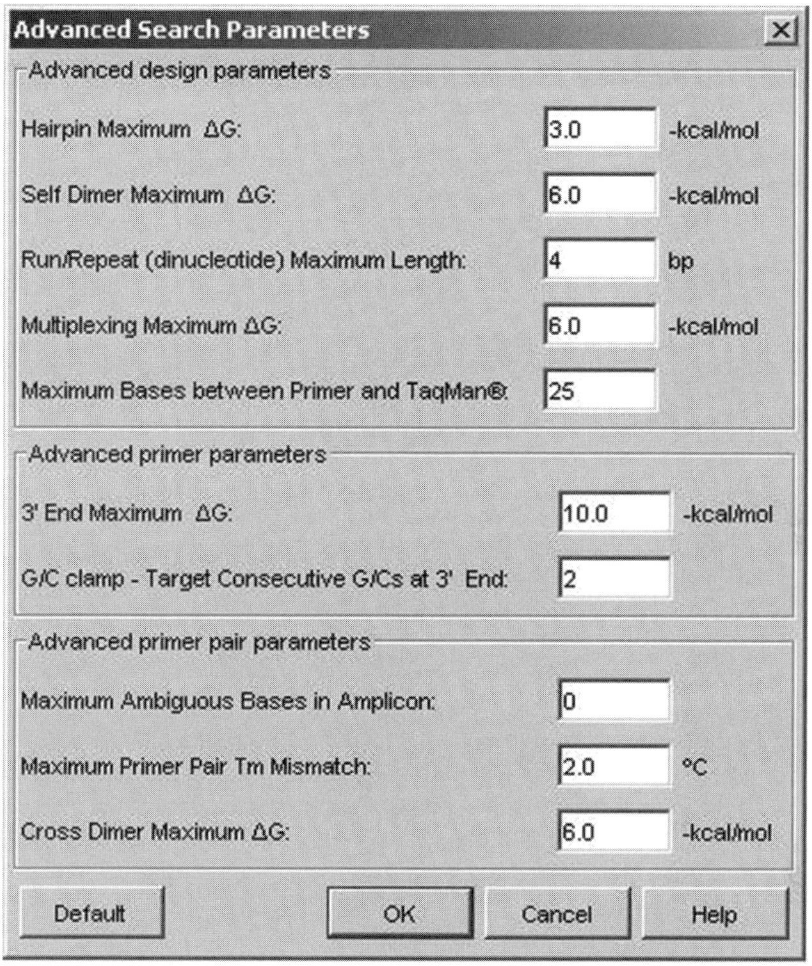

Fig. 8. Advanced search parameters for TaqMan® probe search.

can BLAST target sequences against all publicly available databases at NCBI including the nr, eukaryotic or microbial databases, or against custom databases on a local server. AlleleID automatically interprets BLAST results and avoids all regions of a sequence that show substantial homologies with other sequences in a database. Avoid cross-homology function is one of the most powerful features. It identifies regions of each sequence that show cross homology to other sequences and automatically avoids these regions while searching for primers and probes. As shown in **Fig. 9**, the avoid cross-homology window shows the interpretation of the BLAST search. The E-value represents the

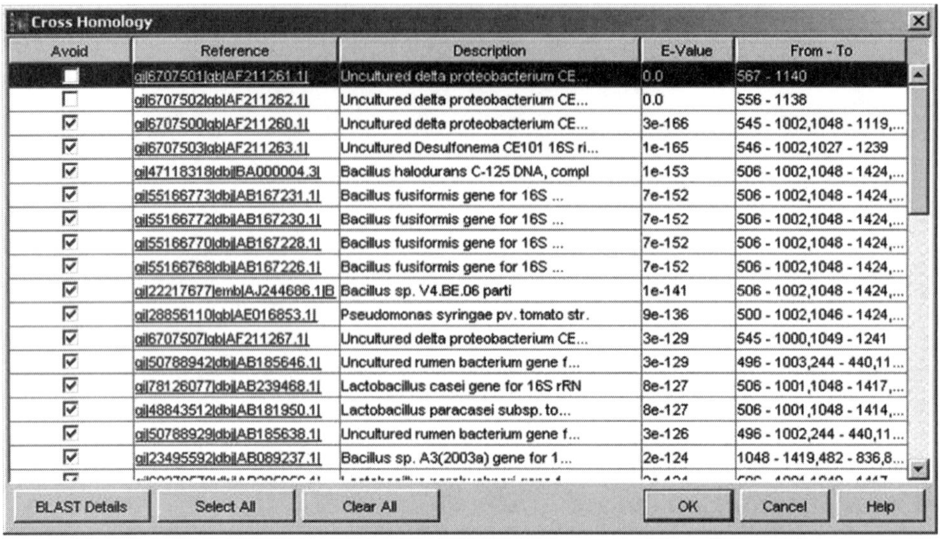

Cross Homology

Avoid	Reference	Description	E-Value	From - To
☐	gi|6707501|gb|AF211261.1|	Uncultured delta proteobacterium CE...	0.0	567 - 1140
☐	gi|6707502|gb|AF211262.1|	Uncultured delta proteobacterium CE...	0.0	556 - 1138
☑	gi|6707500|gb|AF211260.1|	Uncultured delta proteobacterium CE...	3e-166	545 - 1002,1048 - 1119,...
☑	gi|6707503|gb|AF211263.1|	Uncultured Desulfonema CE101 16S ri...	1e-165	546 - 1002,1027 - 1239
☑	gi|47118318|dbj|BA000004.3|	Bacillus halodurans C-125 DNA, compl	1e-153	506 - 1002,1048 - 1424,...
☑	gi|55166773|dbj|AB167231.1|	Bacillus fusiformis gene for 16S ...	7e-152	506 - 1002,1048 - 1424,...
☑	gi|55166772|dbj|AB167230.1|	Bacillus fusiformis gene for 16S ...	7e-152	506 - 1002,1048 - 1424,...
☑	gi|55166770|dbj|AB167228.1|	Bacillus fusiformis gene for 16S ...	7e-152	506 - 1002,1048 - 1424,...
☑	gi|55166768|dbj|AB167226.1|	Bacillus fusiformis gene for 16S ...	7e-152	506 - 1002,1048 - 1424,...
☑	gi|22217677|emb|AJ244686.1|B	Bacillus sp. V4.BE.06 parti	1e-141	506 - 1002,1048 - 1424,...
☑	gi|28856110|gb|AE016853.1|	Pseudomonas syringae pv. tomato str.	9e-136	500 - 1002,1046 - 1424,...
☑	gi|6707507|gb|AF211267.1|	Uncultured delta proteobacterium CE...	3e-129	545 - 1000,1049 - 1241
☑	gi|50788942|dbj|AB185646.1|	Uncultured rumen bacterium gene f...	3e-129	496 - 1003,244 - 440,11...
☑	gi|78126077|dbj|AB239468.1|	Lactobacillus casei gene for 16S rRN	8e-127	506 - 1001,1048 - 1417,...
☑	gi|48843512|dbj|AB181950.1|	Lactobacillus paracasei subsp. to...	8e-127	506 - 1001,1048 - 1414,...
☑	gi|50788929|dbj|AB185638.1|	Uncultured rumen bacterium gene f...	3e-126	496 - 1002,244 - 440,11...
☑	gi|23495592|dbj|AB089237.1|	Bacillus sp. A3(2003a) gene for 1...	2e-124	1048 - 1419,482 - 836,8...

BLAST Details | Select All | Clear All | OK | Cancel | Help

Fig. 9. View Cross-Homology Window showing the sequence regions that will be avoided to do cross homology with other sequences in the target database.

significance of each match. The “From–To” column shows the region of the template homologous to database sequences. Clicking the accession number opens its database record for a detailed look. All of the marked regions will be avoided during the primer and probe design searches. The researcher can override any of these recommendations by manually checking or unchecking the boxes.

Often, a researcher has a pre-designed assay from the literature. AlleleID can check these pre-designed assays for cross-searching by BLAST searching any specified primers and probes against any target database. In addition, all secondary structures for the primers and probe are shown.

6. Conclusions

Pathogen detection and identification assay design has always been a challenge. Although traditional methods available can detect and identify pathogens, they are time consuming and often have poor accuracy. Molecular methods i.e., sequence-based assays for detecting and identification of pathogens are rapid, have high sensitivity, and high accuracy. Design of primers and probes for pathogen identification requires sophisticated software capable of handling multiple sequences, database cross-homology identification, and primer/probe thermodynamic considerations simultaneously. AlleleID has been optimized for the design of pathogen and bacterial identification systems.

7. References

1. Briese, P., Palacios, G., Kokoris, M., Jabado, O., Liu, Z., Renwick, N., Kapoor, V., Casas, I., Pozo, F., Limberger, R. and Lipkin, W. (2005) Diagnostic system for rapid and sensitive differential detection of pathogens. *Emerg Infect Dis* 11(2), 310–313.
2. Guilbaud, M., Coppet, P., Bourion, F., Rachman, C. and Dousset, X. (2005) Quantitative detection of Listeria monocytogenes in biofilms by real-time PCR. *Appl Environ Microbiol*, 2190–2194.
3. Higgins, D. G. and Sharp, P. M. (1998) CLUSTAL: a package for performing multiple sequence alignment on a microcomputer. *Gene* 73, 237–244.
4. Holland, P. M., Abramson, R. D., Watson, R. and Gelfand D. H. (1991) Detection of specific polymerase chain reaction product by utilizing $5'-3'$ exonuclease activity of Thermus aquaticus DNA polymerase. *Proc Natl Acad Sci USA* 88, 7276–7280.
5. Verweij, J. J., Blange R. A., Templeton, K., Schinkel, J., Brienen E. A., van Rooyen, M. A, et al. (2004) Simultaneous detection of *Entamoeba histolytica, Giardia lamblia*, and *Cryptosporidium parvum* in fecal samples by using multiplex real-time PCR. *J Clin Microbiol* 42, 1220–1223.
6. Vet J. A., Majithia, A. R., Marras, S.A., Tyagi, S., Dube, S., Poiesz, B.J., et al. (1999) Multiplex detection of four pathogenic retroviruses using molecular beacons. *Proc Natl Acad Sci USA* 96, 6394–6399.
7. Salvatore, M., Kramer F. R. and Tyagi, S. (1999) Multiplex detection of single-nucleotide variations using molecular beacons. *Genet Anal* 14, 151–156.
8. Matveeva, O. V., Foley, B. T., Nemtsov, V. A., et al. (2004) Identification of regions in multiple sequence alignments thermodynamically suitable for targeting by consensus oligonucleotides: application to HIV genome. *BMC Bioinformatics* 5(1), 44–51.
9. Relman, D. A. (1998) Detection and identification of previously unrecognized microbial pathogens. *Emerg Infect dis* 4(3), 382–389.

D

Long PCR Primer Design

18

Designing Primers for Whole Genome PCR Scanning Using the Software Package GenoFrag

A Software Package for the Design of Primers Dedicated to Whole-Genome Scanning by LR-PCR

Nouri Ben Zakour and Yves Le Loir

Summary

Whole-genome polymerase chain reaction (PCR) scanning (WGPS) is based on the PCR amplification of small-sized chromosomes (e.g., bacterial chromosomes) by long-range PCR with a set of primers designed using a reference strain and applied to amplify several other strains. Such an approach of genome variability has specific requirements for the selection of primers and the design of primer pairs for the optimal coverage of the chromosome. To facilitate such analysis, we have developed GenoFrag, a software package for the design of primers optimized for whole-genome scanning by long-range PCR. GenoFrag works in a two-step procedure: first, a list of primers is selected according to the basic criteria, and second, the list of primer candidates is used for the coverage of the whole chromosome. These two steps are presented here with a part of the algorithm scripts developed for this software. Examples of what can be done using GenoFrag are illustrated by results obtained from the online version of the software. GenoFrag has already been validated in long-range (LR)-PCR experiment on several bacterial species. It is a robust and reliable tool for primer design for WGPS.

Key Words: Whole-genome PCR scanning; LR-PCR; primer design; combinatorial computation.

1. Introduction

Increasing numbers of complete genome sequences for prokaryotic organisms are now available. More than 330 bacterial species have been completely sequenced (genome online database, http://www.genomesonline.org/,

From: *Methods in Molecular Biology, vol. 402: PCR Primer Design*
Edited by: A. Yuryev © Humana Press, Totowa, NJ

April 2007), and several strains have been sequenced in some species. This wealth of data allows comparisons between whole genomes, a powerful and accurate approach to genome diversity. However, when whole-genome sequence is available for only one or a few strains in a given species, some other means have to be developed to gather some data on genome diversity. The whole-genome scanning using long-range polymerase chain reaction (LR-PCR) is one of these means. This approach is based on comparative analysis of the whole-genome structure of different strains of the same species, as determined by whole-genome amplification using the LR-PCR technique. It was successfully used to study genome diversity in enterohemorrhagic *Escherichia coli* O157 ***(1)***, *Bacillus licheniformis* ATCC 14580 ***(2)***, and *Streptococcus pyogenes* ***(3)***. Its robustness for studying genome variability in other species was also demonstrated for *Staphylococcus aureus* ***(4)*** and several species of lactic acid bacteria ***(5)***.

Before 2004, several bioinformatics tools were available to design and to test the robustness of primers, for example, Primer3 ***(6)***, PRIDE ***(7)***, or PRIMO ***(8)***. However, these tools did not satisfy the specific requirements of a whole-genome PCR scanning (WGPS) project, that is, the selection of primer pairs covering the whole genome and allowing segmentation of the genome into fragments whose length and overlap could be set by the users. There was no publicly available software that could process a whole-genome sequence to design primers taking into account WGPS specificities except one commercial software, the Vector NTI package ***(9)***. The long PCR primer design of this software was based on an algorithm using a rollback approach, able to select primer pairs corresponding to consecutive overlapping fragments. However, a primer pair selection by rollback approach has been proved to be quite time-consuming in the case of long sequences such as genome sequences. Thus, we developed the GenoFrag software package, a tool dedicated to the WGPS, which is based on a powerful algorithm allowing an efficient selection of primers for the amplification of overlapping fragments.

1.1. Principle of Whole-Genome PCR Scanning

WGPS takes advantage of a unique (or a few) genome sequence(s) available for a given species to get data about the genome variability of several other strains of this species. Briefly, a whole-genome sequence is taken as a reference to design a set of primers used to amplify the whole chromosome in DNA fragments of ~ 10 kb that overlap on ~ 1 kb (*see* **Fig. 1**). A 10-kb size allows routine LR-PCR experiments without major difficulties, and size variations around 10 kb can easily be visualized. The same set of primers is then used for

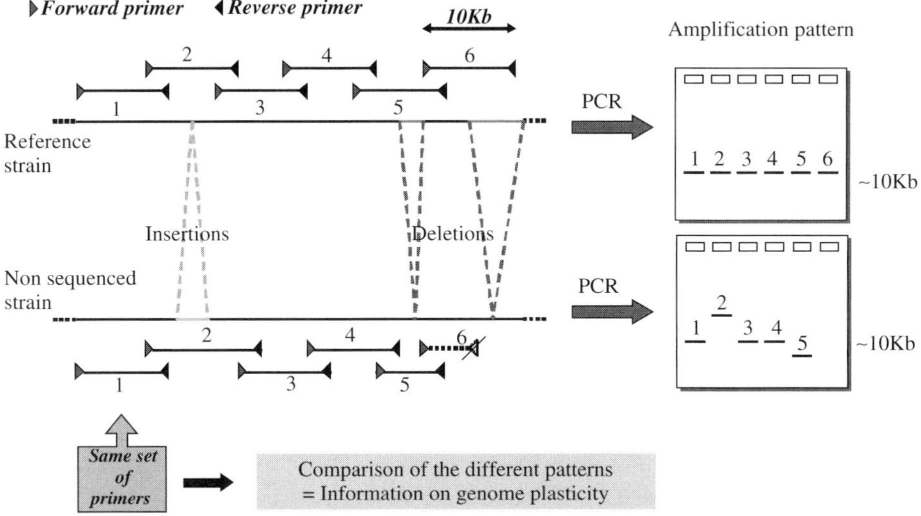

Fig. 1. Schematic view of the whole-genome polymerase chain reaction scanning (WGPS) principle.

LR-PCR experiments on several other (non-sequenced) strains. The LR-PCR products are analyzed after electrophoresis, and the resulting pattern of each strain is compared to that of the reference strain. The size variations in PCR products can result from the deletion or insertion of some DNA in between the two primers of a given pair. The absence of a PCR product may result from a deletion that included the hybridization site of one (or both of the) primer(s), from a nucleotide divergence at the hybridization site of the primer(s). It may also be due to the insertion of a large DNA segment that renders the PCR technically unfeasible.

Comparison of the WGPS patterns enables a rapid detection of the variable regions in the chromosome. As the gene content of each product is known in a reference strain, the gene content of the constant (non-variable) regions of the chromosome and some of the variations observed can be readily identified.

1.2. Requirements of LR-PCR

As WGPS is based on the amplification by LR-PCR of DNA fragments on a whole-chromosome DNA template, the primer pairs have to be designed such that they are highly specific (to avoid parasitic PCR products that impair the result analysis). Moreover, in many cases, the LR-PCR conditions include a concomitant annealing and elongation, which are both, carried out at the same temperature (around 68 °C). This imposes some specific requirements with regard to the thermodynamic stability of primer-template duplexes. The criteria for primer selection are discussed below. Some of the codes developed for these different steps are given in **Subheading 3** of this chapter.

Both the length and G + C content of primers are pre-established in GenoFrag by default, but users can modify the values. Thermodynamic stability of the primer-template duplex has to be fixed according to the physical conditions required for LR-PCR. A primer length of 25-mer is fixed as a default value (the 25-mer size was chosen according to Rychlik, *(10)*). Thermodynamic stability has to be identical for all primers designed by GenoFrag to facilitate large-scale PCR. The Tm value can be calculated using the rules devised by of Suggs et al. *(11)*.

To increase specificity, GenoFrag favors primers that exhibit a higher degree of free energy (ΔG) in the 5′-primer extremity than in the 3′-primer extremity. The nearest-neighbor method was used to calculate the thermodynamic stability of hybrids *(12)*. Five bases at each extremity are considered: GC clamp in 5′ and lower stability in 3′ are favored because PCR yields dramatically decrease with high ΔG in the 3′-extremity *(10)*. To avoid non-specific annealing of primers, GenoFrag eliminates sequences that contain words of five or more identical and consecutive bases *(10)*.

Putative hairpin formation is checked, and GenoFrag rejects primers when they are likely to form hairpin structures with a stem of four nucleotides (minimum) and a loop of four nucleotides (minimum). These default values were fixed in line with the suggestion made by Blommers et al. *(13)*.

Self-complementarity is a common criterion for primer selection. Here, a first selection step eliminates primers with regard to their overall self-complementarity. A second step focuses on the 3′-extremity and further eliminates primers that present self-complementarity but with more discriminative values. Each step requires the implementation of computation involving the nearest-neighbor method *(12)*.

One critical point in the WGPS results is the number of PCR products. Only one product is expected for a given primer pair, and the presence of parasitic PCR products would lead to misleading or unexplainable results. It is thus very important to avoid any secondary binding sites within a range of length that could be PCR amplified under LR-PCR conditions and thus give rise to non-specific PCR products (additional bands that are smaller or slightly longer than those expected). This criterion is particularly important during the early cycles of PCR amplification because the non-specific PCR fragment can become the predominant template for the remainder of the reaction. The methods employed to eliminate candidates with secondary binding sites involved the use of alignment algorithms: a candidate is rejected when a putative annealing site is found with minimum values (set by default) of 17 matches with one gap allowed. Secondary binding sites are searched within a limited range of the template sequence corresponding to twice the amplicon size. Indeed, the parameters used for LR-PCR allow amplification of fragments shorter or slightly longer than those expected (e.g., for an expected amplicon of 10 kb, a range of 0 to ~ 15 kb could be amplified in the reaction in the event of non-specific annealing).

Partial complementarity between two primers for a given pair may interfere with annealing. If complementarity occurs at the 3′-end of the primers, primer dimer formation may take place and will prevent formation of the desired product (i.e., hybridization of the primers with the template) through competition. The methods used here to evaluate inter-primer complementarity are similar to that used to calculate self-complementarity.

1.3. GenoFrag Can be Used Through a WWW Interface or As a Standalone Version

An online version of GenoFrag is hosted on the West Genopole bioinformatics server: http://genoweb.univ-rennes1.fr/Serveur-GPO/outils_acces.php3?id_syndic=89&lang=en. GenoFrag is also available by request from the authors. It can be run under UNIX, Windows, or LINUX environments. The source code has been written in C *(14)*, Python (http://www.python.org), and Perl (http://www.perl.com) (with Bioperl modules *(15)*). In practice, two programs need to be executed sequentially. The first one generates primers, whereas the second searches for an optimum set of amplicons. Each program takes as its parameter a file describing the different physical values.

2. GenoFrag from the End-User Point of View

In this section, we first describe the different operations the users have to perform on the genome sequence to identify the so-called forbidden regions (see subheading 2.1). This sequence treatment must be performed before the submission to GenoFrag. Then, we focus primarily on the WWW interface for a step-by-step presentation of the software capacities.

2.1. Sequence Treatment Before Submission to GenoFrag

Genome sequence can be used to perform a search of exact and degenerate repeated regions using *REPuter (16)* or BLASTn *(17)*. Using these outputs, users can precisely localize redundant sequences with or without biological significance and their coordinates. This preliminary analysis allows GenoFrag to define genome regions that will not be taken into account for primer design, so as to avoid ambiguous PCR results. These regions are hereinafter referred to as forbidden regions: these include numerous repetitive sequences without known biological significance. Other forbidden regions include short and redundant mobile genetic elements such as tranposons or insertion sequences. These short mobile genetic elements are included in amplified regions. This is particularly interesting because these mobile genetic elements are often involved in genome plasticity *(18,19)*. Sequences of ribosomal RNA (that are in several copies on the chromosome) and other repeated sequences are not used for primer design but are retained in amplified regions. On the contrary, mobile genetic elements too large to be included in an amplicon (e.g., prophages or pathogenicity islands that can be above 20 kb long) are usually unique or present at low copy levels in the chromosome. They may be absent from strains other than the reference strain or present elsewhere on the chromosome. They may also be subject to internal rearrangements from a strain to another, thus participating in genome plasticity *(20,21)*. GenoFrag uses these sequences to generate primer pairs. It should be noted that users can also submit these large genetic elements independently to GenoFrag for the design of specific primer sets (e.g., giving smaller amplicons) to focus on their intrinsic plasticity. The sequence treatments define forbidden regions, which are identified by their coordinates on the reference chromosome. These data are submitted to GenoFrag for primer design.

2.2. Graphical Interface and Step-by-Step Procedure

When users enter the GenoFrag web page, a short introduction explains the principle of the software and indicates where and what users have to input (*see* **Fig. 2**). Users can then enter into the primer selection process. At this step, the inputs are the genome sequence and a suite of parameters (*see* **Fig. 3**). Each parameter has a default value that was set for the analysis of bacterial genomes with a low G+C content. The output of this step is a list of primers in a FASTA format as shown in the example given in **Fig. 4**. The start- or end-status and the coordinates of each primer are designated. Primers are listed in their order of appearance in the genome sequence selected. This primer list is the input of the fragmentation step (*see* **Fig. 5**). At this step, users have to provide some other parameters such as the genome length, a list of the forbidden regions (designated "prohibited areas" as a text file), and they have to choose SITA or SPP program for the chromosome coverage. A figure illustrates what are the different parameters related to (*see* **Fig. 6**). For example, the starting zone length corresponds to the length of sequence (starting from the start index) where GenoFrag will select the forward primer of the very first primer pair for the fragmentation. If default parameters are chosen, GenoFrag may select the

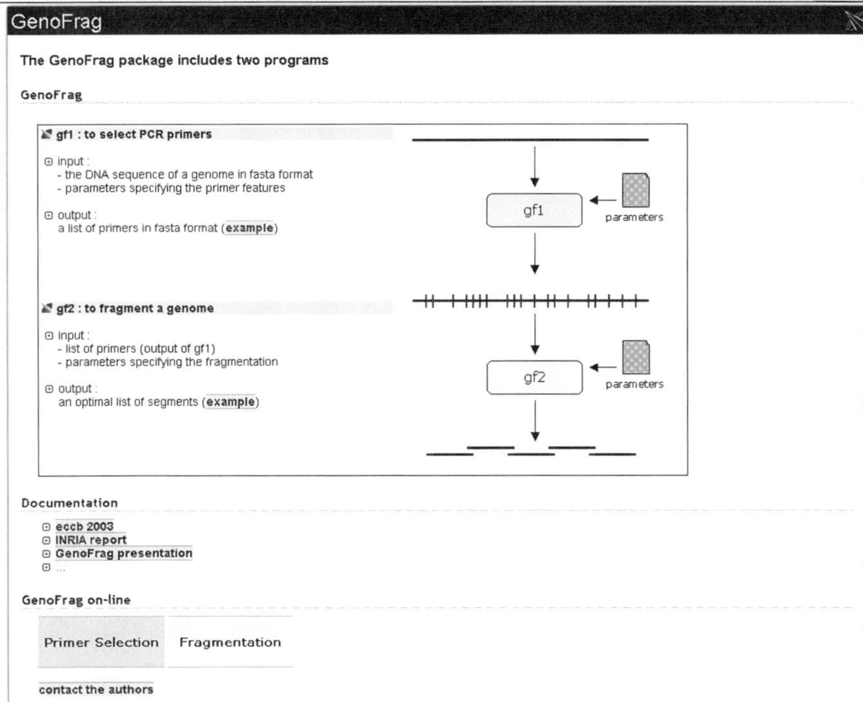

Fig. 2. Graphical interface for the explanations of the GenoFrag principle the online version.

first forward primer in a range of 5 kb. If it does not find a forward primer for an optimal coverage in this range, there will be no solution found. The same principle is applied for the end of the sequence to be covered.

The output of this step, and the final result of GenoFrag use, is a list of primer pairs as shown in **Fig. 7**. A table gives the features on the fragments that will be generated (coordinates of the start-point and end-point, fragment length, and overlap length). A second table gives the primer sequence.

At each step, users can get some help by clicking on the help link (see **Figs. 3** and **5**).

2.3. What to do When No Solution is Found

In the first version of GenoFrag, when no solution was found, there was no message indicating what and where was the problem, although this situation could be easily encountered because some bacterial genomes present AT-rich

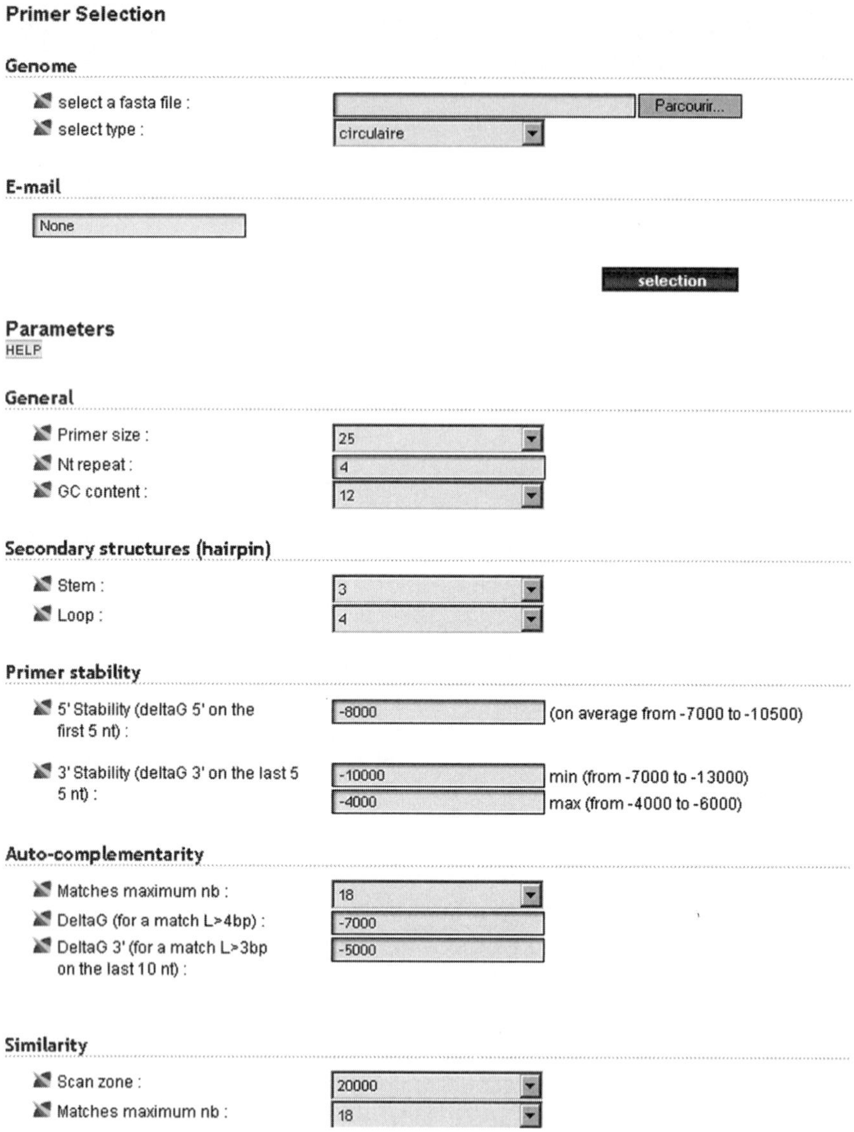

Primer Selection

Genome

select a fasta file : [ ] Parcourir...
select type : circulaire

E-mail

None

selection

Parameters
HELP

General

Primer size : 25
Nt repeat : 4
GC content : 12

Secondary structures (hairpin)

Stem : 3
Loop : 4

Primer stability

5' Stability (deltaG 5' on the first 5 nt) : -8000 (on average from -7000 to -10500)

3' Stability (deltaG 3' on the last 5 5 nt) : -10000 min (from -7000 to -13000)
-4000 max (from -4000 to -6000)

Auto-complementarity

Matches maximum nb : 18
DeltaG (for a match L>4bp) : -7000
DeltaG 3' (for a match L>3bp on the last 10 nt) : -5000

Similarity

Scan zone : 20000
Matches maximum nb : 18

Fig. 3. Graphical interface to set the parameters for the primers selection step.

regions or chromosomal portions with a high density of repeated sequences. These can lead to a local scarcity in primer candidates with regard to the parameters set by the users for the primer selection. In some of these cases, GenoFrag cannot find a solution when the coverage process enters these particular regions

GenoFrag

Outputs

Primer Selection

The output of primer selection is a list of primers in fasta format:

```
> 11660 11684 start primer
GCGTATGCGTGATTGGTTAGAAGGA
> 12001 12025 end primer
GGCTCTGGACTAAGAAATGCTGGTA
> 12002 12026 end primer
TGGCTCTGGACTAAGAAATGCTGGT
> 12097 12121 start primer
CGCATCCAGCGAGTGTTGATTCTAT
> 12316 12340 start primer
TGCCATCCATTACACATGACCGTCA
> 12317 12341 start primer
GCCATCCATTACACATGACCGTCAA
> 13584 13608 start primer
GCAGGATCAGCAGGTAGAGATACAA
> 13850 13874 end primer
CGCTTGGAAATCCGTACAGTTTGAG
> 13896 13920 start primer
AAGCGTGACAAAGCAGCTAAACCAG
> 13897 13921 start primer
AGCGTGACAAAGCAGCTAAACCAGA
> 13898 13922 start primer
GCGTGACAAAGCAGCTAAACCAGAA
> 13939 13963 end primer
CGTCCAACTGCTAAACCACTACCAT
> 14708 14732 start primer
GTGGCAAAGGAGGTAATTGAGATGG
> 14752 14776 end primer
TGGGATACACTCTTGTACGCCTTGT
> 14753 14777 end primer
TTGGGATACACTCTTGTACGCCTTG
> 14754 14778 start primer
AAGGCGTACAAGAGTGTATCCCAAC
> 14755 14779 start primer
AGGCGTACAAGAGTGTATCCCAACA
> 14756 14780 start primer
GGCGTACAAGAGTGTATCCCAACAT
> 15194 15218 end primer
ACCCTAGCGTTTGCGGATTTGAGAT
> 15195 15219 end primer
AACCCTAGCGTTTGCGGATTTGAGA
```

Fig. 4. Example of output after primer selection.

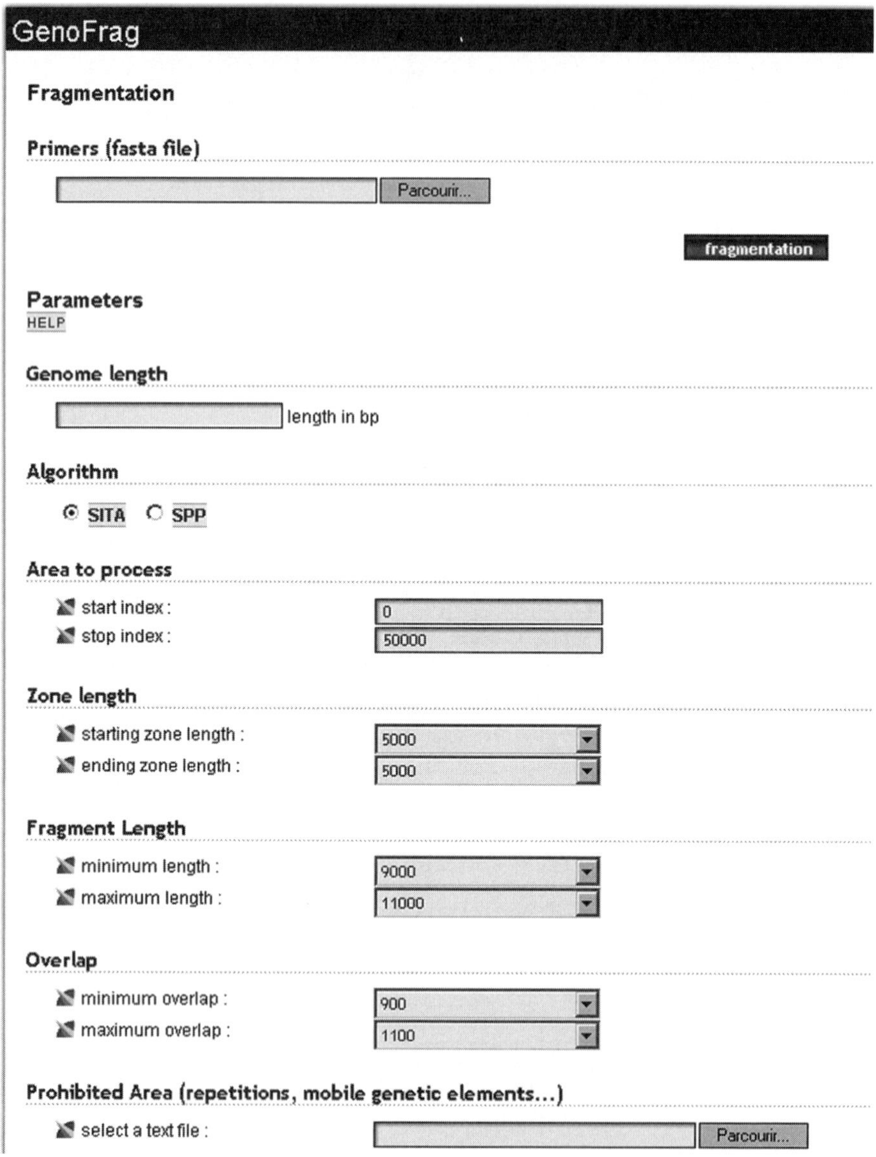

Fig. 5. Graphical interface to set the parameters for the fragmentation step.

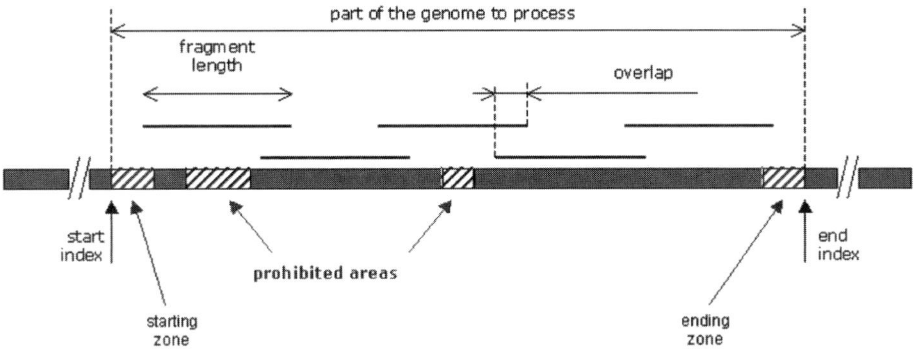

Fig. 6. Explicative figure of the different parameters that can be set by the users.

Fragmentation

The output of fragmentation is an array of segments:

Optimal solution

FRAGMENT	START	END	LENGTH	OVERLAP
0	1264	11314	10051	0
1	10044	20182	10139	1271
2	19261	29346	10086	922

```
Summary :
Max Error : 88
Max Overlap : 1271
Length of the non-covered start zone : 1265
Length of the non-coverd end zone : 4225
```

List of Amplicons

Start Primer	End Primer	Length
TTTCGAGTGATCGACCACCAAAGGA	GAGCCTTGTTGTGGGATTATCGGTA	10051
GCACCTAAACGATCAACTGCTTCAC	CGCCACCAAGTGCTTCCATTGTTAA	10139
CCGATGGAGAAACGTACTCGTGTAA	GCACCATCTGTGGGCTTAATGTCAT	10086

Computation time: 0.02 s

Fig. 7. Example of output after fragmentation. Primer sequences are oriented $5' \rightarrow 3'$. General features of the fragmentation are summarized. Max error gives the spread between the shortest and the longest fragment of the solution found. Max overlap gives the size of the longest overlap between two consecutive fragments. The non-covered zones require the design of a primer pair dedicated to the coverage of the ori region (when GenoFrag is run on a circular chromosome sequence).

with a low density of primer candidates. Nevertheless, in most cases encountered, GenoFrag could find a solution for an optimal coverage for the rest of the chromosome. This drawback has been recently corrected in a newly released online version of the software. A graphical interface now focuses on the region where the problem is and maps the forward and reverse primers. This rapidly gives a view of the primer density and allows the user to change the parameters used for primer selection. Forward primers and reverse primers are indeed indicated on the graph by a blue bar and a red one, respectively. Users can modify parameters (e.g., relax the size of the overlap) by simply clicking on the corresponding + and − buttons above the graph.

Once the problem is found, two solutions can be envisioned. The regions that are AT-rich or rich in repeated sequences can be considered as forbidden regions. An alternative is to relax the parameters for the primer selection, so that there are more candidates available for the coverage. However, relaxing these parameters would lead to lower primer specificity. To avoid this, we suggest to increase the size of the overlap between two consecutive segments. In most cases, it is sufficient to find a good coverage solution without lowering the hybridization specificity.

3. GenoFrag from the Bioinformatic Perspective

An overview of the GenoFrag software package is shown in **Fig. 8**. It has two principal parts: the generation of primers and the segmentation of the genome.

3.1. Generation of Primers

This software program identifies all primers suitable for LR-PCR. It acts as a sequential pipeline of seven filters. All potential k-mers (default value = 25) are considered and enter filter 1. Each filter yields only those oligonucleotides satisfying specific constraints set by the user. To limit the computation time, filters with the highest selectivity are the first to be activated:

- Filter 1 selects oligonucleotides according to their G + C content.
- Filter 2 removes oligonucleotides with N consecutive identical nucleotides.
- Filter 3 removes oligonucleotides with hairpin loops.
- Filter 4 selects oligonucleotides according to thermodynamic stability constraints.
- Filter 5 tests both overall and 3′-extremity self-complementarity of oligonucleotides.
- Filter 6 checks that no other binding sites exist in the neighboring sequence.
- Filter 7 (optional) compares the primer list with other genome sequences available. If sequences other than that of the reference strain are available, the user can ask GenoFrag to perform a BLASTn *(17)* search with these genome sequences against

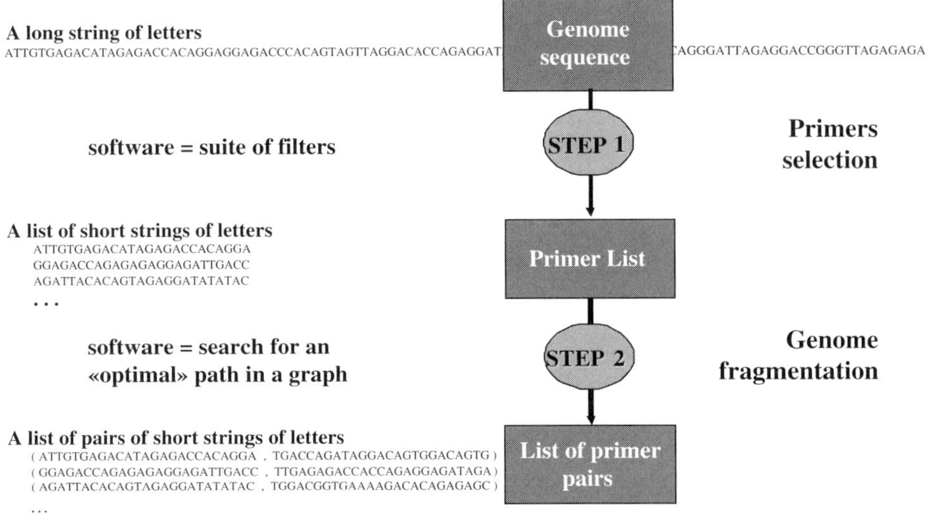

Fig. 8. Schematic view of the two-step process in GenoFrag software.

the primer list. This option allows GenoFrag to reject primers that may not produce PCR amplification because of sequence divergence.

All oligonucleotides passing successfully through the filters are proposed as primers for LR-PCR. In addition, they are labeled with their position in the genome and with their ability to start or end an amplicon. Each of the six first filters relies on common algorithms that are widely used and described elsewhere ***(6,10)***. The seventh filter required some development. It has been written in perl, and the code is as follows:

```
#!/usr/local/bin/perl -w
# this program runs a BLAST search using Bioperl, interprets
the results
# to retrieve only the sequences corresponding to criteria
# and produce a file in fasta format with the selected
primers
#
# localization of directory containing blastall program

BEGIN {$ENV{BLASTDIR} = 'blastall _directory_path';}

use Bio::Tools::Run::StandAloneBlast;
```

```
use Bio::SearchIO;
use Bio::SeqIO;
use Bio::Seq;
# process Command line arguments

if (scalar @ARGV ! = 5) {
    $usage = "Enter command as follow perl
/PATH/parser_test.pl <QUERY> <DATABASE> <LENGTH PRIMER>
<BLAST OUTPUT FILE> <FASTA OUTPUT FILE> ";
     print "Wrong number of arguments \n";
     die "$usage \n";
}

# read in parameters from the command line

$input_seq = $ARGV[0];
$db = $ARGV[1];
$length_primer = $ARGV[2];
$out=$ARGV[3];
$fastout=$ARGV[4];

#################
# BLAST process #
#################

# format the database file for fastacmd treatment
@table = 'cat $input_seq';
$table = @table;
$input_seq = ~ s/\.[a-z]*/\.tmp/;

for (my $i=0 ; $i < $table ; $i+=2)
{

    $table[$i] =~  s/>\s/>lcl\|p/;
}

open (FDB, ">>$input_seq") or die "can't open fichier FDB :$?
$input_seq";
print FDB @table;
close FDB;

# call formatdb using system
```

```
system "blastall_directory_path/formatdb -i
$input_seq -p F -o T");

# read in the sequence using the Bioperl SeqIO;
$in = Bio::SeqIO->new(-file => $db, '-format' => 'Fasta');

# load the sequence into a Bio::Seq object
$seq   = $in->next_seq;

# prepare the parameter list for BLAST
# query = genome file and bank = primers db

$program = 'blastn';
$cutoff = 10;
@params = ('program' => $program, 'database'=> $input_seq,
        'outfile' => $out, 'e' => $cutoff, 'v'=> 500000,
'b'=> 500000, 'F' => F);

# create the Standalone Blast object with the
parameter list
$blast = Bio::Tools::Run::StandAloneBlast->new(@params);

# run BLAST
$blast_report = $blast->blastall($seq);

#################
# BLAST parsing #
#################

#  parse the blast report
my $result = $blast_report->next_result;

# return the best hsp by hit considering conditional
criteria
@tab_inter_rej=();
$return="\ n";
while (my $hit=$result->next_hit()) {

for (my $hsp=$hit->next_hsp()) {

        # conditional criteria to be satisfied by the hsp to be
rejected
```

```
        # and return the corresponding database entry in a
table
    if ($hsp->end('hit') < ($length_primer-1)||$hsp-
>length() <($length_primer-4)|| $hsp->frac_identical() <0.90
|| $hsp->gaps()>= 2) {

           $acc_reject = $hit->accession();
        $acc_reject=$acc_reject. $return;
           push (@tab_inter_rej, $acc_reject);

        }
    }
}
(…) Accession entries filtering and fasta file
creation not shown
```

3.2. Segmentation of the Genome

This second software program aims to provide a list of amplicons ensuring optimum coverage of the whole genome, or part of the genome, from the set of primers previously generated. Constraints are the minimum and maximum length of amplicons, and the minimum and maximum overlaps allowed. If, for the sake of simplicity, we assume that a solution is made up of a list of N amplicons and that each of the amplicons can occupy only P different locations, then the number of possibilities is equal to P^N. Finding the best option when N is large is clearly a combinatorial problem: computing of all the possibilities is untenable. We have developed computational optimization methods to solve this problem within a reasonable time interval. Two solutions to this problem have been implemented: Shortest Path Problem (SPP) and Single Traverse Algorithm (SITA) (*see* **Fig. 9**). SPP looks for an optimum list of amplicons whose sizes are as close as possible to an ideal length. Under the second solution, the ideal length is not required a priori but is computed by the SITA in such a way that the best segmentation is that which provides homogeneous amplicon sizes and minimizes the difference between the shortest and the longest amplicons. For both solutions, we have (1) formulated a suitable combinatorial optimization model and (2) programmed a dedicated graph algorithm to solve these models (the mathematical analysis described in **ref.** ***22***). Both programs provide a list of primer pairs as output. The computation time for processing a sequence of 1 Mb ranges from 1 to 2 min on a 1.5 GHz PC, depending on the number of primers selected during the first step. Of course, the higher the number of primers, the longer the computation time.

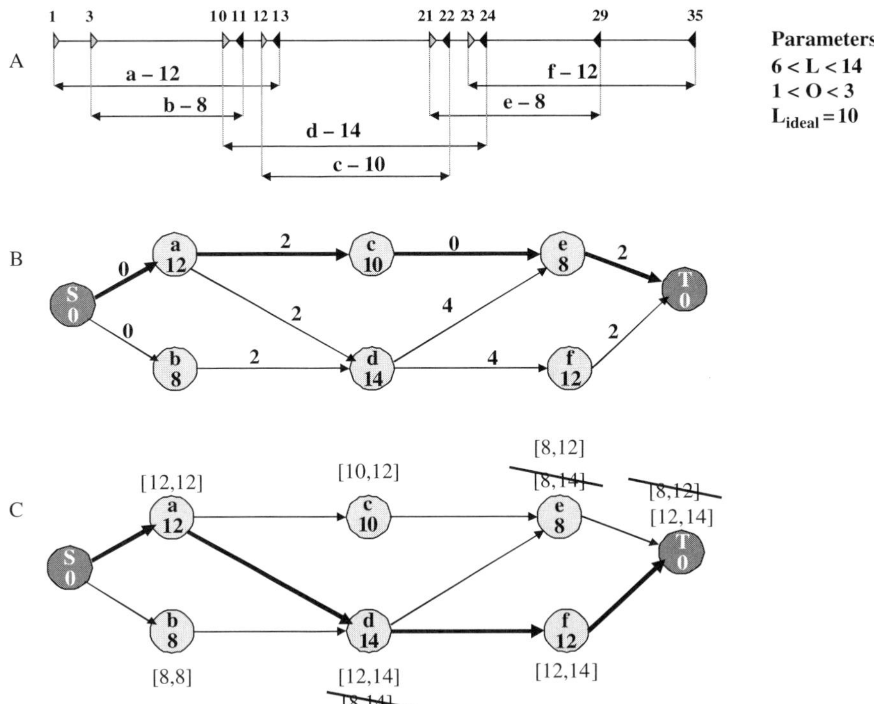

Fig. 9. Example of a 35-kb region coverage using SPP or SITA programs. **(A)** Schematic view of the 35-kb DNA region with coordinates (in kb) of the forward (white) and reverse (black) primers. Each putative fragment is indicated by a letter followed by the fragment size. The set of parameters is given on the right: fragment length (L) can range from 6 to 14 kb, overlap (O) size from 1 to 3 kb. The ideal length (SPP program) is then 10 kb. **(B)** Graph representation of the path chosen through SPP program. The graph has got a virtual start (S) and a terminus (T) node. Each fragment is represented by a node on a path (name and length of the fragment are given). A link between two consecutive nodes is possible when the parameters are respected (e.g., no link between b and c because they do not overlap). The difference between the fragment length and the ideal length is mentioned on each link. SPP chooses the path (in bold) that minimizes the sum of these differences. The average size of the fragments on the path is close to the ideal size but the spread between the longest and the shortest fragment of this solution is up to 4 kb. **(C)** Graphical representation of the path chosen through SITA program. In SITA, at each node of the path, the length of the longest and the shortest fragment encountered on the path is given. Each time several solutions are available, SITA keeps the node that minimizes the range between these two features. In the solution found (in bold), the average size of the fragments is not the ideal one (12.66 versus 10 kb) but the spread is reduced down to 2 kb. This gives more homogenous reference patterns.

3.3. Efficiency Considerations

The execution time for the "generation of primers" program to produce a list of all potential primers is around 40 s in the case of a genome 2.8 Mb long (corresponding to the size of the *S. aureus* chromosome) and using the default parameters (*see* **Fig. 3**). This time may vary slightly and linearly, depending on the stringency of the parameters when tested on a 1.6 GHz Linux machine. To evaluate and compare the performance of the SPP and SITA algorithms, both were run on randomly generated genomes of increasing length (where primers are uniformly distributed over the segment) but of a fixed primer density for each curve. Computational experiments performed on a Pentium 4 (1.6 GHz) machine under Linux revealed a linear behavior compared to the genome length *(4)*.

4. Future Developments

GenoFrag is constantly evolving because we modify and enhance the software in response to the users comments or suggestions. We are aware that sequence treatment before submission to GenoFrag can be a heavy task for the users and that it requires a good knowledge of the genome. We are currently working on a version 2 of GenoFrag that will avoid most of the sequence treatment required to detect the so-called forbidden regions. Furthermore, the new version will take account of the whole-genome sequence to search for secondary hybridization sites. These secondary hybridization sites are currently sought on a given interval (that is believed to be amplifiable in the LR-PCR conditions we use and give parasitic PCR products). A GenoFrag 2 version will therefore be developed. It will be based on the evaluation of the thermodynamic stability of the primer-template duplexes.

Acknowledgments

The development of GenoFrag was possible thanks to the help of Symbiose team at IRISA (Rennes, France). Rumen Andonov, Philippe Veber, Nicolas Yanev, and Dominique Lavenier were responsible for the development of the segmentation program. The GenoFrag WWW interface is supported and hosted by Ouest Genopole (http://genoweb.univ-rennes1.fr/Serveur-GPO/outils_acces.php3?id_syndic=89&lang=en). The authors thank Dr. S. D. Ehrlich and Dr. Alexei Sorokin (INRA Jouy en Josas, France) for constructive discussions at the beginning of this work and for their kind help and advice. Nouri Ben Zakour received financial support in the form of a PhD grant from the French Ministry of Research and Education.

References

1. Ohnishi, M., Terajima, J., Kurokawa, K., Nakayama, K., Murata, T., Tamura, K., Ogura, Y., Watanabe, H., and Hayashi, T. (2002) *Proc Natl Acad Sci USA* 99, 17043–8.
2. Lapidus, A., Galleron, N., Andersen, J. T., Jorgensen, P. L., Ehrlich, S. D., and Sorokin, A. (2002) *FEMS Microbiol Lett* 209, 23–30.
3. Beres, S. B., Sylva, G. L., Sturdevant, D. E., Granville, C. N., Liu, M., Ricklefs, S. M., Whitney, A. R., Parkins, L. D., Hoe, N. P., Adams, G. J., Low, D. E., DeLeo, F. R., McGeer, A., and Musser, J. M. (2004) *Proc Natl Acad Sci USA* 101, 11833–8.
4. Ben Zakour, N., Gautier, M., Andonov, R., Lavenier, D., Cochet, M. F., Veber, P., Sorokin, A., and Le Loir, Y. (2004) *Nucleic Acids Res* 32, 17–24.
5. Ben Zakour, N., Grimaldi, C., Gautier, M., Langella, P., Azevedo, V., Maguin, E., and Le Loir, Y. (2005) *Res Microbiol* 157, 386–94.
6. Rozen, S., and Skaletsky, H. (2000) *Methods Mol Biol* 132, 365–86.
7. Haas, S., Vingron, M., Poustka, A., and Wiemann, S. (1998) *Nucleic Acids Res* 26, 3006–12.
8. Li, P., Kupfer, K. C., Davies, C. J., Burbee, D., Evans, G. A., and Garner, H. R. (1997) *Genomics* 40, 476–85.
9. Gorelenkov, V., Antipov, A., Lejnine, S., Daraselia, N., and Yuryev, A. (2001) *Biotechniques* 31, 1326–30.
10. Rychlik, W. (1995) *Mol Biotechnol* 3, 129–34.
11. Suggs, S. V., Hirose, T., Myoke, E. H., Kawashima, M. J., Johnson, K. I., and Wallace, R. B. (1981) In Brown, D. D. (ed.), *ICN-UCLA Symposium for Developmental Biology Using Purified Gene*. Academic press, New York, 683–93.
12. Breslauer, K. J., Frank, R., Blocker, H., and Marky, L. A. (1986) *Proc Natl Acad Sci USA* 83, 3746–50.
13. Blommers, M. J., Walters, J. A., Haasnoot, C. A., Aelen, J. M., van der Marel, G. A., van Boom, J. H., and Hilbers, C. W. (1989) *Biochemistry* 28, 7491–8.
14. Kernighan, B. W., and Ritchie, D. M. (1988) The C Programming Language, Prentice Hall Professional Technical Reference.
15. Stajich, J. E., Block, D., Boulez, K., Brenner, S. E., Chervitz, S. A., Dagdigian, C., Fuellen, G., Gilbert, J. G. R., Korf, I., Lapp, H., Lehvaslaiho, H., Matsalla, C., Mungall, C. J., Osborne, B. I., Pocock, M. R., Schattner, P., Senger, M., Stein, L. D., Stupka, E., Wilkinson, M. D., and Birney, E. (2002) *Genome Res* 12, 1611–8.
16. Kurtz, S., Choudhuri, J. V., Ohlebusch, E., Schleiermacher, C., Stoye, J., and Giegerich, R. (2001) *Nucleic Acids Res* 29, 4633–42.
17. Altschul, S. F., Gish, W., Miller, W., Myers, E. W., and Lipman, D. J. (1990) *J Mol Biol* 215, 403–10.
18. Kreiswirth, B. N., Lutwick, S. M., Chapnick, E. K., Gradon, J. D., Lutwick, L. I., Sepkowitz, D. V., Eisner, W., and Levi, M. H. (1995) *Microb Drug Resist* 1, 307–13.

19. Schneider, D., Duperchy, E., Depeyrot, J., Coursange, E., Lenski, R., and Blot, M. (2002) *BMC Microbiol* 2, 18.
20. Yarwood, J. M., McCormick, J. K., Paustian, M. L., Orwin, P. M., Kapur, V., and Schlievert, P. M. (2002) *J Biol Chem* 277, 13138–47.
21. Ubeda, C., Tormo, M. A., Cucarella, C., Trotonda, P., Foster, T. J., Lasa, I., and Penades, J. R. (2003) *Mol Microbiol* 49, 193–210.
22. Andonov, R., Lavenier, D., Veber, P., and Yanev, N. (2005) *Concurr Comput: Pract Exper* 17, 1657–68.

E

DNA Methylation Mapping

19

Designing PCR Primer for DNA Methylation Mapping

Long-Cheng Li

Summary

DNA methylation is an epigenetic mechanism of gene regulation, and aberrant methylation has been associated with various types of diseases, especially cancers. Detection of DNA methylation has thus become an important approach for studying gene regulation and has potential diagnostic application. Bisulfite-conversion-based PCR methods, such as bisulfite-sequencing PCR (BSP) and methylation-specific PCR (MSP), remain the most commonly used techniques for methylation detection. Primer design for this type of PCR is challenging because of the extreme DNA sequence composition after bisulfite modification and the special constraints on the primers and their location on the DNA template. To facilitate methylation detection, a primer design program called MethPrimer has been developed specifically for bisulfite-conversion-based PCR. MethPrimer accepts a DNA sequence as input, performs a digital bisulfite conversion of the input sequence, and then picks primers on the converted sequence. Results of primer selection are delivered through a Web browser in text and graphic views. This chapter discusses the process of using MethPrimer to design BSP and MSP primers.

Key Words: DNA methylation; PCR; primer design; bisulfite; MSP; BSP.

1. Introduction

Cytosine methylation is recognized as an important mechanism of epigenetic regulation of genomic function and plays important roles in diverse biological processes including embryogenesis *(1)*, genomic imprinting *(2)*, X-chromosome inactivation *(3)*, and cancer *(4–7)*. Methylation of cytosines is most likely to be restricted to CpG dinucleotides in many higher eukaryotic genome, where both cytosine residues on the opposite strands are methylated *(8)*. Compared with other dinucleotides, CpG dinucleotides are underrepresented in vertebrate DNA

From: *Methods in Molecular Biology, vol. 402: PCR Primer Design*
Edited by: A. Yuryev © Humana Press, Totowa, NJ

except in clusters known as CpG islands. CpG islands are usually unmethylated and often linked to promoter regions of genes ***(9)***. The accepted definition of a CpG island is a region of DNA greater than 200 bp, with a guanine and cytosine content above 0.5 and an observed/expected (Obs/Exp) CpG ratio above 0.6 ***(10)***. Approximately, 60% of all human genes contain CpG islands at their 5′-ends ***(11)***.

Mapping of methylation patterns in CpG islands has become an important tool for understanding both normal and aberrant gene expression. In addition, current research suggests that DNA methylation may be a promising tumor biomarker and detection of methylated DNA may become a routine diagnostic procedure in a clinical setting. Numerous techniques have been designed for the mapping of cytosine methylation, with bisulfite-conversion-based methods among the most widely used. The advantage of the procedure is that it allows for the rapid identification of 5-methyl cytosines (m5Cs) in any sequence context. The bisulfite reaction was first described in the early 1970s ***(12, 13)*** and was used by Frommer et al. ***(14)*** in 1992 to distinguish between cytosines and 5mCs in DNA. In this reaction, DNA is first denatured to create single-stranded DNA, which is further treated with sodium bisulfite to convert cytosine residues to uracil under conditions whereby 5mCs remains essentially non-reactive. The DNA sequence under investigation is then amplified by PCR with primers specific for bisulfite-modified DNA ***(15)***. Since the first description of the bisulfite reaction for studying 5mCs, many methods based on the same principle have been developed including bisulfite-sequencing PCR (BSP) ***(15)*** and methylation-specific PCR (MSP) ***(16)***. All methods share the same procedure of initially modifying DNA with sodium bisulfite followed by PCR amplification with primers specific for the modified DNA (*see* **Fig. 1**).

BSP is performed using bisulfite-modified DNA as the template using primers that selectively amplify bisulfite-modified DNA and contain no CpG sites so that both methylated and unmethylated DNA are amplified. The resulting PCR product can then be used in the following three ways: (1) direct sequencing to examine strand-specific methylation for all population of molecules in the DNA sample; (2) cloning and sequencing to study the methylation status of individual molecules; and (3) digestion with restriction enzymes to examine methylation of particular CpG sites recognized by the enzymes.

MSP needs two pairs of primers, one specific for modified and methylated DNA (M pair) and the other specific for modified and unmethylated DNA (U pair). For each sample to be studied, two PCRs are performed with each

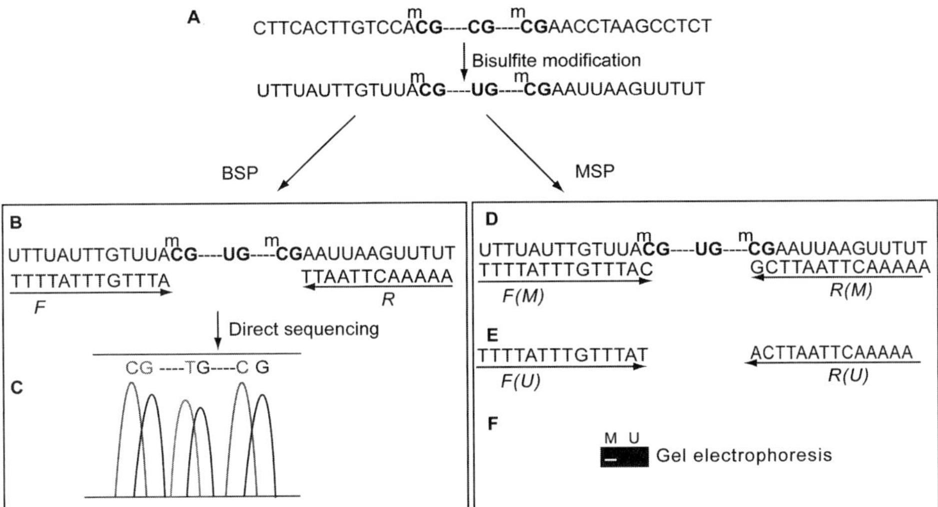

Fig. 1. Bisulfite-conversion-based PCRs. DNA is first denatured to create single-stranded DNA, which is further treated with sodium bisulfite to convert all cytosines except 5-methyl cytosines (m5Cs) to uracil (**A**). The modified DNA is amplified with two different PCR methods: bisulfite-sequencing PCR (BSP) and methylation-specific PCR (MSP). In BSP, DNA is amplified using primers that contain non-CpG "C"s so that all DNA molecules are amplified regardless of their methylation states (**B**). The resulting PCR product is sequenced directly or after cloning (**C**). In MSP, DNA is amplified with two pairs of primers with one pair specific for methylated DNA (**D**) and the other for unmethylated DNA (**E**). The resulting PCR products are separated by agarose gel electrophoresis and visualized by ethidium bromide staining (**F**).

pair of primers. Amplification with the M pair indicates methylation of CpG site(s) within the primer sequences, the U pair no methylation, and both pairs partial methylation.

2. System and Methods

A complete primer design process using the MethPrimer program involves two major phases. The first is the user phase in which a user retrieves a DNA sequence, feeds the sequence into the program, and selects appropriate parameters. When the user clicks the submit button, the program phase is initiated in which it first performs a digital bisulfite conversion of the input sequence, predicts CpG islands, picks primers based on input parameters, prepares data for output, and displays the primer design results to the user.

2.1. User Input

The MethPrimer program can be accessed from any computer with an Internet connection and a Web browser at the URL http://www.urogene.org/methprimer. When the program is launched, it requires a DNA sequence as input from a user and provides default values for all other parameters and options.

2.1.1. Sequence Retrieval

To make methylation mapping biologically meaningful, it is very important to examine areas where DNA methylation may have an impact on gene transcription. In this regard, DNA methylation mapping focuses on the 5′-regulatory sequence or the promoter region of genes where CpG islands are frequently found. However, sometimes CpG islands can extend well into the first exon and intron, so that these areas can also be a target of methylation study. Although the mRNA sequences of many genes are well annotated and can be easily found in major databases such as GenBank at National Center for Biotechnology Information (NCBI), the retrieval of 5′-regulatory sequences may not be a trivial task. Promoter sequences can be retrieved from various sources such as GenBank, Ensembl, and UCSC Genome Bioinformatics site (*see* **Note 1**) and pasted into the sequence box on the MethPrimer input page. MethPrimer accepts original DNA sequences in any format, which means that an input sequence should not be converted by changing "C"s to "T"s.

2.1.2. Parameter Selection

As shown in **Fig. 2**, MethPrimer offers many options to users.

2.1.2.1. Type of PCR

Users must specify the type of PCR primers to be designed. By default, the program will design BSP primers.

2.1.2.2. Use CpG Island Prediction for Primer Selection?

For all primer design tasks, the program will scan, by default, the input sequence for CpG islands. If a user wishes to design primers on CpG islands only, this selection should be checked.

2.1.2.3. Optional General Parameters for Primer Selection

- Sequence name: Users can give a name to the input sequence. If the input sequence is in FASTA format, the description line in the FASTA format will be used as the sequence name.

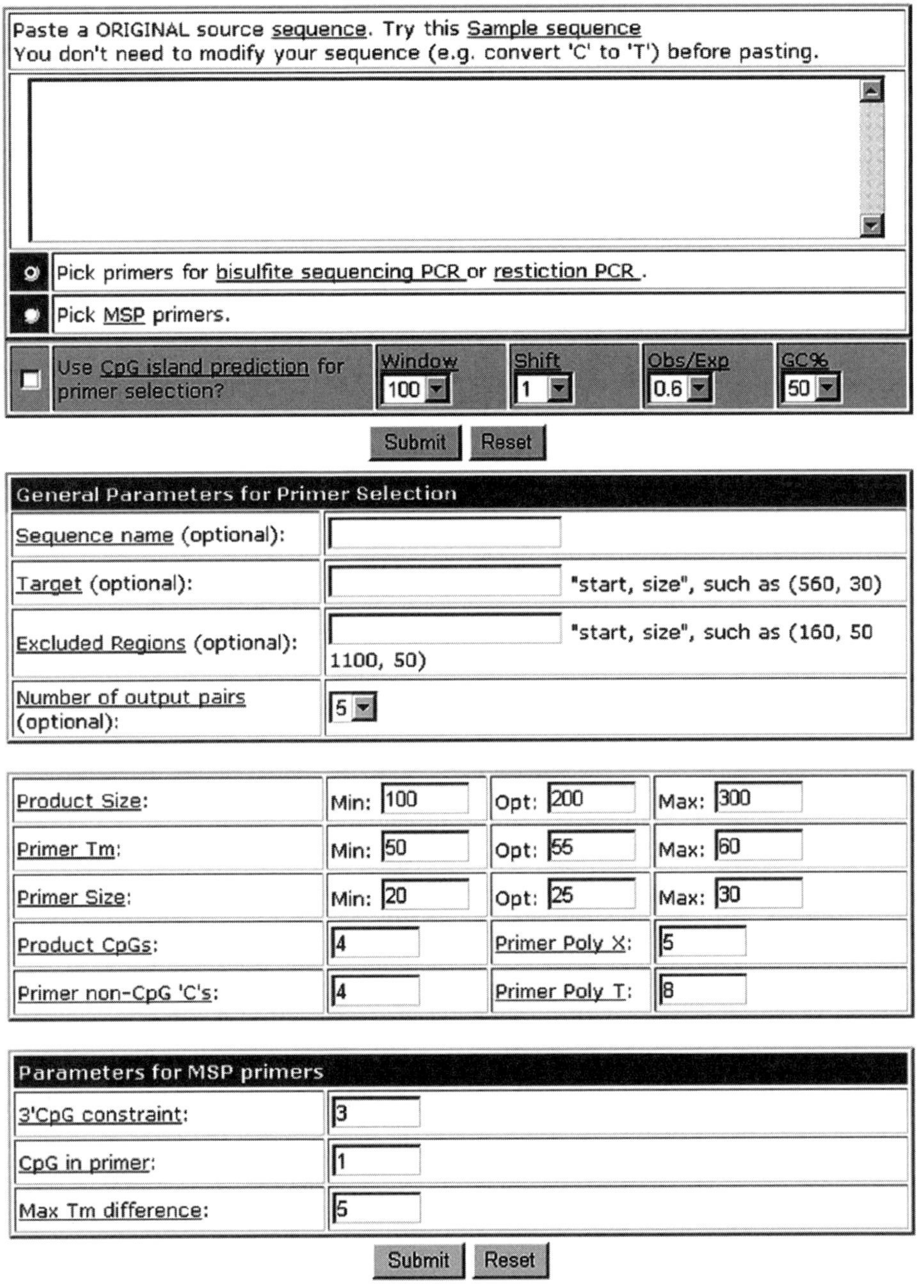

Paste a ORIGINAL source sequence. Try this Sample sequence
You don't need to modify your sequence (e.g. convert 'C' to 'T') before pasting.

Pick primers for bisulfite sequencing PCR or restiction PCR .

Pick MSP primers.

	Use CpG island prediction for primer selection?	Window 100	Shift 1	Obs/Exp 0.6	GC% 50

Submit Reset

General Parameters for Primer Selection	
Sequence name (optional):	
Target (optional):	"start, size", such as (560, 30)
Excluded Regions (optional):	"start, size", such as (160, 50 1100, 50)
Number of output pairs (optional):	5

Product Size:	Min: 100	Opt: 200	Max: 300
Primer Tm:	Min: 50	Opt: 55	Max: 60
Primer Size:	Min: 20	Opt: 25	Max: 30
Product CpGs:	4	Primer Poly X:	5
Primer non-CpG 'C's:	4	Primer Poly T:	8

Parameters for MSP primers	
3'CpG constraint:	3
CpG in primer:	1
Max Tm difference:	5

Submit Reset

Fig. 2. User interface of the MethPrimer program.

- Target: It is a single region in the sequence that the amplicon should cover. By giving a target, the program can be forced to pick primers surrounding the target. Target can be specified by providing a starting position from the beginning of the input sequence and the length from the starting position. For example, "350, 10" means the amplicon should at least cover the region from position 350 to 360. The target should not be bigger than the maximum product size (see 2.1.2.4). Target can also be specified by marking input sequence with " [ ]," e.g.,...ATCT[...CCGT...]ATCT.... If no target is defined, any primers that satisfy other criteria may be chosen regardless of their location.
- Excluded regions: One or more than one region that should be excluded for primer selection. It can be defined by numbers such as "160, 50 1100, 50," which means that the regions from 160 to 210 and from 1100 to 1150 should be excluded.
- Number of output pairs: It is the number of good primers to be returned by the program, which by default is 5. Because primer pairs returned are in order of overall quality from highest to lowest, the pair at the top of the list should be first considered by users.

2.1.2.4. Parameters in the Following Section Are Bisulfite PCR specific and Are Common to Both BSP and MSP

- Product size: Users can specify minimum, optimum, and maximum product size; the default is 100, 200, and 300 bp respectively. Product size for bisulfite PCR should be kept under 400 bp because of severe DNA degradation during bisulfite modification (*see* **Note 2**).
- Primer melting temperature (T_m): Owing to bisulfite conversion of non-CpG "C"s to "T," the primer T_m value is usually lower than that for standard PCR with the default ranging from 50 to 65 °C and 55 °C being the optimum.
- Primer size: Primer size for bisulfite PCR is usually longer than that for standard PCR because of the lower sequence complexity of the modified sequence, with the default ranging from 20 to 30 nucleotides (nt) and 25 nt being the optimum (*see* **Note 3**).
- Product CpG: The number of CpG sites in the amplicon. This parameter is more important for BSP than for MSP, because this forces the program to pick primers surrounding CpG-dense regions that may be of interest.
- Primer non-CpG "C"s: The number of non-CpG "C"s in a primer. To bias against unmodified or incompletely modified DNA, primers should be picked from a region that has enough numbers of non-CpG "C"s in the original sequence ***(16)***. Primers with more "C"s will be preferred and receive higher weighing scores (*see* **Note 4**).
- Primer Poly X: It is the maximum allowable length of a mononucleotide repeat in a primer sequence. The default number is 5.
- Primer Poly T: "T" repeats are treated differently as other nucleotides, because all non-CpG "C"s are treated as "T" during primer selection. By default, a maximal number of eight consecutive "T"s is allowed.

2.1.2.5. Parameters for MSP Primers

An MSP experiment usually requires two pairs of primers with one specific for the methylated template and the other for the unmethylated template. The following options are for MSP primer design.

- 3′-CpG constraint: For maximum discrimination between methylated and unmethylated alleles, primers should contain at least one CpG site at the very 3′-end. Users can specify the maximum distance from the "C" in the CpG dinucleotide to the very 3′-end of the primer. By default, this value is set to 3, which means that among the last three bases in the primer, at least one of them should be a CpG "C."
- Number of CpG's in a primer: Other than the CpG site(s) at the very 3′-end, more CpG sites in a primer sequence are preferred (*see* **Note 5**). The default value is 1.
- Max T_m difference: Two sets of primers should preferably have similar product T_m values. This constraint will produce primers so that two PCR reactions can be carried out in a single PCR machine using the same cycling conditions. By default, this parameter is set to 5 °C.

2.2. MethPrimer Backend

MethPrimer first takes an input sequence that can be in any format (no sequence editing is needed before input) and internally converts non-CpG "C"s in the sequence to "T." The resulting sequence is called a "modified sequence." For MSP primers, in addition to the "modified sequence," another version called the "unmethylated sequence" is also stored internally. The program will shift one base at a time from the beginning of the input sequence using a window size ranging from the minimum to maximum primer size and test every substring of the input sequence in the window against the various parameters. It will then match two primers (forward and reverse) to make a primer pair. Lastly, the program will output the best primers that meet all the constraints.

2.2.1. CpG Island Prediction

CpG islands are predicted using a simple sliding window algorithm. The algorithm slides across the sequence at a specified shift value, examining the GC content and the Obs/Exp ratio in a window size defined by users ***(10)***. By default, a CpG island is defined as a DNA stretch at least 200 bp long with a GC content > 50% and an Obs/Exp ratio of CpG dinucleotides > 0.6 ***(10)***. If users choose to pick primers using the predicted CpG islands as the target region, the following rules are applied:

1. If more than one island is found, any of the predicted islands will be a potential target region for amplification.
2. If the size of a CpG island is smaller than the minimum product size, the primer pair should span the whole island.
3. If the size of a CpG island is greater than the maximum product size, the primer pair should be within the island.
4. If the size of a CpG island is between the minimum and maximum product size, the primer pair should cover at least two-thirds of the island.

2.2.2. General Parameter Calculations

MethPrimer uses Primer3's algorithms and code to compute primer self-annealing, self-end annealing, pair complementarity, GC content, and T_m *(17)*.

2.2.3. Picking Forward and Reverse Primers

MethPrimer tests every possible substring in the input sequence for parameters including T_m, self-annealing, self-end annealing, GC content, length, poly X, poly "T" non-CpG "C"s, and CpG "C"s (for MSP) and assigns them a weighted score for every parameter. The program then sorts the primers by the sum of weighted scores.

2.2.4. Picking Primer Pairs

To make a primer pair, the program matches a forward and reverse primer starting with those with a high score and checks their product size, difference in T_m values, pair complementarity, number of product CpGs, and product T_m and assigns a weighted score for each criterion for each pair. The resulting pairs are sorted by the overall scores.

2.2.5. Picking MSP Primer Sets

An MSP primer set consists of an M and a U pair. MethPrimer will match high-score M pairs and U pairs of a set by checking whether both pairs cover the same CpG sites and have similar product T_m values. For example, if the sequence for a forward primer in an M pair is ATTTAGTTT*CG*TTTAAGGTT*CG*A, the forward primer in the U pair must also contain the two CpG sites (italicized) as the M pair, although they may be of different size (*see* **Note 6**).

2.2.6. Prepare Data for Output

MethPrimer uses a Perl module, the GD module (http://stein.cshl.org/WWW/software/GD) to generate a PNG image for each input sequence.

An image map is also generated and embedded into HTML code using a Perl script for each image to display text explanation as tool tips for each element in the image.

3. Program Output

On the output page, the top is a graphic view in which the input sequence is drawn proportionally on the *x*-axis, percentage of GC content is plotted as a curved line against the *y*-axis, and CpG islands are drawn as a light-blue area (*see* **Fig. 3**). Under the *x*-axis, individual CpG sites are drawn as gold-colored bars. Primer pairs are drawn as square boxes connected by a line. When the mouse cursor is over the input sequence, CpG islands, individual primers and primer pairs, and a text tool tip will show up displaying information related to those elements in the graph such as the position of a CpG island and a primer, the PCR product size of a primer pair, etc. Clicking on the graphic elements will lead to their corresponding part in the text output.

Below the graph are the primer picking results. The first few items are the input sequence name, length of the sequence, and CpG island prediction results followed by the primer picking results. In this section, primer pairs are ordered by their overall quality scores from the highest to the lowest. Further down are the original input sequence aligned with the bisulfite-modified sequence and the best primer pair. In the alignment, non-CpG "C"s are indicated by ":", CpG sites by "++", and other bases by "|".

4. Tips on Using MethPrimer

4.1. Narrow Down the Sequence of Interest

It is not a good idea to use, for example, a 5-kp promoter sequence and use whatever primers the program picks for methylation mapping, because the program will only return high-score primers without consideration of their biological context. It is the responsibility of the users to define and specify where to look for primers so that the results of methylation mapping can answer their biological questions. This can be best done by specifying a target region or even better by using a shorter input sequence containing the important regions of interest.

4.2. No Primers Were Found for My Sequence, What's Wrong?

Designing primers for bisulfite-conversion-based methylation PCR is very different from that for standard PCR, because, initially, the input sequence is modified to convert non-CpG "C"s to "T," resulting in low GC content with

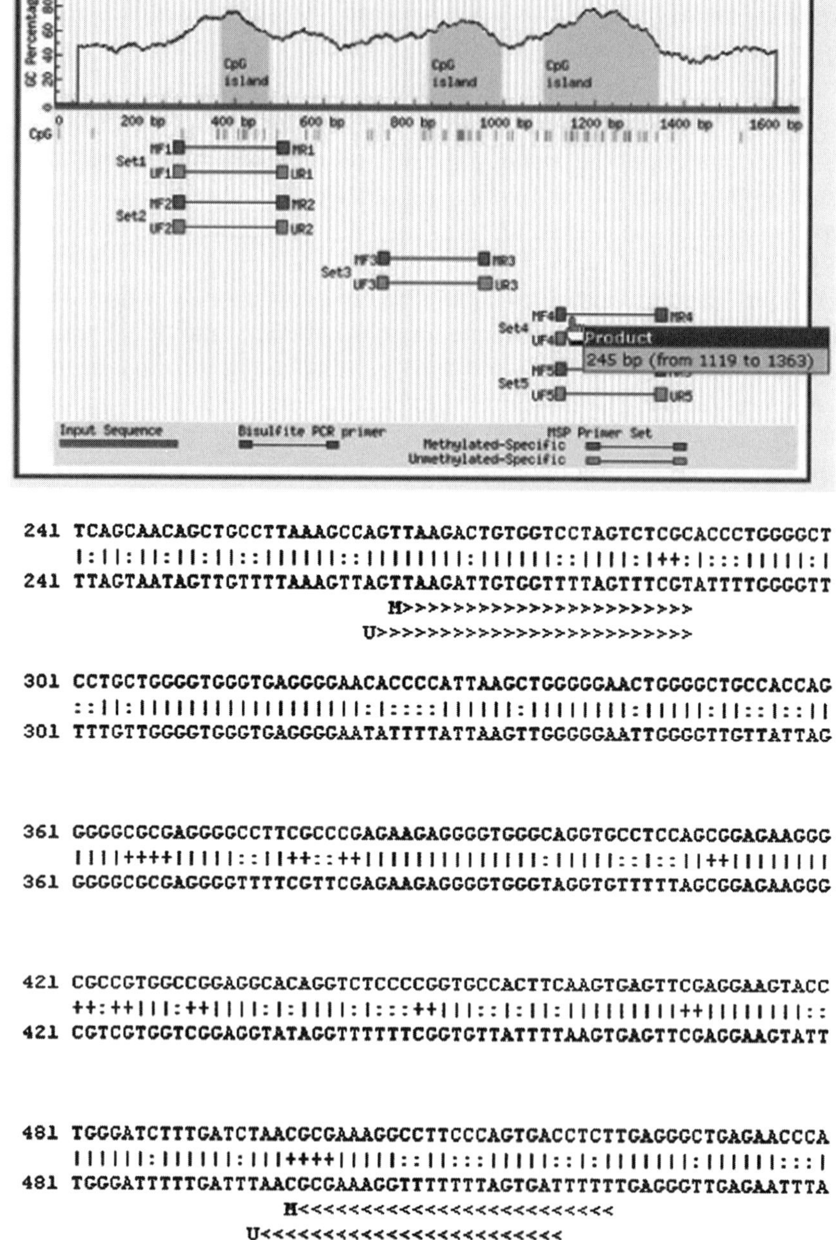
GC Percentage
0 20 40 60 80
CpG island
CpG island
CpG island
0 200 bp 400 bp 600 bp 800 bp 1000 bp 1200 bp 1400 bp 1600 bp
CpG
Set1 MF1 MR1 UF1 UR1
Set2 MF2 MR2 UF2 UR2
Set3 MF3 MR3 UF3 UR3
Set4 MF4 MR4 UF4
Product
245 bp (from 1119 to 1363)
Set5 MF5 UF5 UR5
Input Sequence
Bisulfite PCR primer
MSP Primer Set
Methylated-Specific
Unmethylated-Specific
241 TCAGCAACAGCTGCCTTAAAGCCAGTTAAGACTGTGGTCCTAGTCTCGCACCCTGGGCT
|:||:||:||:||::||||||::||||||||:||||||::||||:|++:|:::||||||:|
241 TTAGTAATAGTTGTTTTAAAGTTAGTTAAGATTGTGGTTTTAGTTTCGTATTTTGGGTT
M>>>>>>>>>>>>>>>>>>>>>>>>
U>>>>>>>>>>>>>>>>>>>>>>>>>>
301 CCTGCTGGGGTGGGTGAGGGGAACACCCCATTAAGCTGGGGGAACTGGGGCTGCCACCAG
::||:||||||||||||||||||:|::::||||||:|||||||||:|||||:||::|::||
301 TTTGTTGGGGTGGGTGAGGGGAATATTTTATTAAGTTGGGGGAATTGGGGTTGTTATTAG
361 GGGGCGCGAGGGGCCTTCGCCCGAGAAGAGGGGTGGGCAGGTGCCTCCAGCGGAGAAGGG
||||++++|||||::||++::++||||||||||||||:|||||::|::||++||||||||
361 GGGGCGCGAGGGGTTTTCGTTCGAGAAGAGGGGTGGGTAGGTGTTTTTAGCGGAGAAGGG
421 CGCCGTGGCCGGAGGCACAGGTCTCCCCGGTGCCACTTCAAGTGAGTTCGAGGAAGTACC
++:++|||:++||||:|:||||:|:::++|||::|:||:|||||||||++||||||||::
421 CGTCGTGGTCGGAGGTATAGGTTTTTTCGGTGTTATTTTAAGTGAGTTCGAGGAAGTATT
481 TGGGATCTTTGATCTAACGCGAAAGGCCTTCCCAGTGACCTCTTGAGGGCTGAGAACCCA
||||||:||||||:|||++++|||||::||:::|||||::|:|||||||:||||||:::|
481 TGGGATTTTTGATTTAACGCGAAAGGTTTTTTAGTGATTTTTTGAGGGTTGAGAATTTA
M<<<<<<<<<<<<<<<<<<<<<<<<<<<
U<<<<<<<<<<<<<<<<<<<<<<<<<<

Fig. 3. Example of a methylation-specific PCR (MSP) primer design result by MethPrimer.

long stretches of "T"s in the input sequence, which is not ideal for primer selection. Also, many extra constraints are applied to the primer selection in addition to those for standard PCR. For example, bisulfite sequencing or restriction PCR primers should contain no CpG sites in their sequence but should span a certain number of CpG sites, whereas for MSP primers, at least one CpG site is required in their sequence. Therefore, it is not uncommon to get no primer hit for your sequence.

4.3. What Should I Do If No Primers Are Found for My Sequence?

There are several measures that can be taken to force the program to pick primers without significantly sacrificing primer quality. (1) Try to relax some parameters such as the number of CpGs in the product (for BSP) and in the primers (for MSP), the number of non-CpG "C"s in the primer (for both BSP and MSP), the T_m difference between two sets of primers (for MSP), and the 3′-end CpG constraint (for MSP). (2) Try to specify a target region of interest if you have not done so.

Notes

1. The primary source for gene regulatory sequences or promoters is the Entrez Nucleotides database (GenBank) at NCBI at the URL http://www.ncbi.nlm.nih.gov/entrez/query.fcgi?db=Nucleotide. Promoter sequences in the GenBank are deposited by researchers, and many of them are experimentally characterized. However, not every gene promoter sequence has been identified and deposited in the GenBank, and moreover, the Entrez search tool is not very efficient. In these cases, using the 5′-flanking sequence of a gene from genome databases for bisulfite PCR primer design is an alternative, because promoter sequences are usually located immediately upstream of the transcriptional start site. The Ensembl database (http://www.ensembl.org) provides the most convenient tool for 5′-sequence retrieval. On the Ensembl GeneView page of a particular gene, click the "Genomic Sequence" link on the left side, and the 5′-flanking sequence will be displayed along with the exons and introns of the gene.
2. The bisulfite reaction converts cytosines to uracils, resulting in a low GC content with long stretches of "T"s in the sequence, and also causes undesired DNA strand breakage. Loss of DNA during the subsequent purification step is another concern especially when studying microdissected DNA samples. All these factors pose challenges to downstream PCR applications and should be taken into consideration when designing primers for these PCRs.
3. Another rule that differs from standard PCR is primer length. Bisulfite-conversion-based PCRs generally require longer primers. Primers with a length of approximately 30 bp usually yield successful results *(15)*. The reason is that bisulfite modification considerably decreases the GC content of DNA templates, producing

long stretches of "T"s in the sequence, thus making it difficult to pick primers with acceptable T_m values or stability. On the contrary, in order to discriminate between modified DNA and unmodified or incompletely modified DNA, an adequate number of "C"s are required in the primers, making the job of picking stable primers more demanding. Thus, to achieve better duplex stability, choosing longer primers is necessary as the T_m of DNA also depends on its length ***(18)***. In practice, the size of primers for these PCRs usually ranges from 20 to 30 bp ***(5,16,19)***. In MethPrimer, 20–30 bp is set as the default range of primer size with 25 bp being the optimal size.

4. Incomplete bisulfite conversion of DNA in methylation detection is sometimes a concern ***(15)*** and results in a false high representation of methylation levels in the DNA samples studied. Thus, selective amplification of only completely modified DNA is important.
5. There are pros and cons regarding (1) having both the forward and reverse primer contain a CpG site and (2) having them contain multiple CpG sites. Neighboring CpG sites are not always equally methylated. If both forward and reverse primers contain a CpG site and one of the sites on a DNA molecule is not methylated, the primer pair may not be able to amplify this DNA molecule. On the contrary, if only one primer contains a CpG site, each PCR can only detect the methylation state of one site. For the second issue, multiple CpG sites in a primer will give the primer better specificity than one CpG site, but again, if these sites are not equally methylated, it is unpredictable as to whether the primer will bind to the DNA or not.
6. This constraint is necessary because nearby CpG sites are not always equally methylated ***(5)***. If two pairs of primers do not anneal to the same CpG sites, PCR results from the primers may not truly reflect the DNA methylation status of the sample studied. However, primers in an M pair and a U pair may not span the exact same sequence and may vary in start position or length. Usually primers in a U pair are longer than those in an M pair to compensate for the low T_m in the U primers caused by the replacement of the 5mCs by "T."

References

1. Monk, M., Boubelik, M., and Lehnert, S. (1987) Temporal and regional changes in DNA methylation in the embryonic, extraembryonic and germ cell lineages during mouse embryo development. *Development* **99**, 371–82.
2. Singer-Sam, J., and Riggs, A. D. (1993) X chromosome inactivation and DNA methylation. *EXS* **64**, 358–84.
3. Li, E., Beard, C., and Jaenisch, R. (1993) Role for DNA methylation in genomic imprinting. *Nature* **366**, 362–5.
4. Baylin, S. B., Herman, J. G., Graff, J. R., Vertino, P. M., and Issa, J. P. (1998) Alterations in DNA methylation: a fundamental aspect of neoplasia. *Adv Cancer Res* **72**, 141–96.

5. Li, L. C., Chui, R., Nakajima, K., Oh, B. R., Au, H. C., and Dahiya, R. (2000) Frequent methylation of estrogen receptor in prostate cancer: correlation with tumor progression. *Cancer Res* **60**, 702–6.
6. Li, L. C., Zhao, H., Nakajima, K., Oh, B. R., Filho, L. A., Carroll, P., and Dahiya, R. (2001) Methylation of the E-cadherin gene promoter correlates with progression of prostate cancer. *J Urol* **166**, 705–9.
7. Nojima, D., Nakajima, K., Li, L. C., Franks, J., Ribeiro-Filho, L., Ishii, N., and Dahiya, R. (2001) CpG methylation of promoter region inactivates E-cadherin gene in renal cell carcinoma. *Mol Carcinog* **32**, 19–27.
8. Paulsen, M., and Ferguson-Smith, A. C. (2001) DNA methylation in genomic imprinting, development, and disease. *J Pathol* **195**, 97–110.
9. Antequera, F., and Bird, A. (1993) CpG islands. *EXS* **64**, 169–85.
10. Gardiner-Garden, M., and Frommer, M. (1987) CpG islands in vertebrate genomes. *J Mol Biol* **196**, 261–82.
11. Larsen, F., Gundersen, G., Lopez, R., and Prydz, H. (1992) CpG islands as gene markers in the human genome. *Genomics* **13**, 1095–107.
12. Hayatsu, H., Wataya, Y., Kai, K., and Iida, S. (1970) Reaction of sodium bisulfite with uracil, cytosine, and their derivatives. *Biochemistry* **9**, 2858–65.
13. Shapiro, R., and Weisgras, J. M. (1970) Bisulfite-catalyzed transamination of cytosine and cytidine. *Biochem Biophys Res Commun* **40**, 839–43.
14. Frommer, M., McDonald, L. E., Millar, D. S., Collis, C. M., Watt, F., Grigg, G. W., Molloy, P. L., and Paul, C. L. (1992) A genomic sequencing protocol that yields a positive display of 5-methylcytosine residues in individual DNA strands. *Proc Natl Acad Sci USA* **89**, 1827–31.
15. Clark, S. J., Harrison, J., Paul, C. L., and Frommer, M. (1994) High sensitivity mapping of methylated cytosines. *Nucleic Acids Res* **22**, 2990–7.
16. Herman, J. G., Graff, J. R., Myohanen, S., Nelkin, B. D., and Baylin, S. B. (1996) Methylation-specific PCR: a novel PCR assay for methylation status of CpG islands. *Proc Natl Acad Sci USA* **93**, 9821–6.
17. Rozen, S., and Skaletsky, H. (2000) Primer3 on the WWW for general users and for biologist programmers. *Methods Mol Biol* **132**, 365–86.
18. Rychlik, W. (2000) Primer selection and design for polymerase chain reaction, in *The Nucleic Acid Protocols Handbook* (Rapley, R., Ed.), pp. 581–88, Humana Press, Inc., Totowa, NJ.
19. Graff, J. R., Herman, J. G., Myohanen, S., Baylin, S. B., and Vertino, P. M. (1997) Mapping patterns of CpG island methylation in normal and neoplastic cells implicates both upstream and downstream regions in de novo methylation. *J Biol Chem* **272**, 22322–9.

20

BiSearch

ePCR Tool for Native or Bisulfite-Treated Genomic Template

Tamás Arányi and Gábor E. Tusnády

Summary

The design of adequate primers for polymerase chain reaction (PCR) is sometimes a difficult task. This is the case when either the target sequence harbors unusual nucleotide motifs or the template is complex. Unusual nucleotide motifs can be repeat elements, whereas complex templates are targets for mispriming and alternative amplification products. Such examples are GC-rich native or bisulfite-treated genomic DNA sequences. Bisulfite treatment leads to the specific conversion of non-methylated cytosines to uracyls. This is the key step of bisulfite genomic sequencing, widely used to determine DNA methylation of a sequence. Here, we describe BiSearch Web server (http://bisearch.enzim.hu), a primer design software created for designing primers to amplify such target sequences. Furthermore, we developed a unique post-design primer analysis module, to carry out genome wide searches to identify genomic mispriming sites and to test by electronic (in silico) PCR (ePCR) for alternative PCR products. This option is currently available on four native or bisulfite-treated mammalian genomes.

Key Words: Primer design; DNA methylation; methylation-specific PCR (MSP); epigenetic control; sequence alignment; mispriming; non-specific amplification; primer validation; genomic DNA.

1. Introduction

Polymerase chain reaction (PCR) is a widely used technique to amplify nucleic acids. One of the most important steps of the setting up of a PCR experiment is the design of the oligonucleotide primers used in the amplification

From: *Methods in Molecular Biology, vol. 402: PCR Primer Design*
Edited by: A. Yuryev © Humana Press, Totowa, NJ

reaction. During the primer design, the user has to face various constraints, because a great number of parameters determine the quality of a unique primer and that of a primer pair. The importance of the individual parameters depends on the aim of the specific PCR amplifications.

The purpose of the PCR experiments can be extremely diverse. An amplification reaction can be a qualitative or a quantitative analysis; the template can be vector DNA, cDNA, or much more complex such as human genomic DNA or chemically modified DNA. The objective may be the amplification of a single unique sequence or the simultaneous amplification of a high number of more or less different templates. Thus, the quality of the templates essentially determines the strategies to adopt while designing the optimal primer pairs to be used in a study.

Here, we describe the BiSearch Web server that was created to design primers for the amplification of single templates ***(1)***. A characteristic feature of the software is its composition of several independent submodules, which give rise to two separate modules such as primer design and post-design analysis by genome search. The primer design web tool can be utilized for any DNA template; however, it is especially suited for primer design for chemically modified DNA used to determine the CpG methylation profile of a DNA template by the bisulfite genomic sequencing technique ***(2,3)***. The principle of this technique is the chemical conversion of non-methylated cytosines to uracyls. The methyl-cytosines are resistant to the treatment; hence, the methylation profile of a sequence is determined by the amplification of the bisulfite-treated template.

Owing to the independence of the submodules, BiSearch can be used also for primer scoring or to determine the Tm of a primer. The unique feature of the software is the possibility of a post-design primer analysis for genomic mispriming and alternative PCR products. This option can be used for both bisulfite-treated and native genomic templates.

The unique format of the BiSearch software is an online tool at the http://bisearch.enzim.hu home page freely available for academic researchers. BiSearch is an interactive tool characterized by two general features, making it user friendly. First, all parameters used in the various applications are modifiable, although default parameters are provided.

A practical guide to the BiSearch Web server is proposed in the following sections. First, the primer design options will be described. This section is followed by the description of the tools for post-design primer analysis. Finally, some technical details of the Web interface conclude the chapter.

2. Primer Design

2.1. General Considerations

2.1.1. Input of a Target Sequence

The BiSearch software proposes three primer design options *(1)*. Primers can be designed for native or bisulfite-treated DNA sequences *(2,3)* after clicking on the "Primer design" menu point. In some (generally medical) cases, the amplification of only the methylated template is needed after the bisulfite treatment of the sample. This approach is called MSP *(4)*. Primer design for these reactions is proposed under the "MSP design" menu point.

In each case, the next step is the input of the target sequence (*see* **Fig. 1**). For each primer design option, the original (native) sequence should be used. The sequence is entered in plain text format. The form is not case sensitive but accepts only the four DNA nucleotides [characters other than the four standard nucleotides (A, C, G, and T) are ignored]. The maximal PCR product length (default value is 400 bp) and the approximate primer locations have to be indicated (*see* **Fig. 1**). If primer location is not indicated, then the forward and reverse primers will be designed automatically in the first and the second halves of the template sequences, respectively. Default values for the beginning and the end of the forward and reverse primers are 0, $n/2$, $n/2$, and n, respectively where n designs the input template length in nucleotides.

2.1.2. Parameters (see **Fig. 2**)

The primer design and selection is based on the calculation of various parameters *(5)*. These include the self and pair (end)annealing of the primers, their melting temperature (Tm), GC content, and length. Tm is a crucially important parameter especially because the small Tm difference of the primers is one of the clues to an efficient PCR amplification. Therefore, the maximum allowable Tm difference between the forward and reverse primers is also adjustable. The software calculates the Tm with the nearest-neighbor method, according to the equation published previously by Wetmur and Sninsky *(6)*.

For each parameter, optimum, minimum, and maximum values are set (*see* **Fig. 2**). Individual preferences can be set by the modification of these parameters or of their weight. The preferences can be saved by the user's browser, or the default parameters can be restored at any time. Primers with one or more parameters outside of the range determined by the minimum and maximum values are not considered. The difference of a calculated value from the optimum is taken into account according to its predefined weight. The

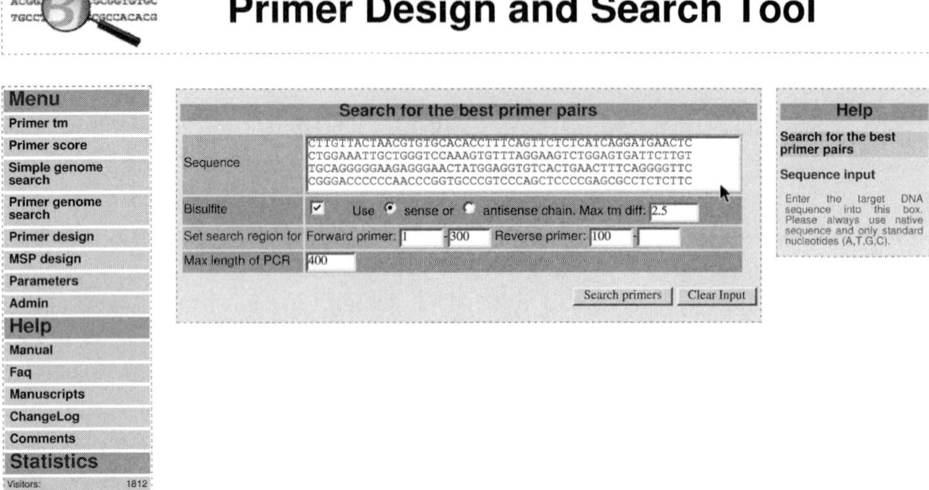

Fig. 1. Target sequence input datasheet. The native genomic sequence is entered in a plain text format in the "Sequence" box. Only A, T, G, and C nucleotides are considered for primer design by the algorithm. Bisulfite sense or antisense chain has to be selected if the template is bisulfite treated. Bisulfite treatment is carried out automatically by the software on the native sequence. Approximate primer locations and maximal product length have to be adjusted on the same input datasheet before primer design.

selected oligonucleotides are scored according to the parameters using simple weighted sums.

As a general observation, some PCR primer pairs do not amplify the target template as efficiently as it is predicted even if they were designed by a very efficient software. Unfortunately, with one parameter set, most of the primer design software packages propose a list of only very similar, almost identical primer pairs. If the best primer pair do not amplify the target, then it is difficult to find alternative primers for the same purpose by the same software. Therefore, to overcome this inconvenience, BiSearch software proposes only significantly different primer pairs. Two primers are considered by the software to be significantly different from each other when the location of either their

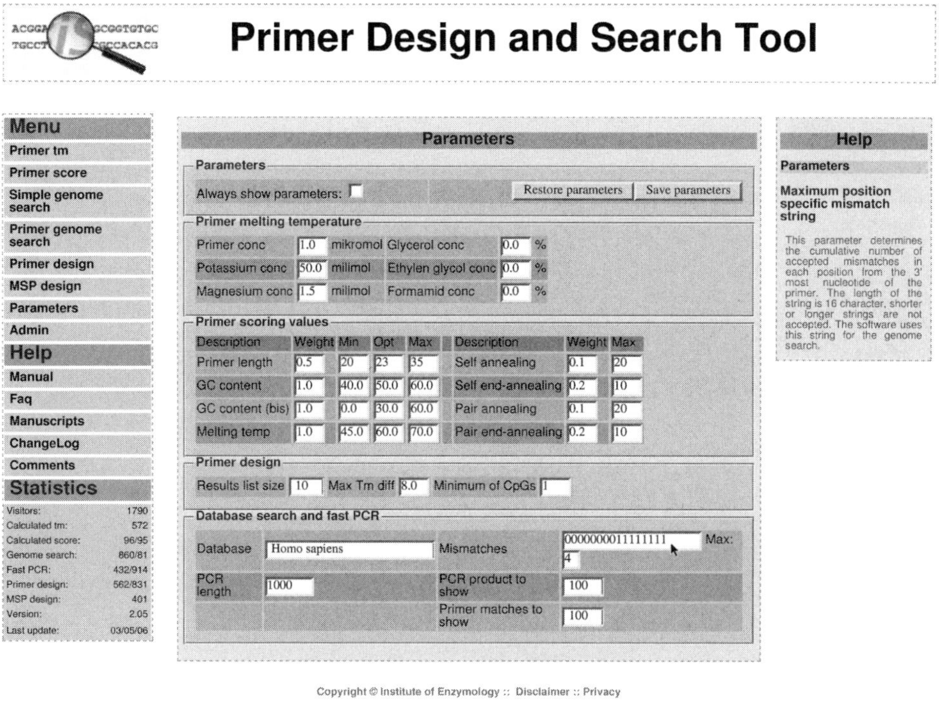

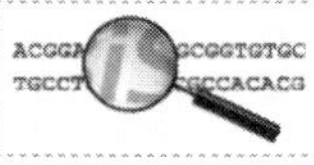

Fig. 2. Parameters. The parameter table indicates the default values used for primer design and post-design analysis. The mouse is over the mismatch string parameter.

5′ or their 3′ ends has at least three bases different. Two primer pairs are significantly different from each other when at least one of the primers of a pair is significantly different from the corresponding member of the other pair. Only the primer pair with the best score is proposed from not significantly different primer pairs.

2.1.3. Output Format

The results of primer design are summarized in a table in which the primer pairs are listed according to their scores (*see* **Fig. 3**). The table also includes the calculated values for the different parameters. When moving the mouse over the different values, a small window appears to illustrate the corresponding results. The PCR amplicon is also shown with the primers underlined and the CpG dinucleotides highlighted. The results are also summarized in a printable window, which appears after clicking on the details. The fast PCR (FPCR)

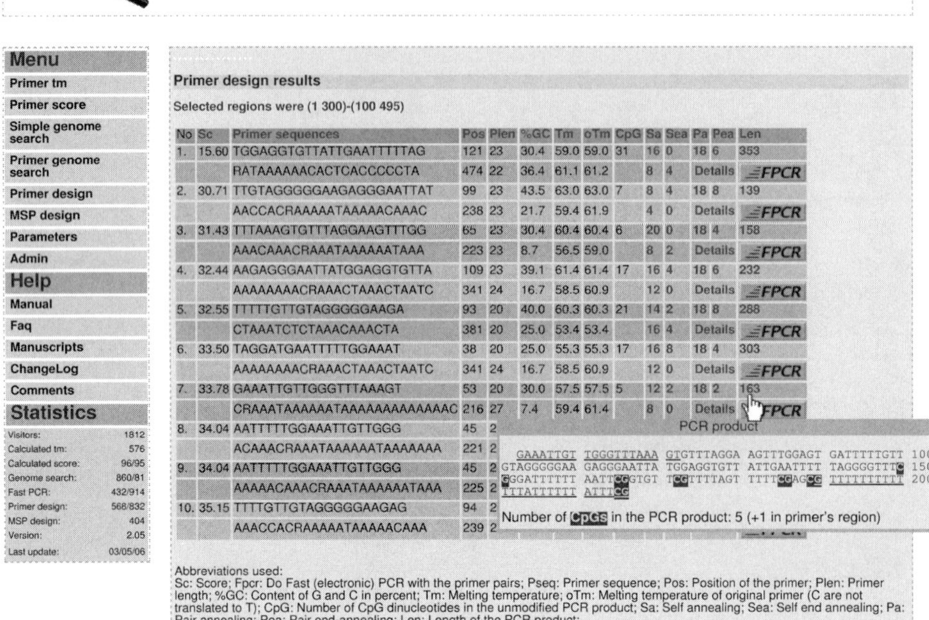

Fig. 3. The results of primer design are visualized in a tabular format. Primer pairs are ranked according to their scores. The best primer pair has the lowest score. Important details (primer length, *T*m, product length, etc.) are shown in the different columns (the abbreviations are listed below the table). The sequence of the PCR product with highlighted CG dinucleotides appears in a new window when the mouse is over the "product length" box. A printable window with all these details appears after double-clicking on the "Details".

button (absent from the MSP primer design result sheet) serves to start the post-design primer analysis (*see* **Subheading 3.**).

2.2. *Primer Design for Native DNA Sequences*

Details of primer design previously described correspond to all three kinds of primer design proposed by BiSearch. There are no specific general considerations applicable to the native DNA sequences. This is the only application when "easy" templates (vectors, cDNA) are frequently used for primer selection. The design of primers, which efficiently amplify the target molecules, is generally easy in these cases.

However, we emphasize that BiSearch is dedicated for primer design for more complex templates, such as mammalian genomic DNA. Owing to the quality of sample or to obtain efficient amplification of these templates, sometimes various chemicals are added to the PCR, which may modify the *T*m of the primers. BiSearch software can take into account the presence of glycerol, ethylene glycol, or formamid when calculating the melting temperature of the primers.

Some users filter genomic DNA for repeat sequences before designing primers for a target sequence. We suggest not to pre-filter the template because it may lower the efficiency of primer design, whereas the highly efficient post-design primer analysis of BiSearch reveals any potential cross-amplification PCR product.

2.3. Primer Design for Bisulfite-Treated DNA Sequences

The most frequent covalent modification of the mammalian genomic DNA is the methylation of cytosines at the C5 position *(7)*. This occurs almost exclusively in the context of CpG dinucleotides. Methylation of DNA has important physiological and pathological consequences. Therefore, it is interesting to determine the methylation profile of some target DNA sequences in various tissue samples. The most widespread technique, which offers the most detailed methylation analysis, is based on the bisulfite treatment of the genomic DNA *(2,3)* (*see* **Fig. 4**). This treatment transforms all the non-methylated cytosines to uracyls, whereas the methylated cytosines remain unchanged. The methylation profile is determined by the PCR amplification of the treated template followed by the sequencing of the amplicon. During these steps, the uracyls are transformed to thymines, and their presence in the final sequences indicates the absence of DNA methylation at the specific cytosines in the original DNA molecule.

Primer design for PCR amplification of a bisulfite-treated sequence is a difficult task, because the template has several new features, which distinguish it from the native DNA molecules. The DNA sequence after the treatment is characterized by a very high percentage of T. This sequence redundancy causes frequent primer dimer formation, low efficiency of amplification because of long T stretches and high frequency of contaminating PCR products. The objective of amelioration of the primer design process for these templates and especially the concern to develop a rapid in silico post-design primer analysis to avoid the contaminating PCR products led us to create BiSearch.

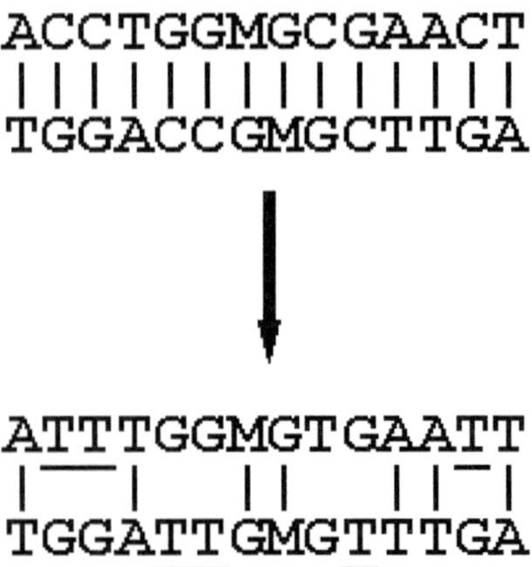

Fig. 4. Bisulfite conversion of DNA. A short double-stranded DNA sequence before (upper part) and after (lower part) bisulfite treatment. Non-methylated cytosines of the original sequence are transformed to uridines and detected as thymines (underlined in the lower sequence) after amplification. Methyl-cytosine (M) is resistant to the treatment. The lower panel illustrates that bisulfite conversion leads to the loss of complementarity between the two strands of the DNA molecule.

Although the sequence of the target DNA molecule is modified by the bisulfite treatment, the original, native sequence should be used as input for primer design (*see* **Fig. 1**). If the primer design option for bisulfite-treated sequence was chosen, then the software carries out an in silico bisulfite treatment, notes the localization of CpG dinucleotides, and changes all C to T. The Cs of the original CG dinucleotides are taken into account during primer design both in their methylated and in their non-methylated forms.

The software may propose primers overlapping with CpG dinucleotides of the original sequence. These cytosines can be either methylated or unmethylated in the target molecule, thus to avoid any PCR bias *(8)* BiSearch proposes a degenerate primer able to anneal to both target sequences. A degenerate primer contains a C/T (Y) or an A/G (R) on the sense and antisense strands, respectively, at the Cs of the original CG dinucleotides. However, the melting temperatures of the methylated and unmethylated sequences are different, which leads to a potential bias during the amplification reaction. Therefore, the maximum allowable *T*m difference between the methylated and unmethylated forms of a primer has to be low. This parameter can be determined on the target DNA input datasheet (*see* **Fig. 1**).

As, because of the bisulfite treatment, the originally double-stranded, self-complementary DNA molecules loose the complementarity of the two strands and are therefore no more double-stranded (*see* **Fig. 4**), the original sense and antisense strands can and should be amplified separately with two different primer pairs. BiSearch proposes to design primers for the amplification of bisulfite-treated sense or antisense strand (*see* **Fig. 1**).

The bisulfite treatment of DNA provokes strand brakes, which leads to an approximate upper limit of PCR products around 400–500 bp *(9)*. The maximal allowable limit for the PCR product is thus 400 bp by default, a parameter to be determined also on the target DNA input datasheet (*see* **Fig. 1**).

Some further parameters have to be also considered during primer design for the amplification of bisulfite-treated DNA (*see* **Fig. 2** "Primer design"). Generally, these sequences are amplified to allow the analysis of DNA methylation of a target gene. It is thus important that the amplicon include some CpG dinucleotides. The lowest acceptable number of CpG dinucleotides in the amplicon can be determined by the user. In the result table the window for the amplicon shows a PCR product corresponding to an originally fully methylated sequence after bisulfite treatment with hoghlighted CpG dinucleotides. The number of CpGs in the amplicon is indicated below the sequence of the PCR product (*see* **Fig. 3**).

Finally, there is one more parameter specifically used for primer design to amplify bisulfite-treated target molecules. The GC content of these sequences importantly drops. In general, the target sequences are originally highly GC rich, with 60% (or higher) GC content. Approximately, the half of the GC content is C (30%). Owing to the treatment, they disappear; therefore, default parameter for GC content of a treated molecule is between 0 and 60% with the optimum at 30%.

2.4. Primer Design for Methylation-Specific PCR

DNA methylation influences several nuclear processes such as chromatin structure modification, X chromosome inactivation, genomic imprinting, and transcriptional regulation *(7)*. All these roles of the methylation of DNA are linked and somehow lead to or are related to gene silencing. Although the listed examples are physiologic ones, gene silencing has a great impact on the molecular pathomechanism of several disorders and especially on cancerogenesis *(10)*. Cancerous cells are characterized by an imbalance of local and global genomic methylation, leading to the silencing of tumor suppressor genes by acquiring DNA methylation at gene regulatory regions and activation of silenced genes by loss of methylation. Therefore, the analysis of DNA methylation profiles in regulatory regions of various target genes in patients and controls has an important role both in understanding molecular carcinogenesis and in diagnostics.

DNA methylation analysis by bisulfite genomic sequencing to determine the precise methylation profile of a target sequence in diagnostics is very time-consuming and expensive. Alternative approaches were thus developed. One of them is MSP ***(4)***, which amplifies selectively the methylated alleles after bisulfite treatment of the extracted DNA, because, according to our current knowledge in most of the cases, the detection of hypermethylation has the diagnostic value ***(11)***.

The "MSP design" menu point of BiSearch was created to design primers for the amplification of a given template by MSP. The considerations for MSP primer design are essentially the same as for any other bisulfite-treated sequence and were discussed in section 2.3. Here, we focus only on the differing aspects.

The principle of MSP technique is based on the high *T*m value difference between the non-converted (methylated) and the converted (unmethylated) strands. This is due to the occurrence of Cs instead of Ts in the methylated targets (*see* **Fig. 4**). The selective amplification of methylated over unmethylated bisulfite-treated templates is achieved by primer pairs designed to anneal to non-converted (methylated) cytosines of CpG dinucleotides. PCR amplification is carried out at high annealing temperature, which does not allow mispriming on the unmethylated sequence.

The input of sequence in native format is followed by in silico bisulfite treatment. However, only the non-CpG cytosines are transformed, and all the others remain non-converted as if they were methylated. Only primers annealing to the methylated (non-converted) sequence are proposed. For each primer, the software calculates the *T*m value for both a converted and a non-converted template. In contrast to the previous section, where the unbiased amplification of both methylated and unmethylated alleles was the objective, here the aim is the specific amplification of the methylated template. Although the maximum *T*m difference between methylated and unmethylated primers was set to a low value (2.5 °C) for the "classical" bisulfite-treated sequences (*see* **Fig. 1**), here the software calculates with a high minimal *T*m (8 °C) difference (*see* **Fig. 2** "Primer design"). Moreover, only those primer pairs are accepted where at least one or both (if higher stringency is needed) of the primers has (a non-converted) C at the 3′-end. This set up allows the specific amplification of methylated alleles.

3. Post-Design Primer Analysis

As mentioned earlier, BiSearch is composed of two independent modules. The first one is dedicated to primer design (described here above), whereas the second one was created for post-design primer analysis. It is generally observed that some PCR primer pairs do not amplify the target template as efficiently as predicted by the primer design algorithm. One of the reasons for this empirical fact is the binding of the selected primers to sequences other than targeted. More the template is complex more often this problem occurs. Mammalian genomic DNA is a prototype of such complex templates and binding of selected primers to similar or identical sequences other than targeted occurs (more or less) frequently. Furthermore, this mispriming may occur at sites located in proximity to each other on the opposite strands in a compatible distribution, allowing the amplification of a spurious PCR product called "contaminating" amplicon. Owing to the duplication of the genome (complementarity of the strands is lost) and the high redundancy of the sequence (T-richness especially in the target sequences) after bisulfite treatment ***(2,3)***, mispriming and misamplification happen much more often. Therefore, post-design primer analysis for genomic mispriming and more importantly for contaminating PCR products is crucially important for genomic and especially bisulfite-treated templates. The unique post-design primer analyses functions of BiSearch were created to meet these needs.

Nevertheless, one can ask why to use a new search engine instead of BLAST ***(12)***. There are two simple answers: (1) BLAST has no bisulfite-converted database; therefore, it is impossible to search with primers designed for bisulfite-treated templates. However, such a database would still give only partial solutions to the needs, as it is exemplified by the methblast Web site (http://medgen.ugent.be/methBLAST/). (2) BLAST software do not contain a final analysis, which would allow easy, user-friendly and fast identification of potential PCR amplicons.

There is also a more complex answer to the above question. BiSearch was developed in view of post-design primer analysis, whereas the development of blast followed a different logic. BiSearch allows the use of mismatches and requires more stringent 3′- than 5′-end homology between the query sequence (primer) and the hits. These principles are essentially important for primer design but are not useful for DNA sequence identification. This is well illustrated with the difference of search algorithms. BLAST requires the presence of at least one "word" (sequence length) to identify a hit. This word of at least 11 nucleotides should be present in the hits; otherwise, the algorithm does not recognize the sequence as a hit. Thus, there are some mismatch sequences (e.g., a primer of 21-base long with a mismatch at the eleventh position), which are not recognized by BLAST, although BiSearch recognizes them easily in a couple of seconds and can be accepted as a highly homologous primer, which can potentially amplify contaminating PCR products.

In the next paragraphs, we describe in detail the post-design primer analysis module of BiSearch. The function of this module is the analysis of primers and primer pairs for genomic mispriming sites and contaminating PCR products. Currently, these applications are available on four mammalian genomes (human, mouse, rat, and chimpanzee) both with and without in silico bisulfite treatments. In silico bisulfite-treated genomes were prepared by changing on both strands all Cs to Ts, assuming that they were completely unmethylated (*see* **Fig. 4**). This creates two genomes, because the two strands both give rise to a new genome. Therefore, the imaginary new antisense strands (they will exist only upon PCR amplification) were also created in silico to allow the post-design primer analysis.

The search function of the post-design module of BiSearch analyzes the potential genomic binding sites by a binary hash function on a 16-base long string. This 16-base long string corresponds to the 16 3′ most nucleotides of the primer. The rest of the primer is not taken into account during the search for the genomic annealing sites. However, the entire primers are analyzed for PCR amplicons (see below).

First, all four genomes were indexed according to 16-base long strings found in their sequences. The position list of each individual string is stored on the hard disk in a way, which allows a quick search for the potential primer binding sites. The post-design primer analysis consists of the identification of genomic locations of the 16-base long string corresponding to the 16 3′ most nucleotides of the query sequence (primer). It has to be noted that only exact matches can be identified with this approach. Therefore, to allow the analysis of genomic binding sites with mismatches in the primer sequence, all theoretical 16-base long mismatch strings of a primer are generated in the first step of the analysis. The mismatch strings are the 16-base long sequences, which correspond to the cumulative number of mismatches defined by the user in the 16 3′ most nucleotides of the primer. The mismatch strings are then sorted according to their hash index.

The mismatch string can be adjusted by the user (*see* **Fig. 2** "Database search and fast PCR"). When the user accepts only complete matches, then the value of all 16 positions should be zero. The default parameter is "0000000011111111." This means that, after the search, all genomic loci will be considered as hits, which have no mismatches in the 8 3′ most nucleotides, and there is maximum one mismatch in the next eight nucleotides. The default parameter results in 24 supplementary oligonucleotide sequences (three for each of the eight positions where the value is 1).

The search time depends on the number of generated mismatch strings. However, because the main time-consuming step in the algorithm is to find the appropriate position list on the hard disk, strings having position lists close together can be searched/found more quickly. Therefore, sorting the oligos according to their hash index speeds up the search algorithm. Moreover, searching the genomic localization of several primers (mismatch strings) is faster if they are analyzed simultaneously. This is the reason of the roughly identical search times for a unique primer and a primer pair.

The search functions are very fast. The analysis of the binding sites of a single oligonucleotide with the default parameters in a "native" (without bisulfite treatment) genome is only 20 s long. The same search on the bisulfite-treated genomes takes approximately 30 s. The output format of the results of a simple genome search with a single oligonucleotide is the listing of the genomic hits with their precise location and an alignment between the query and the hit sequences (*see* **Fig. 5**). When the number of hits is higher than a limit set before the search, then only the number of hits is indicated in the report. If the search is carried out on a bisulfite-treated genome, the results are separately shown for the (original) sense and antisense strands. It is important to note that these searches ignore the 5′ part of the oligonucleotide between the 17th nucleotide from the 3′-end and the first (5′ most) nucleotide of the sequence.

An even more important function of the software is the "primer genome search" function. The main interest of this function is the analysis of the primers for genomic binding sites and especially for potential (contaminating) PCR products. The principle of this part of the algorithm is a refined "simple genome search." In the first step, both primers are analyzed by the "simple genome search," then in a second step, the locations of the hits are further analyzed for appropriate distribution and distance of the primers for potential PCR products. This search includes two additional parameters (*see* **Fig. 2** "Database search and fast PCR"). The first one is the maximal theoretical length of PCR products (default is 1 kb). Primer hits at longer distance from each other will not be taken into account for the analysis of potential PCR products. The second additional parameter is the maximal acceptable number of mismatches in the primers.

During this further analysis, those well-oriented hits are further filtered, which are not more distanced from each other than the maximal acceptable

Primer Design and Search Tool

Menu
Primer tm
Primer score
Simple genome search
Primer genome search
Primer design
MSP design
Parameters
Admin

Help
Manual
Faq
Manuscripts
ChangeLog
Comments

Statistics

Visitors:	1790
Calculated tm:	572
Calculated score:	96/95
Genome search:	859/81
Fast PCR:	432/913
Primer design:	562/831
MSP design:	401
Version:	2.05
Last update:	03/05/06

Results of your search

6 matches were found based on the 3' 16mer oligo search.

```
Chromosome 1
        Query:            1 GCGATGTAGTTCTGCT 16
                              || |||||||||||||
        Sbjct:     49834110 GCTATGTAGTTCTGCT 49834094   Frame: -1

Chromosome 5
        Query:            1 GCGATGTAGTTCTGCT 16
                              | ||||||||||||||
        Sbjct:    172403028 GAGATGTAGTTCTGCT 172403044  Frame: 1

Chromosome 12
        Query:            1 GCGATGTAGTTCTGCT 16
                              || |||||||||||||
        Sbjct:     37516157 GCTATGTAGTTCTGCT 37516173   Frame: 1

Chromosome 14
        Query:            1 GCGATGTAGTTCTGCT 16
                              | ||||||||||||||
        Sbjct:    106199540 GTGATGTAGTTCTGCT 106199524  Frame: -1

Chromosome 22
        Query:            1 GCGATGTAGTTCTGCT 16
                              || |||||||||||||
        Sbjct:     20502897 GCCATGTAGTTCTGCT 20502913   Frame: 1

Chromosome X
        Query:            1 GCGATGTAGTTCTGCT 16
                              | ||||||||||||||
        Sbjct:    108384822 GAGATGTAGTTCTGCT 108384838  Frame: 1
```

Elapsed time: 16s

Fig. 5. Output format of the results of a "Simple genome search." Default mismatch string was used to analyze the potential primer binding sites on the native human genomic sequence. Chromosomal locations of the genomic hits are indicated. The query sequence (16 3′ most nucleotides of the primer) is aligned with each hit.

PCR amplicon length indicated as the first additional parameter. These data are obtained after the analysis of the 16 3′ most nucleotides of the primer pair. During the final analysis, the entire primers are considered and compared to the sequences found at the locations identified by the "simple genome search." Only those hits are accepted as PCR amplicons where the number of maximal allowable mismatches in the primers is not exceeded.

Interestingly, these run last generally the same time as mentioned previously for simple genome searches. In the output results, first, the total number and the sequence data of the potential PCR products are reported (*see* **Fig. 6**). In the PCR products, the color-coded entire length of the primers is shown, and the mismatches are highlighted. If the number of potential PCR products exceeds a limit, then only their total number is reported. Finally, the report is finished by

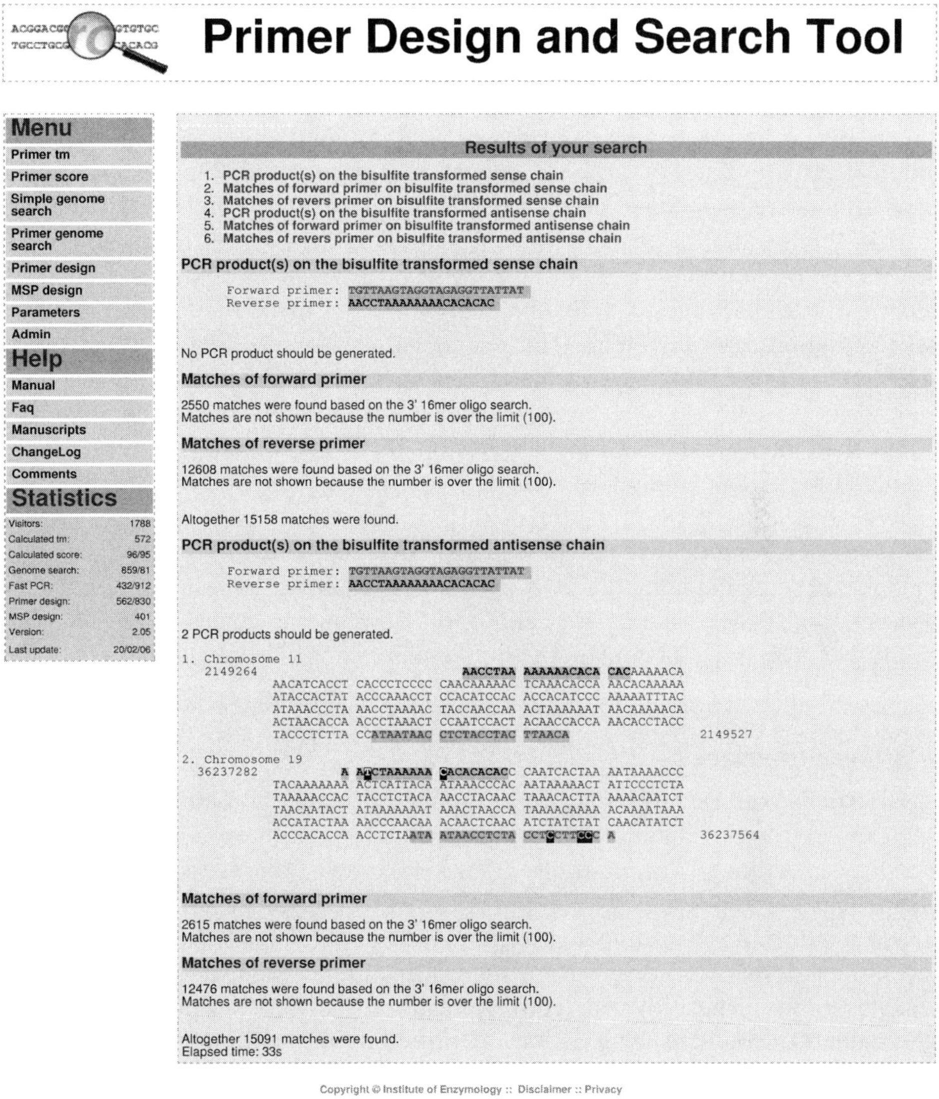

Primer Design and Search Tool

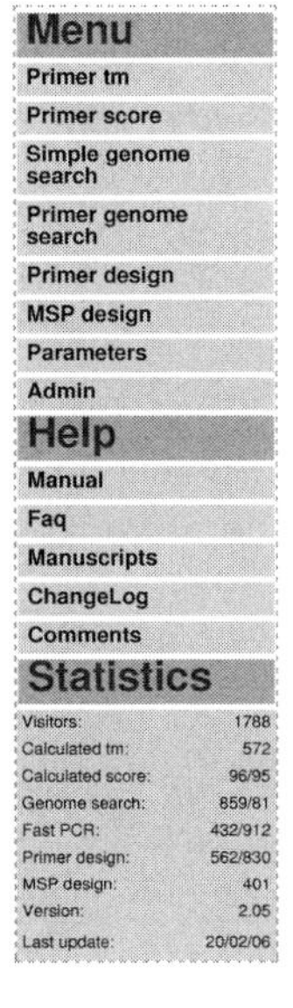

Results of your search

1. PCR product(s) on the bisulfite transformed sense chain
2. Matches of forward primer on bisulfite transformed sense chain
3. Matches of revers primer on bisulfite transformed sense chain
4. PCR product(s) on the bisulfite transformed antisense chain
5. Matches of forward primer on bisulfite transformed antisense chain
6. Matches of revers primer on bisulfite transformed antisense chain

PCR product(s) on the bisulfite transformed sense chain

```
Forward primer: TGTTAAGTAGGTAGAGGTTATTAT
Reverse primer: AACCTAAAAAAAACACACAC
```

No PCR product should be generated.

Matches of forward primer

2550 matches were found based on the 3' 16mer oligo search.
Matches are not shown because the number is over the limit (100).

Matches of reverse primer

12608 matches were found based on the 3' 16mer oligo search.
Matches are not shown because the number is over the limit (100).

Altogether 15158 matches were found.

PCR product(s) on the bisulfite transformed antisense chain

```
Forward primer: TGTTAAGTAGGTAGAGGTTATTAT
Reverse primer: AACCTAAAAAAAACACACAC
```

2 PCR products should be generated.

```
1. Chromosome 11
  2149264                         AACCTAA AAAAAACACA CACAAAAACA
          AACATCACCT CACCCTCCCC CAACAAAAAC TCAAACACCA AACACAAAAA
          ATACCACTAT ACCCAAACCT CCACATCCAC ACCACATCCC AAAAATTTAC
          ATAAACCCTA AACCTAAAAC TACCAACCAA ACTAAAAAAT AACAAAAACA
          ACTAACACCA ACCCTAAACT CCAATCCACT ACAACCACCA AACACCTACC
          TACCCTCTTA CCATAATAAC CTCTACCTAC TTAACA                 2149527

2. Chromosome 19
 36237282        A ATCTAAAAAA CACACACACC CAATCACTAA AATAAAACCC
          TACAAAAAAA ACTCATTACA ATAAACCCAC AATAAAAACT ATTCCCTCTA
          TAAAAACCAC TACCTCTACA AACCTACAAC TAAACACTTA AAAACAATCT
          TAACAATACT ATAAAAAAAT AAACTAACCA TAAAACAAAA ACAAAATAAA
          ACCATACTAA AACCCAACAA ACAACTCAAC ATCTATCTAT CAACATATCT
          ACCCACACCA ACCTCTAATA ATAACCTCTA CCTCCTTCCC A           36237564
```

Matches of forward primer

2615 matches were found based on the 3' 16mer oligo search.
Matches are not shown because the number is over the limit (100).

Matches of reverse primer

12476 matches were found based on the 3' 16mer oligo search.
Matches are not shown because the number is over the limit (100).

Altogether 15091 matches were found.
Elapsed time: 33s

Fig. 6. Output format of the results of a "Primer genome search." Default mismatch string was used to analyze the native human genomic sequence for potential primer binding sites and polymerase chain reaction (PCR) products. Only the number of genomic hits for each primer is indicated, because their total number was superior to the limit. PCR products with highlighted mismatches in the primers are also shown with the genomic locations.

the data of genomic (mis)priming sites. Results from bisulfite-treated templates are given first for the original sense then for the antisense strands.

This "primer genome search" menu point is directly linked to the primer design module of BiSearch. The result table of primer design contains a link called FPCR for each primer pair (*see* **Fig. 3**). A double-click on this link automatically initiates a primer genome search run with the corresponding primer pair on the genome indicated in the parameter table and using bisulfite-treated genome if the primers were designed to a bisulfite-treated template. The FPCR option is not available for MSP primers. Primer pairs designed by other software packages can also be analyzed when using the "primer genome search" option from the menu. The search starts after selecting the appropriate search parameters (genome, mismatch string, maximal number of mismatches, and bisulfite treatment, if necessary). Degenerate primers cannot be analyzed, others than the four standard nucleotides (A, T, G, and C) are ignored. Primers designed by MSP can also be tested by this function. However, the test will analyze the primers as unmethylated (C/T converted) on a bisulfite-treated unmethylated genome, which will lead to an overestimation of both the genomic binding sites and the potential PCR products. Indeed, the genomic binding sites and potential PCR products are valid only if the hit sequences are resistant to bisulfite conversion and contain methyl-cytosines at the CG sites corresponding to the MSP primers.

4. WWW Interface

BiSearch Web server is designed to be fast and user friendly. This is achieved by the combination of various techniques and programing languages. The user interface is written in php (PHP: Hypertext Preprocessor); the HTML (Hypertext Markup Language) output is paired by CSS (Cascading Style Sheet) to be esthetic and brushed up javascript to be user friendly. The main programs, the primer design and post-design algorithms are written in C and C++ to ensure rapidity. The programs communicate by the PHP scripts through CGI (Common Gateway Interface).

The different options and modules of BiSearch Web server are structured by the main menu located on the left side of the pages. The usage of the software is facilitated by a dynamic help. Indeed, if the cursor is over an input field, a pull-down explanatory help box automatically appears. Finally, a third feature of the Web server completes the user-friendly characteristics. The user's contact is permanent to BiSearch Web server during primer design and genome searches. This is because of the continuous exchange of information between the user and the server. This programing technique ensures that the connection remains

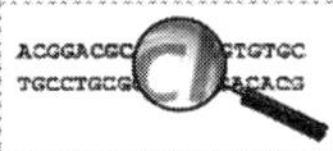

Fig. 7. Running job. The contact between the user and BiSearch Web server is intact even during longer calculations. This allows a feedback from the server on the remaining part of the current task.

intact even during long calculation. This continuous contact makes possible the feedback from the server, which is demonstrated by the follow-up of the primer design calculations and genome searches (*see* **Fig. 7**). In conclusion, BiSearch Web server is a useful tool in primer design for complex matrices and especially for bisulfite-treated genomes. The post-design primer analyses for genomic mispriming sites and alternative PCR products are unique features of the software. These characteristics combined with the rapidity and facility of use make BiSearch more and more popular.

References

1. Tusnady, G. E., Simon, I., Varadi, A. & Aranyi, T. (2005). BiSearch: primer-design and search tool for PCR on bisulfite-treated genomes. *Nucleic Acids Res* **33**, e9.
2. Frommer, M., McDonald, L. E., Millar, D. S., Collis, C. M., Watt, F., Grigg, G. W., Molloy, P. L. & Paul, C. L. (1992). A genomic sequencing protocol that yields a positive display of 5-methylcytosine residues in individual DNA strands. *Proc Natl Acad Sci USA* **89**, 1827–1831.
3. Clark, S. J., Harrison, J., Paul, C. L. & Frommer, M. (1994). High sensitivity mapping of methylated cytosines. *Nucleic Acids Res* **22**, 2990–2997.
4. Herman, J. G., Graff, J. R., Myohanen, S., Nelkin, B. D. & Baylin, S. B. (1996). Methylation-specific PCR: a novel PCR assay for methylation status of CpG islands. *Proc Natl Acad Sci USA* **93**, 9821–9826.
5. Kampke, T., Kieninger, M. & Mecklenburg, M. (2001). Efficient primer design algorithms. *Bioinformatics* **17**, 214–225.
6. Wetmur, J. & Sninsky, J. (1995). Nucleic acid hybridization and unconventional bases. In *PCR Strategies* (Innis, M., Gelfand, D., Sninsky, J., eds), pp. 69–83 Academic Press.
7. Bird, A. (2002). DNA methylation patterns and epigenetic memory. *Genes Dev* **16**, 6–21.
8. Warnecke, P. M., Stirzaker, C., Melki, J. R., Millar, D. S., Paul, C. L. & Clark, S. J. (1997). Detection and measurement of PCR bias in quantitative methylation analysis of bisulphite-treated DNA. *Nucleic Acids Res* **25**, 4422–4426.

9. Grunau, C., Clark, S. J. & Rosenthal, A. (2001). Bisulfite genomic sequencing: systematic investigation of critical experimental parameters. *Nucleic Acids Res* **29**, E65–65.
10. Robertson, K. D. & Wolffe, A. P. (2000). DNA methylation in health and disease. *Nat Rev Genet* **1**, 11–19.
11. Laird, P. W. (2005). Cancer epigenetics. *Hum Mol Genet* **14**(Spec No 1), R65–76.
12. Altschul, S. F., Gish, W., Miller, W., Myers, E. W. & Lipman, D. J. (1990). Basic local alignment search tool. *J Mol Biol* **215**, 403–410.

21

Graphical Design of Primers with PerlPrimer

Owen Marshall

Summary

PerlPrimer is a cross-platform application for the design of standard PCR, bisulfite PCR, real-time PCR, and sequencing primers. The program combines accurate primer-design algorithms with powerful interfaces to commonly used Internet databases, such as sequence retrieval from the Ensembl genome databases and the ability to BLAST search primer pairs. The use of PerlPrimer for designing primers is described, together with a synopsis of the primer search and primer-dimer algorithms.

Key Words: PCR; bisulfite PCR; real-time PCR; QPCR; sequencing.

1. Introduction

Accurate design of primers is essential for effective PCR reactions. Several significant open-source applications to design primers have been written previously—most notably the application Primer3 *(1)*. However, a graphical utility for end-users that interfaces directly with popular Internet databases has been lacking.

PerlPrimer is a cross-platform application that provides a graphical user interface to easily design primers for standard PCR, real-time PCR (QPCR), bisulfite PCR, and sequencing reactions. The aim of PerlPrimer is to give an investigator maximum control over primer design, whilst at the same time simplifying the process as much as possible. Users can retrieve gene sequences from the Ensembl database *(2)*, for example, or run BLAST queries on any primer pair with a single click.

From: *Methods in Molecular Biology, vol. 402: PCR Primer Design*
Edited by: A. Yuryev © Humana Press, Totowa, NJ

1.1. Obtaining PerlPrimer

PerlPrimer is open-source software released under the GNU Public License; it may be freely downloaded from http://perlprimer.sf.net. The software is designed to be cross-platform and has been written in Perl using the Perl/Tk toolkit. As such, the software will run under any system with Perl and Perl/Tk installed, including Linux, Microsoft Windows, and MacOSX (provided an X-server is also present). In addition, for real-time PCR primer design, PerlPrimer uses the open-source application Spidey *(3)* to find intron/exon boundaries.

A software installer is provided for Microsoft Windows systems which contains a standalone executable version of PerlPrimer built using the Perl Archiving ToolKit, PAR (http://par.perl.org) and a Windows binary of the Spidey executable.

2. Methods

2.1. Graphical Design of Primers

2.1.1. Common Interface Elements

PerlPrimer uses a tabbed interface format, having a separate tab for each primer design method and a final tab to display details of any selected primer pair. The structure of each primer design section is kept the same—options are presented at the top of the page, followed by entry boxes for the DNA sequence to be analyzed (*see* **Fig. 1**). The results are displayed below, in a table that lists sorted primer pairs and on a graphical representation of the DNA sequence where the selected primer pair is drawn. A detailed alignment of the DNA sequence with the selected primer pair may be displayed by right-clicking on the graphical DNA representation (*see* **Fig. 2**).

The graphical representation of the DNA sequence also displays the regions used to search for primers. These are defined by an outer range limit (represented in orange) and an inner range limit (represented in light blue). Although the automated functions provided by the program mean that it is generally not necessary to manipulate these by hand, the ranges may be adjusted by using the mouse or by manually entering the base-pair numbers in the "Amplified range" section. By clicking or dragging on the DNA graphic, the inner range settings may be adjusted using the left mouse button, and the outer range values may be adjusted using the middle mouse button (or pressing "ctrl" together with the left mouse button).

Details (including thermodynamic calculations and primer-dimers) for any primer pair may be displayed by double-clicking on a primer pair in the table, which switches the interface to the "Primers" page. From here, the user can

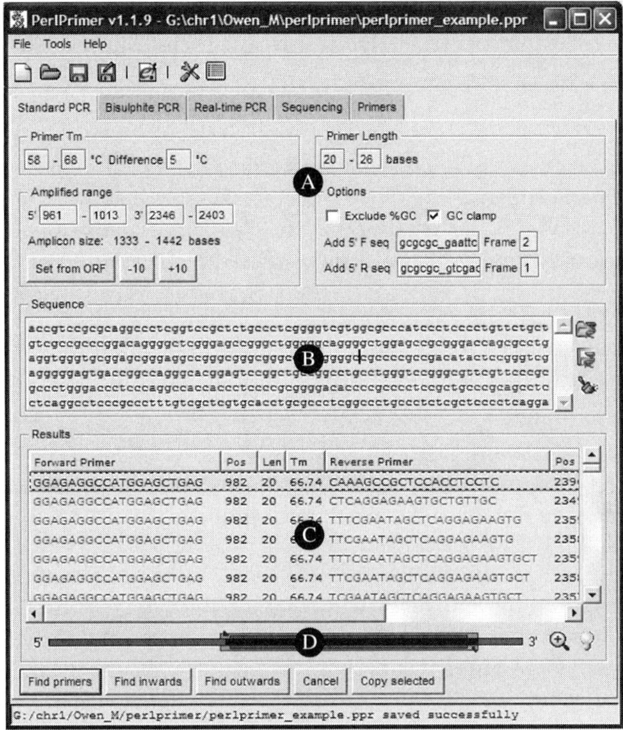

Fig. 1. The PerlPrimer main window showing the basic interface elements: **(A)** common options, **(B)** the sequence entry box, **(C)** the table of matching primer pairs, and **(D)** the graphical representation of the DNA sequence, search ranges, and primers.

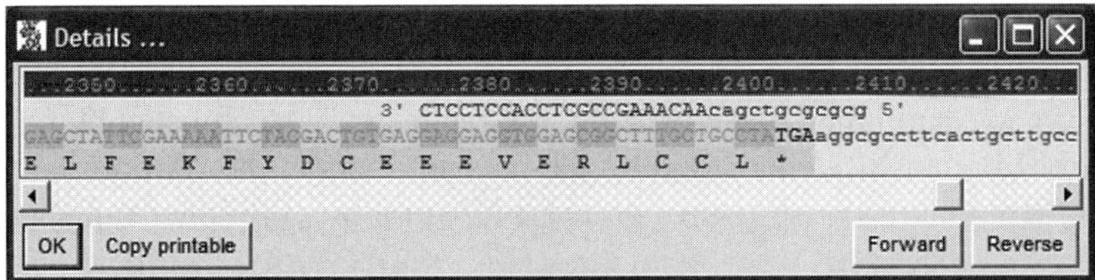

Fig. 2. The detailed DNA sequence view (in Standard PCR mode), illustrating a reverse primer with restriction enzyme site and spacer matched against the DNA sequence with codons highlighted and an open-reading frame (ORF) translation below.

choose to run a BLAST search *(4)* using (by default) the National Center for Biotechnology Information (NCBI) server or a local BLAST server (set in the "Preferences" dialogue box).

Beneath the "Results" section, several buttons provide access to common functions:

- Find primers: find primers using the set parameters (note that this will initially always return 0 primer pairs as the inner and outer ranges are identical—to automatically find primers around a gene or CpG island, use the "Find inwards" or "Find outwards" buttons below).
- Find inwards: repeatedly reduce the inner range by 10-bp steps on each side until primer pairs are found.
- Find outwards: repeatedly increase the outer range by 10-bp steps on each side until primer pairs are found.
- Cancel: cancel the running task.
- Copy selected: copy the selected primers to the clipboard in tab-delimited format (suitable for pasting into a spreadsheet application).

Finally, at the top of the interface, a toolbar provides access to many of the menu functions, including the ability to open and save project files, generate a report file of the selected primer pair aligned with the DNA sequence, and the ability to retrieve sequence information from the Ensembl database *(2)* (*see* **Subheading 2.1.2.**).

Access to the program Preferences is also provided in the toolbar, where many fine adjustments to primer selection parameters may be made.

2.1.2. Opening DNA Sequences

PerlPrimer provides a number of options for opening DNA sequences in plain text or FASTA format. A user can open local sequence files by using the "Open" command from the "File" menu and toolbar or the "Open DNA Sequence" button beside the Sequence entry box. The sequence may also be pasted directly from the system clipboard into the entry box. If opening a FASTA sequence, the program will automatically take the FASTA header information as the working title for the project.

An alternative, and perhaps the most useful method of entering sequence data, is to use the "Retrieve gene from Ensembl" function (accessible through the "Tools" menu or the toolbar), which provides access to the Ensembl project *(2)* (*see* **Note 1**). Using this option, a user can search for a gene sequence (the cDNA, genomic, coding or 5′- or 3′-untranslated regions) in any of the Ensembl genomic databases. If a search is successful, the alternative transcripts for the gene are listed and may be selected by the user. Although a vital part of the functionality of the program, this feature is extremely useful for finding QPCR primers (*see* **Subheading 2.1.5.**), as both the cDNA and genomic sequences for a gene transcript are automatically retrieved together.

2.1.3. Finding Standard PCR primers

The "Standard PCR" tab is appropriate for finding PCR primers for any sequence, but is especially useful when cloning genes into expression vectors through PCR. PerlPrimer includes code to automatically find the largest open-reading frame (ORF) in the entered sequence and will automatically set the search range to the ORF boundaries. Additionally, restriction enzyme sites can be added to primers, with the option of keeping the amplified ORF in frame with the expression vector if cloning into a fusion protein system.

Finding PCR primers around a gene using PerlPrimer would use the following method:

1. Retrieve the gene of interest from the Ensembl database, or open or paste a DNA sequence into the sequence box.
2. If the PCR product is intended to encompass the entire ORF, including the stop codon, click on the "Find outwards" button (*see* **Note 2**). Alternatively, if the PCR produce is designed to be cloned into an expression vector as both an N-terminal and C-terminal fusion protein, select the "Find primers for cloning" option from the "Tools" menu (*see* **Note 3**).
3. PerlPrimer will automatically widen the search range until suitable primers are found or until the end of the DNA sequence is reached. Primer pairs that match the search criteria are displayed in the Results table, sorted by extensible primer-dimer $\Delta G°37$ (*see* **Note 4**).
4. Primers may be sorted by any criteria by right-clicking on the Results table—it may be preferable here to sort by amplicon size to obtain primers that are closest to the original ORF.
5. Right-clicking on the graphical representation of the sequence will display a detailed alignment of the DNA sequence together with the primer pair (*see* **Fig. 2**). A translation of the largest ORF found in the sequence is shown below the sequence, with codons highlighted.
6. Double-clicking a primer pair (*see* **Note 5**) will switch to the Primers tab, where detailed thermodynamic parameters are displayed for each primer and the most stable primer-dimers are displayed graphically.
7. The Primers tab also provides the ability to BLAST search each primer using the NCBI server or a local BLAST server—it is highly recommended that primer pairs be checked to ensure their relative uniqueness.
8. If the amplified PCR product is intended to be cloned into a vector, the user can add a 5′-spacer and a restriction enzyme site to each primer using the "Add cloning sequences" menu option (*see* **Note 6**). If the PCR product is to be cloned as a

fusion protein, adding the correct frame for the forward and reverse primers will cause adenine bases to be automatically added between the restriction site and the sequence to ensure the amplified ORF remains in frame with the reporter gene (*see* **Note 7**).

2.1.4. Finding Bisulfite PCR Primers

This page is specifically for the design of primers to be used in methylation-specific PCR reactions, based on the parameters discussed by Warnecke et al. *(5)*. The program will automatically detect CpG islands in a sequence using the algorithm of Gardiner-Garden and Frommer *(6)* (*see* **Note 8**), and the amplified range is by default set to encompass any CpG islands found in the sequence. PerlPrimer designs primers against a bisulfite-converted DNA template and provides the option to only design primers that contain converted cytosine bases at the 3′-end, thereby allowing specific amplification of the converted DNA in a reaction. By default, PerlPrimer will also avoid primers that contain CpG dinucleotides due to the methylation-specific conversion of these residues.

The following method describes the design of primers for bisulfite PCR:

1. Open or paste the relevant DNA sequence containing the CpG island of interest into the Sequence box.
2. Select "Set from CpG island" from the "Amplified range" options—PerlPrimer will search for any CpG islands in the sequence and set the range to encompass all islands found (*see* **Note 9**).
3. Depending on the nature of the CpG island, choose either the "Find inwards" or "Find outwards" functions (*see* **Note 2**) to search for primers around the start and end point of the island.
4. Selecting a primer pair in the Results box will display each primer in the status bar, with cytosine bases colored red and CpG dinucleotides colored blue—this allows a quick assessment of how well each primer will selectively amplify the bisulfite-converted template (*see* **Note 10**).
5. Primer pairs may be assessed by following **steps 4–7** in **Subheading 2.1.3**. The detailed DNA sequence representation in this method will display CpG dinucleotides highlighted in red, with the CpG islands present marked by a blue box beneath the sequence.
6. If nested or semi-nested primers are desired, the primer search range may be set to a particular primer pair by right-clicking on the desired pair in the Results table and choosing the menu option "Set range from selected." Then by using the "Find inwards" function, the resulting primers found will be nested within the original primer pair.

2.1.5. Finding Real-Time PCR Primers

One of the more powerful features of PerlPrimer is the design of SYBR Green compatible QPCR primers. Here the ability to retrieve sequences from Ensembl is especially useful, as both the genomic and mRNA sequences from the database are automatically retrieved for any gene. The application Spidey *(3)* is used to accurately find intron/exon boundaries, and finding QPCR primers is in most cases a simple, two-step procedure.

Using the default settings, PerlPrimer searches for small (100–300 bp) amplicons that span an intron/exon boundary and possess at least one primer that hybridizes across an intron/exon boundary. Finding real-time primers for a particular gene would therefore use the following method:

1. Retrieve the gene of interest from the Ensembl database, or open or paste the relevant genomic and mRNA sequences into the sequence boxes.
2. If the mRNA transcript lacks introns, primers may still be designed by unchecking the "Span intron/exon boundary" and the "Overlap intron/exon boundary" options in the Options box (*see* **Note 11**).
3. If only primers amplifying across specific exons are desired, this may be set in the Amplicon size box. Limiting exons may be entered manually or may be selected by clicking on the graphical representation of the DNA sequence (*see* **Note 12**).
4. Click on the "Find primers" button—PerlPrimer will now automatically calculate intron/exon boundaries and search for appropriate primers.
5. Primers may be analyzed by following **steps 4–7** in **Subheading 2.1.3**. The detailed DNA sequence representation in this method will display the intron/exon boundaries in the sequence highlighted in red.

2.1.6. Finding Sequencing Primers

PerlPrimer also provides a function for quickly finding evenly spaced single forward primers for sequencing reactions. Primers may be searched across the entire input sequence, or the user can choose to limit this search to an ORF, if present. Again, the process is almost entirely automatic, generally requiring the user to simply click the "Find Primers" button to be provided with a list of appropriate single primers, as detailed below:

1. Retrieve the gene of interest from the Ensembl database, or open or paste the relevant genomic and mRNA sequences into the sequence boxes.
2. Select the desired distance between primers in the Spacing/Coverage box (*see* **Note 13**).
3. If the only region of interest is the ORF, click the "Set range from ORF" button in the Spacing/Coverage box.

4. Options are also provided to limit the amount of secondary structure present in primers (*see* **Note 14**) if required.
5. Click on the "Find primers" button to search for sequencing primers in the sequence—a list of sequencing primers to cover the region of interest will be generated.
6. If the program fails to find a contiguous primer set, slight adjustments to the primer spacing parameters, maximum allowable secondary structure parameters, or GC-content exclusion may be required.

2.2. Using Other Applications with PerlPrimer

Whilst PerlPrimer attempts to provide built-in solutions for many common PCR primer design conditions, it is also possible to use the primer searching capabilities of the software with other applications. PerlPrimer establishes a listening socket on local port 2500 (which may be modified in the program Preferences), allowing any application to send data in the form of a modified FASTA format. Upon receiving data, PerlPrimer will automatically search for suitable primers (although this behavior may be changed in the program preferences).

The syntax of the data format is identical to a normal FASTA sequence, except in the description line. Here, following the sequence name, the line should include the primer search range and the PCR design method to use, as follows:

```
>Name of DNA sequence 5prime_region[?-?]
3prime_region[?-?] page[?]
```

where page 1 = Standard PCR, page 2 = Bisulfite PCR, etc. If the page argument is not given, the Standard PCR page will be used as default.

An example of a program that communicates directly with PerlPrimer can be found in ContigViewer (http://www.atgc.org/Py_ContigViewer/), a program used in finding single-nucleotide polymorphisms (SNPs) and/or insertions/deletions in genome assembly contigs.

2.3. Program Implementation

2.3.1. Primer Searching Techniques

PerlPrimer utilizes a simple primer search algorithm for all PCR methods. To search for primers, PerlPrimer first limits the DNA sequence to the search range and then windows the sequence into primers. Primers are excluded on the basis of nucleotide composition, repetitive nature, and finally melting temperature (T_m). The forward primer set is then searched against the reverse primer set for pairs that meet T_m constraints and, in the case of QPCR, amplicon size constraints. Finally, the primer pairs are analyzed for secondary structure and sorted into the Results table on the basis of extensible primer-dimer stability.

2.3.2. Primer Melting Temperature Calculations

A central aim of PerlPrimer is to design accurate primers. To this end, the program calculates primer T_m through the nearest-neighbor thermodynamic approach ***(7,8)***. To adjust the calculated entropy for the salt conditions of the PCR, PerlPrimer uses the empirical formula derived by von Ahsen et al. ***(9)*** and allows the user to specify the concentration of Mg^{2+}, dNTPs, and primers (set in the program Preferences), or use default, standard PCR conditions. The result is a highly accurate prediction of primer T_m under experimental salt concentrations, giving rise to a maximum yield of product when amplified.

2.3.3. Primer-Dimer Estimation

Elimination of primer-dimer artifacts is important for reliable PCR amplification and essential for accurate QPCR primer design. Particular distinction needs to be made between extensible primer-dimers that have complementarity at one or both 3′-ends and will generate an amplified product, and non-extensible dimers that will merely reduce the available primer population. Whilst both populations of dimers will make the PCR reaction less efficient, the former are especially undesirable in QPCR reactions where they contribute to the total amplified product signal. As a result, PerlPrimer considers each set of primer-dimers separately.

PerlPrimer calculates primer-dimers by using a variation on the standard "sliding" algorithm approach. In this method, a matrix of base-pair compatibility between each primer and itself and between the forward and reverse primers is created. The complementarity of each pairing combination between the two primers is then read from the matrix, with the range of pairing combinations considered being dependant on whether extensible or non-extensible dimers are searched for. In each combination, bonds between single isolated base pairs—which are generally unfavorable to dimer stability—are ignored for algorithm speed considerations. From the complementarity that remains, the approximate $\Delta G°37$ is calculated using the nearest-neighbor approach, incorporating base mismatch data ***(7,10–13)***. Because the $\Delta G°$ nearest-neighbor values are dependent on the entropy of the system, which in turn is dependent on the salt concentration of the PCR reaction, these values are calculated internally by the program and are re-calculated whenever the salt conditions are changed by the user.

The above algorithm has the advantage of being relatively efficient; however, the results obtained are only an approximation of primer-dimer formation. Most significantly, the possibility of unequal internal loops between regions of complementarity within a dimer is not considered. The separate formation of hairpin loops within a primer is also not explicitly calculated, although such hairpin loops may be considered to be equivalent to a subset of the calculated primer homo-dimers.

3. Notes

1. The Ensembl database is accessed through the textview/searchview cgi functions provided at http://www.ensembl.org and sequence data are retrieved through the exportview function.
2. The "Find inwards" and "Find outwards" buttons provide functions that either decrease the inner range (both 5′ and 3′) or increase the outer range (both 5′ and 3′). The region in which primer pairs are searched for is thus incrementally increased until suitable primers are found.
3. The "Find primers for cloning" menu option works on the principal that a user is cloning the mRNA ORF into an expression plasmid to create a fusion protein in both N-terminal and C-terminal orientations. By incrementally increasing the 5′ outer range and incrementally decreasing the 3′ inner range, the program attempts to automatically find forward primers 5′ to the initiating ATG codon and reverse primers 5′ to the stop codon of the largest ORF of the supplied sequence. The resulting PCR product may then be cloned as both an N-terminal and C-terminal fusion protein.

   When using this feature, it is important to note that forward primers will need to be screened manually for stop codons 5′ to the start codon. A simple means of achieving this is to use the detailed DNA view, where translation is provided from the start of the forward primer to the start codon of the ORF.

   If the user simply seeks to clone the ORF into a standard expression vector, the "Find outwards" button is more appropriate (*see* **Note 2**).
4. Extensible primer-dimers are primer-dimers that can be amplified from their 3′-end. Such primer-dimers will contribute to the PCR reaction, creating visible primer-dimer artifacts. In comparison, non-extensible primer-dimers will reduce the amount of available primer in the PCR reaction but will not create a visible product.

   In all cases, the more negative the ΔG°37 for a primer-dimer, the more stable the dimer will be and the more likely it will be to contribute to the reaction.
5. The user can quickly switch between the PCR tab and the Primers tab by using the right and left arrow keys on the keyboard; the up and down arrow keys will change the selected primer pair in the Results table.
6. The list of restriction enzymes displayed is generated from the REBASE project database ***(14)***. By default, PerlPrimer will only list those restriction enzymes that

do not cut the PCR amplicon, and the number of restriction enzymes displayed is further reduced by only listing standard six-base cutters. Both of these options may be changed in the program Preferences.

7. The frame is specified as 0 (in-frame), 1 (one base out-of-frame), or 2 (two bases out-of-frame) and refers to the position of the restriction site relative to the fusion protein ORF.
8. The method for searching for CpG islands can be fine-tuned in the Preferences dialogue. The search window size, minimum island size, minimum (observed CpG)/(expected CpG) ratio, and minimum CG content can all be adjusted here. There is also an option to emulate the popular program "cpgplot" ***(15)***, which uses slightly relaxed parameters to those specified due to fencepost errors.
9. Although PerlPrimer makes no attempt to limit the primer search range, it is generally recommended that the amplicon size for bisulfite sequencing be less than 600 bp due to strand-breaks induced by the conversion process ***(5)***, with an amplicon of 300 bp being optimum. For the sequencing of large CpG islands, therefore, several sets of primers may be required.
10. The more converted cytosine residues present in a primer, the more specific the primer will be for the bisulfite-converted template. Conversely, the less CpG dinucleotides present in a primer, the more accurate the PCR reaction will be regardless of the methylation status of the template.
11. Note that QPCR primers that do not span intron/exon boundaries will also amplify product from any genomic DNA contamination present. It is suggested that RNA preparations used with such primers be treated with RNase-free DNase before reverse transcription and PCR.
12. Primers that are specific to certain exons may be useful in selecting all or only one transcript from a gene that may have multiple transcripts. If cDNA has been generated from poly-adenine primers, it may also be useful to select primers in the 3′-region of large transcripts.
13. The distance between sequencing primers will be dependent on the read-length generally obtained from sequencing and the fold coverage desired. The default values, searching for primers every 500–700 bases, are appropriate for single-fold coverage of sequencing reactions with an average read-length of 800 bp.
14. Secondary structure in sequencing primers will only affect the available primer population—thus the default value of disallowing any primers with secondary structure more stable than $-5\Delta G°37$ is conservative and may be safely reduced if necessary.

References

1. Rozen, S. and Skaletsky, H. (2000) Primer3 on the WWW for general users and for biologist programmers. *Methods Mol Biol* **132**, 365–386.
2. Hubbard, T., Barker, D., Birney, E., Cameron, G., Chen, Y., Clark, L., Cox, T., Cuff, J., Curwen, V., Down, T., Durbin, R., Eyras, E., Gilbert, J., Hammond,

M., Huminiecki, L., Kasprzyk, A., Lehvaslaiho, H., Lijnzaad, P., Melsopp, C., Mongin, E., Pettett, R., Pocock, M., Potter, S., Rust, A., Schmidt, E., Searle, S., Slater, G., Smith, J., Spooner, W., Stabenau, A., Stalker, J., Stupka, E., Ureta-Vidal, A., Vastrik, I., and Clamp, M. (2002) The Ensembl genome database project. *Nucleic Acids Res* **30**, 38–41.

3. Wheelan, S. J., Church, D. M., and Ostell, J. M. (2001) Spidey: a tool for mRNA-to-genomic alignments. *Genome Res* **11**, 1952–1957.
4. Altschul, S. F., Gish, W., Miller, W., Myers, E. W., and Lipman, D. J. (1990) Basic local alignment search tool. *J Mol Biol* **215**, 403–410.
5. Warnecke, P. M., Stirzaker, C., Song, J., Grunau, C., Melki, J. R., and Clark, S. J. (2002) Identification and resolution of artifacts in bisulfite sequencing. *Methods* **27**, 101–107.
6. Gardiner-Garden, M. and Frommer, M. (1987) CpG islands in vertebrate genomes. *J Mol Biol* **196**, 261–282.
7. Allawi, H. T. and SantaLucia, J., Jr (1997) Thermodynamics and NMR of internal G.T mismatches in DNA. *Biochemistry* **36**, 10581–10594.
8. SantaLucia, J., Jr (1998) A unified view of polymer, dumbbell, and oligonucleotide DNA nearest-neighbor thermodynamics. *Proc Natl Acad Sci USA* **95**, 1460–1465.
9. von Ahsen, N., Wittwer, C. T., and Schutz, E. (2001) Oligonucleotide melting temperatures under PCR conditions: nearest-neighbor corrections for Mg(2+), deoxynucleotide triphosphate, and dimethyl sulfoxide concentrations with comparison to alternative empirical formulas. *Clin Chem* **47**, 1956–1961.
10. Allawi, H. T. and SantaLucia, J., Jr (1998a) Nearest neighbor thermodynamic parameters for internal G.A mismatches in DNA. *Biochemistry* **37**, 2170–2179.
11. Allawi, H. T. and SantaLucia, J., Jr (1998b) Thermodynamics of internal C.T mismatches in DNA. *Nucleic Acids Res* **26**, 2694–2701.
12. Allawi, H. T. and SantaLucia, J., Jr (1998c) Nearest-neighbor thermodynamics of internal A.C mismatches in DNA: sequence dependence and pH effects. *Biochemistry* **37**, 9435–9444.
13. Peyret, N., Seneviratne, P. A., Allawi, H. T., and SantaLucia, J., Jr (1999) Nearest-neighbor thermodynamics and NMR of DNA sequences with internal A.A, C.C, G.G, and T.T mismatches. *Biochemistry* **38**, 3468–3477.
14. Roberts, R. J., Vincze, T., Posfai, J., and Macelis, D. (2005) REBASE–restriction enzymes and DNA methyltransferases. *Nucleic Acids Res* **33**, D230–D232.
15. Rice, P., Longden, I., and Bleasby, A. (2000) EMBOSS: the European Molecular Biology Open Software Suite. *Trends Genet* **16**, 276–277.

Index

Q

R

T

Printed in The United States of America